ROUTLEDGE LIBRARY EDITIONS:
SOCIAL AND CULTURAL GEOGRAPHY

Volume 14

THE POWER OF GEOGRAPHY

THE POWER OF GEOGRAPHY

How Territory Shapes
Social Life

Edited by
JENNIFER WOLCH
AND MICHAEL DEAR

LONDON AND NEW YORK

First published in 1989

This edition first published in 2014
by Routledge
2 Park Square, Milton Park, Abingdon, Oxfordshire OX14 4RN
and by Routledge
711 Third Avenue, New York, NY 10017

First issued in paperback 2015

Routledge is an imprint of the Taylor & Francis Group, an informa business

British Library Cataloguing in Publication Data
A catalogue record for this book is available from the British Library

ISBN: 978-0-415-83447-6 (Set)
ISBN 13: 978-1-138-98968-9 (pbk)
ISBN 13: 978-0-415-73504-9 (hbk)

Publisher's Note
The publisher has gone to great lengths to ensure the quality of this reprint but points out that some imperfections in the original copies may be apparent.

Disclaimer
The publisher has made every effort to trace copyright holders and would welcome correspondence from those they have been unable to trace.

THE POWER OF GEOGRAPHY

How Territory Shapes Social Life

EDITED BY

Jennifer Wolch

Michael Dear

University of Southern California,
Los Angeles

Boston
UNWIN HYMAN
London   Sydney   Wellington

Unwin Hyman Inc.,
8 Winchester Place, Winchester, Mass. 01890, USA

Published by the Academic Division of
Unwin Hyman Ltd
15/17 Broadwick Street, London W1V 1FP, UK

Allen & Unwin (Australia) Ltd,
8 Napier Street, North Sydney, NSW 2060, Australia

Allen & Unwin (New Zealand) Ltd in association with the
Port Nicholson Press Ltd,
60 Cambridge Terrace, Wellington, New Zealand

First published in 1989

Library of Congress Cataloging-in-Publication Data
The power of geography: how territory shapes social life/edited by
Jennifer Wolch, Michael Dear.
    p.   cm.
Bibliography: p.
Includes index
ISBN 0-04-445056-7 (alk. paper): £30.00 (Great Britain)
1. Anthropo-geography. 2. Social history. 3. Human territoriality.
I. Wolch, Jennifer R. II. Dear, M. J. (Michael J.)
GF50.P68 1988
304.2–dc19   88-11103

British Library Cataloguing in Publication Data
The power of geography: how territory
    shapes social life.
1. Capitalist societies. Cultural processes.
I. Wolch, Jennifer R. II. Dear, Michael
306'.342
ISBN 0-04-445056-7

Typeset in 10/11pt Bembo by Columns of Reading
and printed in Great Britain by Biddles of Guildford

Preface

This book is about social life under contemporary capitalism. It focuses on the profound way in which geography facilitates and constrains the practice of everyday life, and shows how the routine practices of daily living, within definite territorial limits, are effective in maintaining and reproducing the social order of capitalist society. In short, this book demonstrates how territory shapes social life, and explores some of the consequences of this shaping.

The essays in this volume are a significant departure from earlier studies. Too much of this work has suffered from acute oversimplification. Instead of trying to portray the richness in socio-spatial process, many analysts seem to have preferred the appeal of a simplistic (and hence misleading) metaphor. This criticism applies *inter alia* to Smith's appeal to the "invisible hand" of the market place, to the heavy-handed economic determinism of classical Marxism, and to the reductionism of much mathematical modeling. More recent analysts have attempted to encompass the variety of our emergent social order through a new terminology. They have used such phrases as "late" capitalism, "advanced" capitalism, and "disorganized" capitalism to convey the essence of the evolution toward complexity. These labels might be convenient abbreviations, but they too subsume an immense variety of social, political, and economic processes, including: the penetration of state control in all aspects of human life; world hunger and homelessness; the decline of Fordism and the growth of flexible production systems; the reassertion of local cultures in the face of the global electronic village; the redefined role of nation-states; the nuclear threat; the specter of AIDS; and the violence and corruption underlying much of our everyday lives.

The interaction between society and space is becoming increasingly complex, and we now have to seek new ways to describe the complexities. We can confidently assert that the forms of late-20th century society could not have been predicted by mid-century scholars. Time and space in our society are being stretched into new dimensions. Frederic Jameson has called ours a "postmodern" society, the time-space coordinates of which are so far only dimly perceived. The implication is that *since society is undergoing such a radical reorganization, our social theories require an equivalent revision if they are to provide insight into the emergent world order.*

A principal goal of this book is to contribute toward such a reconstruction of social theory. The general purpose of our effort is to provide a new focus on contemporary capitalist society. At the core of our reconstruction is the concept of society as a "time-space fabric" upon which are engraved the processes of political, economic, and socio-cultural

v

life. Human existence is thus expressed through specific histories and geographies. The task of social theory is to unravel the interactions of time and space in the structuring of people's lives and in the production of human landscapes.

Another distinctive feature of this collection is its substantive focus on the relation between territory and social practice. We are less concerned with the *production-related* aspects of contemporary capitalism. Instead, our emphasis is on what is commonly regarded as the realm of social *reproduction*. In this, we include concerns with gender, family, education, culture and tradition, race and ethnicity. A large and significant proportion of everyday life is engaged in the reproduction sphere. The general purpose and effect of this engagement is to perpetuate and maintain the existing structures of social, political, and economic relations. Discontinuities in the reproduction process are, of course, possible; but such discontinuity requires that human agents interrupt and transcend the inertia implicit in social practice.

An integral source of complexity in contemporary capitalism is the bewildering variety of sites at which the processes of reproduction may occur. They include the home, workplace, classroom, church, voluntary organization, and nation-state. Such complexity is both a source of powerlessness and empowerment. On one hand, the penetration of the reproductive practices is so extensive that a powerful agent can manipulate individual actions in order to pre-empt any attempt at subversion or transcendental social change. On the other hand, power and social control are extensively diffused throughout human organizations. Within the neglected interstices of these organizations, localized expressions of autonomy and change can occur without interference from outside or from above. In this book, we show how territory (which we define as geographically-organized human activity) is a crucial element permitting localized experimentation with the limits of autonomy and with the possibilities of transcendental social change.

The contributors to this collection provide a tentative articulation of a theory of territory and reproduction, and concrete empirical analyses of how social practices in particular places evolve in contemporary capitalism. They are concerned to identify the processes of reproduction; to explain the necessarily *socio-spatial* character of reproduction; to demonstrate the wide range of sites at which reproduction may occur; to explore the diverse outcomes of reproduction; and to examine the conditions under which discontinuities in the reproduction process become possible. Needless to say, we shall not encompass all aspects of this research task within a single volume. Yet our contributors together provide a conclusive demonstration of the territorially-dependent nature of the political, social, and economic spheres of human activity. We regard these essays as a significant step in the general reconstitution of social theory, and, more specifically, in the redefinition of the research agenda in human geography.

JENNIFER WOLCH
Santa Monica, 1988   MICHAEL DEAR

Acknowledgments

Many people helped make this book possible. Most of all, we are greatly indebted to our contributing authors for delivering insightful and beautifully written essays. We would like to thank Michael Storper and Allen J. Scott in particular, for stimulating the discussions which led to this effort.

The University of Southern California povided Wolch with the sabbatical leave during which the ideas for the volume were developed. Virginia Westbrooke expertly handled the correspondence involved in producing the book, and we are most grateful for her assistance.

We are grateful to the following individuals and organisations who have kindly given permission for the reproduction of copyright material (figure numbers in parentheses):

Figure 4.1 reprinted, with permission, from *The Columbus Dispatch*; Chapter 6 © Université Laval; Figures in chapter 12 by permission of the Syndics of Cambridge University Library; Phyllis Moore (15.1, 15.2); Houston Public Library (15.3, 15.4); Harper Leiper Studios (15.5, 15.6); Chapter 15 is an English language version of the article Un aperçu féministe sur la restructuration de l'emploi et sur la gentrification. *Cahiers de Géographie du Quebec* **31**, 83; Chapter 7 contains material originally quoted in Howe, A. & O'Connor 1982. Travel to work and labor force participation of men and women in an Australian metropolitan area. *Professional Geographer* **34**, 50–64.

For Hiram, Frances, and Hilda

Contents

List of tables

Part I

Introduction and overview

1

How territory shapes social life

MICHAEL DEAR and JENNIFER WOLCH

Society and space: an introduction

The journey along Mulholland Drive, atop the Hollywood Hills, provides one of the world's great urban vistas. To the south lies the Los Angeles basin, a glittering carpet. To the north, the San Fernando Valley (still part of the City of Los Angeles) unfolds in an equivalent mass of freeways, office towers, and residential subdivisions. There is probably no other place in North America where such an overpowering expression of the human impact on landscape can be witnessed. And yet, the physical landscape cannot be denied. Even in this region of almost 12 million people, the landscape still contains and molds the city.

In this book, we wish to examine how territory shapes social life, and vice versa. Above all else, the chapters contained in this volume provide a convincing argument for the power of geography in shaping human existence. It is impossible to understand human society without accounting for its geographical underpinnings. In this book, we are in search of a theory of territory or "humanly differentiated geographical space." This search is a direct response to the challenge issued by Scott & Storper in their collection of essays on the geographical anatomy of industrial capitalism. In setting a research agenda, they observe:

> The salient feature of the geographical landscape under capitalism is its status as an assemblage of territorial complexes of human labor and emergent social activity. The configuration of this landscape can be understood at three specific levels of analysis. First, it is constituted out of an overarching system of rules of order rooted in the basic relationships of capitalist society. . . . Secondly, it is the direct manifestation of a set of intricate locational *cum* spatial processes. . . . Thirdly, its immediate phenomenal form consists in a congeries of human communities in which the bases of social reproduction and social action are secured (Scott & Storper 1986, 310).

From this, they derive a principal objective for research in human geography: "to grasp the dynamics of the creation, reproduction and transformation of territorial complexes of human labor and social activity" (Scott & Storper 1986).

The chapters in this collection have been written and assembled to address this challenge. However, our book differs significantly from the previous effort by Scott & Storper. In their volume, the contributors focused almost entirely on the relationships of production, work, and territory. In this book, we focus on *reproduction* and territory. *By this term, we mean to encompass the wide range of social relations and social practices which derive from, and which serve to protect and maintain, the basic structures of capitalist society.* We are not implying a crude economic determinism in this broad statement. We reject the notion that the economic (base) relations are the primary determinants of socio-political (superstructural) relations. Society is much more complex than this. The formation of territorial outcomes is contingent upon the essentially unpredictable interactions of the spatial with the economic *and* the political and social/cultural spheres. Moreover, one of our key objectives is to explore how social practices can transcend existing social arrangements, thus making social *change* a constant part of the everyday life of individuals, communities, and nations.

In sum, this book focuses on the way social life structures territory, and the way territory shapes social life. The interdependencies between these processes – the socio-spatial dialectic – ensure that one cannot be understood without reference to the other. Our emphasis is on the power of geography to affect the social practices of everyday life. We argue that such territory-based practices have the power to protect and maintain (i.e., to reproduce) social relations, but also to transcend these relations to produce significant social change.

No single collection of essays can hope to encompass the entire range of work implied by this research agenda. The remainder of this introductory chapter develops a much simplified framework for the analysis of territory and reproduction. In the next section, some fundamental assumptions concerning territory and reproduction are briefly recounted. Then, our theory of territory and social life is outlined. Finally, an account of the current dynamics of social reproduction in contemporary capitalist society is developed; this provides a framework for understanding the relevance and place of the individual contributions within our general problematic.

Territory and reproduction: conceptual preliminaries

Before we outline our concept of territory and social life, some basic assumptions have to be clarified. These pertain to the two fundamental categories on which our analysis is predicated, namely, territory and reproduction. We shall begin by examining the classical Marxian notion of reproduction in order to demonstrate the major lineage of our approach. However, we shall quickly move from this to a more complex understanding of reproduction, based essentially in the writings of Weber and Foucault *inter alia*. Second, we provide some elementary statements about the nature of our geographical analysis.

The concept of reproduction

As a point of departure, we adopt the classical Marxist notion of reproduction. Marx noted that "every social process of production is, at the same time, a process of reproduction" (1971, vol. 1, 531). Thus any production process not only produces material objects, but also continually reproduces the associated production and distribution relations (Marx 1971, vol. 3, 857). The concept of reproduction implies much more than the mere replication of existing production processes. Reproduction involves several kinds of continuity and discontinuity in social processes, including: a link between individual capitals and economic subjects; a link among the different levels of the social structure, including the noneconomic elements of the production process; and a link between successive historical production processes (Althusser & Balibar 1970, 258–9). Hence, reproduction is the method by which the total social "ensemble," including modes of circulation, distribution, and consumption, is protected and repeated through time.

Reproduction is a dynamic concept, emphasizing historical continuity during periods of transition. In such transitions, reproduction allows for the replacement and transformation of things, but retains the fundamental relationships. In classical theory, the perpetuation of political, legal and other institutions in support of the economic order may be anticipated, as well as the key relations of the economic sphere. But exactly how is the reproduction of social relations secured? The answer has traditionally been sought in the functioning of the "legal-political and ideological superstructure." The state apparatus is a primary institutional manifestation of this superstructure. Other important superstructural elements have included the church, which may now have been superseded by the educational establishment. The role of the family has also been stressed.

The effect of reproduction is to perpetuate the social structures of capitalism. Social relations in capitalist society are constituted through a variety of fundamental work- and nonwork-related mechanisms. The *division of labor* is both a force for consolidation and for fragmentation of class relationships. It favors the formation of classes to the extent that it creates homogeneous groupings. On the other hand, it often implies a specialization of labor functions and, hence, fragmentation within an otherwise homogeneous group. *Authority relations* occur as a hierarchy of command within the productive enterprise, although non-market elements in society are also ordered to sustain the system of production, circulation, and distribution. *Distributive groupings* are those relationships which involve common patterns of consumption of material goods. With their concomitant status implications, these goods act to reinforce the separations initiated by differential work-specific capacities (Giddens 1973, Ch. 6).

The social relations of capitalism invariably take on a geographical expression. This can occur in many guises, as in the cases of the structure of nation-states or ethnically homogeneous neighborhoods. What is often less clear is the precise way in which spatial form is related to social forces.

5

Some researchers have argued that technical progress may reduce the role of space; it is not that space is external to the social structure and unaffected by it, but that its specific importance may be diminishing. Others have criticized the tendency to make a fetish of space in urban research (Castells 1976, 1977). However, *in our analysis, the central significance of space is emphasized.* The organization of space is regarded as a purposeful social product. Hence, it cannot be regarded as a separable structure, with rules of construction and transformation independent of social practice. The organization of space is a vital process in capitalist production and reproduction (Lefebvre 1976).

A concrete example of the socio-spatial dialectic in action may help to illustrate our position. Inequality and poverty are endemic to capitalism, and the facts of geography facilitate their reproduction through generations. Central to this thesis is Hagerstrand's notion of a "daily-life environment," composed of residence and/or workplace, and defined by the physical friction of distance plus the social distance of class. Every social group operates within a typical daily "prism," which, for the disadvantaged, closes into a "prison" of space and resources. Deficiencies in the environment (for instance, limitations on mobility, and the quality of social resources) clearly limit an individual's potential, or market capacity; and poverty limits access to more favorable environments. A self-reinforcing process thus sets in. It is easy to understand how an individual can carry an imprint of a given environment, and how the daily-life environment can act to transmit inequality.

The focus of human geography

Human behavior is expressed through a complex set of social, political, and economic processes which characterize every society to some degree. Time and space define the two dimensions of a fabric upon which the processes of human existence are inscribed. Our objective in human geography is to understand the simultaneity of social, political, and economic life in time and space. But exactly how can we conceptualize the processes and patterns of human life on the time-space tapestry?

Human landscapes are created by knowledgeable actors (or agents) operating within a specific social context (or structure). The structure-agency relationship is mediated by a series of institutional arrangements which both enable and constrain action. Hence, three "levels of analysis" can be identified: structures, institutions, and agents. Structures include the long-term, deep-seated social practices which govern daily life, such as law and the family. Institutions represent the phenomenal forms of structures, including, for example, the state apparatus. And agents are those influential individual human actors who determine the precise, observable outcomes of any social interaction.

It is impossible to predict the exact geographical or social outcome of the interactions between structure, institution, and agency. Although individual activities are framed within a particular structural context, they can also transform the context itself. Moreover, any outcome is necessarily a consequence of the reciprocal relationship between relatively long-term structural forces and the shorter-term routine practices of individual human agents. Economic, political, and social history is therefore time-specific in that these relationships evolve at different temporal rates; but it is also place-specific because these relationships unfold in recognizable "locales" according to the variable logics of spatial diffusion.

Interaction through space is further complicated by the different scales over which human activities operate. Geographical regions, or locales, are defined by physical or human boundaries which delimit fields of process and interaction. In general terms, the processes of social life may operate at macro- or micro-level scales. We may expect the structure-institution-agency sequence to be replicated (sometimes in different ways) at each scale. So, for instance, national urban structure may be the result of the interaction between global capital and labor relations; but local neighborhood structure may also be defined by a capital-labor relation operating in significantly different ways at the community level.

Any locale is, therefore, at once a complex synthesis of objects, patterns, and processes derived from the simultaneous interaction of *different levels of social process*, operating at *varying geographical scales and chronological stages*. It is as though a multi-tiered sequence of events had been telescoped into a single dimension; many levels and scales of process are simply collapsed on to a single territory. But this is precisely the intellectual challenge posed by the "geographical puzzle": to unravel the complex locale into its constituent elements and processes (for further discussion of these points, see Dear 1988).

Territory and social life: outline of a theory

The organization of capitalist society

We shall focus, for the sake of simplicity, solely upon capitalist societies. Such a society is characterized essentially by the institution of commodity production for profit, and by a fundamental cleavage between a class of capitalists and a class of workers.[1] We also assume that we are dealing with "advanced" capitalist society (what some may call "late" capitalism). One of the most significant dimensions of such societies is their organizational complexity. Large bureaucracies are now characteristic of multinational firms and of most state organizations. A highly centralized, extended state apparatus is a dominant element in social relations, which tend increasingly toward corporatist-style arrangements. Capitalist production is increasingly organized on a global scale, which has the effect of allowing remarkably extensive, world-wide adjustments to be made in response to the various crises of capitalist accumulation (for instance, the current era of "post-Fordist" production arrangements). However, somewhat paradoxically, the scale of capitalist mega-organizations also permits a myriad of detailed and varied actions at the local level.

Advanced capitalist societies are conceived as having three spheres of social life. In no particular order, these are: (1) an *economic sphere*, which is characterized by industrialism, and (axiomatically) is organized on capitalist principles; (2) a *political sphere*, which is dominated by the state and its efforts at crisis management, social control, and repression, as well as by the whole panoply of institutions of democracy; and (3) a *social sphere*, which is regarded as the domain of civil society, and is identified separately in order to emphasize the set of social and cultural relations which exist outside the realm of the state or production-determined class cleavages. Each of these three spheres operates autonomously to some degree. No single sphere can ever be entirely independent, and there is a high degree of overlap between the three spheres. Hence, each sphere enjoys only a relative autonomy from the others. The effectiveness of any one sphere at a particular time will depend upon specific local conditions, including the power of the human agents associated with each sphere.

In sum, the social organization of advanced capitalist society is predicated upon three relatively autonomous spheres of social life. Specific time-space configurations (including such diverse outcomes as deindustrialization, disease control, or reading habits) will be related to the dynamics of power and practice within and between the three spheres. The point is not to determine which sphere is pre-eminent in the social practices of advanced capitalism, but to realize that all three spheres are necessarily implicated in the process of reproduction.

Social process

A central dynamic in life is the innate human tendency to strive for security and status, and to protect those gains that have already been achieved. This fundamental human tendency toward self-protection is also projected on to those institutions with which humans identify. This can amount to a joint survival strategy, as when employment will last only as long as the firm is successful. In other cases, identification with a particular cause or belief (as embodied in voluntary or religious organizations) can occur. The net effect of these extensions of human identity on to institutional structures is that people vie for the perpetuation of the institutions as much as for their own self-survival.

Social reproduction (with its emphasis on the perpetuation and maintenance of social relations) occurs when humans are unable to transcend their circumstances, be they personal, institutional, structural or environmental. For instance, a family's life chances may be restricted, so children follow their parents into menial occupations and limited horizons. *Social change* occurs when the reproductive cycle is inhibited. Such discontinuities are possible when social practices overcome the strictures of the time-space prism/prison. This can occur in several ways, as in the case of long-term, non-catastrophic evolution of social relations, or through the impact of the unanticipated side-effects of human action. We also identify a category of *transcendental social action* which we shall define as a purposeful change in the structure of social relations, sufficient to alter the practices of reproduction. Transcendental social change occurs through the autonomous actions of classes and groups within the economic, political, and social spheres. The potential for such change is intimately bound up with breaking the repetitive cycles of reproduction. Central to this dynamic is the relative autonomy of the separate spheres of social life, and the power of key human agents within each sphere.

In the economic sphere, human agents will strive to maintain or improve their material status; political groups will strive to retain or advance the legitimacy of their power bases; and social groups will strive for maintained or improved status (according to specifics of gender, ethnicity, etc.). Diminished status, or frustration in the attempt to improve status, are the primary causes of what we shall term material, legitimacy, or status *crises*. For example, a material crisis can occur when a trade union fails in its attempt to win financial concessions for its members, or when a firm is obliged to declare bankruptcy. A legitimacy crisis can occur whenever a constituency withdraws its support from a political group, be it a formal political party or an informal community group. (Even the threat of withdrawal may be sufficient to induce crisis.)[2] Status crises reflect dissatisfactions and deprivations which pertain to such important categories as gender and race.[3] Status crises develop when individuals, groups, or even nation-states are affronted or diminished (as when a superpower is challenged by a terrorist group or insurgent social movements, and whenever discrimination based on sex, race, etc. is practiced). It must be obvious that the genesis of status crises cannot always be clearly separated from the incidence of material and legitimacy crises. The point is to recognize that pressures for social change can derive from separate spheres with identifiably different logics, but that the outcome in one sphere is intimately bound up with the other two.

Spatial process

Social practices are inherently spatial, at every scale and all sites of human behavior. Geography is thus an important element in social reproduction. The relationship between space and human activity is very subtle, and constantly evolving. In abbreviated terms, we recognize three aspects to the socio-spatial dialectic: (1) *in which social relations are constituted through space*, as when opportunities for natural resource exploitation influence arrangements for production (such as extractive or agricultural activities); (2) *in which social relations are constrained by space*, such as the inertia imposed by an obsolete built environment, or the degree to which the physical environment facilitates or hinders human activity (as in the case of natural hazards); and (3) *in which social relations are mediated by space*, as when the general action of the "friction of distance" facilitates the development of a wide variety of social practices, including the patterns of everyday life, or the development of homogeneous belief systems in geographically isolated communities.

The patterns of human territoriality result from the interaction between the economic/political/social spheres and the constitutive/constraining/ mediative roles of space. It is a far from simple task to disentangle the respective contributions of each component of this socio-spatial dialectic, or to predict particular geographic outcomes. This is because the possibilities and limitations on human behavior are established in socially- and spatially-specific circumstances. Outcomes will vary according to the relative autonomy which characterizes each sphere in a given locale. In addition, a single geographical setting can serve multiple functions, simultaneously possessing constitutive, constraining, and mediating dimensions.

Some of the most basic impetuses to human territorial organization stem from the relations of production, in particular from the division of labor. As Scott & Storper argue, the peculiarities of the labor process have historically given rise to distinctive "territorial production complexes' (1986, Ch. 15). However, many other social and political forces reinforce and extend the geographical manifestations of production, and in so doing, set in motion an autonomous set of social relations. Thus, to extend our example, territorial production complexes can give rise to distinct localized communities of workers with particular class- or community-based political consciousness. These socially homogeneous communities may be reinforced in their separateness by common habits in the consumption of material goods or in cultural practices (these are the distributive groupings mentioned above). Such work/political/cultural stereotypes can develop with remarkable speed and flexibility, and quickly transmit their example to other communities. Perhaps the best instance of this kind of "new reproduction space" is the high-tech Sunbelt community (such as Silicon Valley and Orange County in California). With their combinations of high-waged skilled labor force, affluent life-styles, and (generally) conservative politics, such communities are often held up for admiration as the new archetype of capitalist growth. (The large underclass of low-paid workers needed to service these complexes is usually ignored by its advocates.)

A further impetus to the social construction of territory derives from authority relations. These are the structures of command which maintain order in both work and non-work activities. Almost all organizations adopt some form of hierachical structure of authority, characterized by vertical and horizontal arrangements of power. For instance, the capitalist state apparatus is invariably constituted as a single central authority together with many lower-tier local state organizations. Such arrangements facilitate strong central control over spatially extensive and socially heterogeneous jurisdictions. Private businesses follow similar patterns (although often for different reasons) in their structures of head office and branch plants. In these cases, the purpose of the hierarchical organization is to maintain control and power through the effective delegation of authority over a purposefully-partitioned territory.

It follows from this that reproduction is occurring simultaneously over many different scales of human activity. Life within any locale is, however, a seamless fabric of macro- and micro-scale influences, representing the localized interactions of structure, institution, and human agency. Thus, in their daily-life environment, individuals may be concerned to protect the "personal space" around key job and family relationships, as well as to cope with illness through a regionally-organized health system, plus the longer term, indirect effects of federal tax policy or a multinational-induced restructuring of the workplace. These multiple influences are all experienced within the same locale, even though they are frequently initiated from outside. Social outcomes and social process may therefore *appear* to be locale-specific, even if their genesis is exogenous to that locale.

The specificity and uniqueness of territorial organization in each locale makes it possible to speak of a *crisis of the locale*. Geographically-specific crises can occur if a particular combination of economic, political, or social crises is concentrated in a single locale. An obvious case is the severe deindustrialization of a regional economy which then precipitates particular patterns of political struggle and social strife affecting the region as a whole or selected locales at lower spatial scales (i.e., urban neighborhoods). The effects of industrial decline thus become cumulatively greater than those which can be ascribed to plant closure alone. The locale itself has become problematical.

The appearance of material, legitimacy, status, and locale crises may induce consequent adjustments in the factors causing the crisis. Without such adjustments, the economic/political/social stability of a particular locale is at risk, and the process of social reproduction placed in jeopardy. This situation is complicated when separate locales undertake unilateral action to solve their experience of crisis. The gains of one community may, for instance, be achieved at the expense of others. In addition, a preoccupation with crisis in one's immediate locale can obfuscate understanding of the origins of crisis (as when social forces exogenous to the locale are overlooked or imperfectly perceived). This can have profound ramifications, if political activists attempt to apportion blame for particular injustices.

In summary, the social construction of territory is the complex outcome of a socio-spatial dialectic (Fig. 1.1). In broad terms, society is structured by three spheres of activity (political, social, and economic), which are articulated via a series of mechanisms (including the division of labor, authority relations, and distributive groupings), and which tend to be organized hierarchically. The practice of these social relations is necessarily constituted, constrained, and mediated through space, and will vary according to the geographical scale of operation. The combined effects of society and space produce the structure of the locale, which is the complex outcome of evolution through time and space. The influence and actions of structures, institutions, and agents are experienced and implemented through the locale.

Social change and transcendental social action

To a very large extent, reproduction involves a focus on inertia – such as the constraints imposed by the built environment, the transmission of beliefs across generations, or the perpetuation of loyalties toward an

11

Figure 1.1 The social construction of territory.

SOCIETY	SPACE
SPHERES	
Political	
Social	
Economic	CONSTITUTION
	CONSTRAINT
MECHANISMS	MEDIATION
Division of labor	
Authority relations	
Distributive groupings	SCALE
HIERARCHIES	
Vertical	
Horizontal	

↓
LOCALE

Structure/Institution/Agency

institution. But for reproduction to occur, concrete actions have to be undertaken; reproduction is not a spontaneous event. Hence, we must anticipate that social life will be in a constant state of flux. The processes of reproduction will continue: evolutionary change occurs; the unexpected appears. Even when crises develop, sophisticated mechanisms of crisis management can be brought to bear.

Transcendental social action implies a fundamental discontinuity in the process of reproduction. It is brought about by purposefully-instituted strategies to alter the structure of social relations, if not irrevocably then at least for a very long time. Social change and transcendental social action are possible because the economic, social, and political spheres enjoy varying degrees of relative autonomy. This means that their agents have some power to *initiate* actions independently, and also possess some degree of *immunity* from interference by outside agencies in those actions. Power is distributed imperfectly, almost haphazardly, throughout the three spheres of social practice. Hence, control by any one group will be imperfect and fissures in the reproduction process will open. These fissures can be exploited, control mechanisms perverted, and social practices altered through the exercise of what Foucault calls "micropowers." Autonomy is a flexible concept, with ambiguous boundaries which are

12

always being tested by human actions. Thus the potential for interrupting the repetitions of the reproduction cycle will always be present.

Whenever human actions result in transcendental social change, it will be largely because the ambiguities of power and autonomy in the structure of society have been successfully exploited. These ambiguities are, however, a double-edged sword. They make it easier for powerful individuals and institutions (such as the state) to penetrate and exploit the niches of micropowers in order to affect or inhibit social change. At the same time, they also render it difficult to launch direct attacks on the amorphous macro-structures of power, and they permit more powerful human agents to conceal their efforts to control the practices of everyday life.

Social change and transcendental social action would not be possible if *locales* did not possess some degree of relative autonomy. The structures, institutions, and agents in locales enjoy a degree of relative autonomy from other adjacent, superior, or subordinate locales. For instance, a local state might act unilaterally within its own jurisdiction, and choose to ignore the deleterious effects of its action on a neighboring municipality. Alternatively, a coalition of local capitalists may persuade their municipality to introduce bylaws to offset punitive central state policies, or they may even flaunt those policies. In these ways, micropowers within locales can be manipulated to great effect by knowledgeable individuals, and the autonomy granted by geography is an important factor in the efficacy of their strategies.

We can never be certain about the precise social and geographical outcomes of diverse reproduction processes. Actions aimed at social change can fail. The processes of social reproduction may then continue unaltered. Even then, however, the outcome is unpredictable, because the potential for evolutionary change persists, and because of the presence of unanticipated side-effects. Moreover, people's tolerance for deprivation, poverty, and discrimination is often very high. Even the gravest crisis may fail to provoke action, and the most strongly-felt injustice may survive centuries before it is removed. We simply cannot predict when transcendental social action will be attempted. At some stage, the provocation becomes too great, but even then, the precise chain of events is often ambiguous and outcomes uncertain. For instance, the contemporary black civil rights movement in the US is generally dated from the day that one woman refused to give up her bus seat to a white person in Montgomery, Alabama. However, many such personal protests had occurred previously in Montgomery and elsewhere in the South. Exactly why this particular incident provided the spark remains unclear. But (consistent with our approach) social conditions *at that time and at that place* were sufficient to launch an irrevocable movement for civil liberty.

13

Territory and reproduction in contemporary capitalist society

The pace of social and spatial change has perhaps never been more rapid. The critical dynamics affecting socio-spatial reproduction in contemporary capitalist society involve all forms of social process and locales at all spatial scales. The results are increasing complexity and diversity of social and spatial relations, and a growing diffusion of micropowers amongst economic classes, state agents, and status groupings. The chapters in this book take up the challenge of deciphering this complexity. In what follows, we briefly account for changing patterns of contemporary socio-spatial process, indicating where contributions to the volume illuminate our understanding of territory and social reproduction.

Relations of production have been fundamentally transformed from the high-Fordist period of the 1950s. In the wake of deindustrialization, and the rise of service and high-technology industries, new systems of organizing production have developed to enhance profitability in increasingly competitive global markets. A prominent characteristic of "post-Fordist" production organization is the rise of flexible production arrangements in many industrial sectors and geographic locales. These regimes have a distinct spatial organization, and often operate as distinctive, regional craft communities. New mechanisms of social regulation and control appear to be implicated in the reproduction of these flexible production complexes (see Ch. 2).

Shifting production arrangements are having a profound impact on the everyday lives of workers. Class relations are being restructured as the rates of labor force casualization, marginalization, and peripheralization expand, particularly among (traditionally male) manufacturing workers, and as new groups such as immigrant labor and part-time women workers emerge. Gender relations related to work are also shifting, as job opportunities and labor force participation rates among men and women change. These new class and gender relations are being defined differently in specific geographic locales. For instance, economic structures, local state arrangements for social reproduction, and the spatial organization of residential and employment centers either facilitate or constrain new forms of class consciousness, formation, and behavior (Ch. 5). In many locales, spatial labor markets have developed to take advantage of the fact that many women workers are relatively immobile, both in terms of residential mobility (since they are part of a male-headed household), and access to urban transportation services. Women are much more likely than men to be "captive riders" on public transit, and are apt to have child-care responsibilities. These make employment opportunities closer to home more attractive, so that women constitute a captive labor pool for employment in lower-paid and often sex-segregated occupations (Ch. 7). This reinforces male domination in the workplace and home, and limits economic mobility chances for women.

In response to deep-seated industrial restructuring and autonomous evolution in the political sphere, the capitalist state has undergone dramatic expansion and reorganization. Driven by destabilized geopolitical circumstances, increasing proportions of nation-state budgets have been channeled into defense, leading to the emergence of the "warfare state." Just at a time when industrial restructuring has created massive labor dislocations and hence need for welfare services, the demands of the warfare state have forced reorganization and retrenchment in the welfare state apparatuses. The state's role in social reproduction still occurs largely within the realm of collective consumption, but specific aspects of state provision have changed due to industrial restructuring (Ch. 3). For example, rationalization of production and associated lay-offs have altered the politics of local government, causing resistance to central state policies of welfare cutback and a re-allocation of resources to more vocal class fractions within the community. Furthermore, the many forms of retrenchment of welfare state services have increased the extent to which self-reliance and self-provisioning have replaced some types of state-provided services.

The restructured rules, regulations, and surveillance practices of the contemporary state penetrate evermore deeply into the interstices of everyday life. This has come about, for example, through differential interpretation of central state law by the judiciary and local enforcement agencies, which attempt to meet requirements for uniform treatment while allowing for highly divergent local circumstances (Ch. 8). In addition, state penetration has occurred via the expanded use of alternative mechanisms for service provision, including privatization, and corporatist strategies for coopting the voluntary, nonprofit and cooperative organizations which together constitute the "shadow state" (Ch. 9). Privatization and the use of low-paid voluntary sector workers and volunteers in the social service sector have altered the characteristics of service provision for consumers, and dramatically affected the bargaining power of welfare state workers, who are often unionized. In the case of nonprofit and cooperative housing providers, dependence on state funding has served to integrate coops more closely into the commodified housing delivery within the state's housing resource allocation priorities (Ch. 10).

Welfare state practices also hold possibilities for transcendental social change, however. In the case of legal statutes, the locale may reinterpret central state directives in order to gain autonomy. The provision of decommodified, cooperative housing services has increased as struggles for higher state funding allocations have met with partial success. In the case of service-dependent populations, privatization and voluntarism have accelerated the deinstitutionalization process, bringing dependent populations a mixed blessing. They enjoy greater autonomy from state controls, yet paradoxically endure a greater reliance on the resources of private sector landlords and nonprofit care providers, and on the social networks of isolated and segregated urban environments (Ch. 11). The spatial site of reproduction of dependent populations has thus shifted from 19th-century asylums which were characterized by a diversity of internal spatial arrangements for the treatment and control of dependent persons (Ch. 12).

15

But space nevertheless remains a fundamental element both in the perpetuation of client and service-provider relations, and in creating opportunities for greater client autonomy.

The precise impacts of economic and social restructuring are clearly contingent upon the circumstances of individual locales, but there are many commonalities of social practice and spatial process. One dominant effect at the urban scale has been the displacement of class-based struggles to conflicts around consumption, status, and family (Ch. 4). The rise of turf or territorial politics has occurred as the long-term investment of both capital and labor in their communities has caused disputes over the distribution of local collective consumption resources. These disputes reflect either class-based divisions, or competition among distributive groupings. In the first instance, business interests (typically aided by the local state) attempt to channel funds into infrastructural and other investments intended to enhance the business climate and site values, thus augmenting their accumulation chances. Neighborhood groups, on the other hand, favor spending on human and urban services designed to improve community value (part of labor's "consumption fund"). In the second case, status groups located in different urban neighborhoods wage a defensive turf politics designed to capture a larger share of collective consumption resources, and thus protect and enhance their status and life chances. Typically, affluent residents in urban communities battle the influx of any land uses seen as threatening, including low-cost housing and facilities for service-dependent populations such as the mentally disabled (Ch. 14).

The status groupings involved in turf politics are diverse, and they have complicated relationships to class categories. For instance, homeowners are one status group frequently assumed (by virtue of their ownership status) to be incorporated into the capitalist social order. However, their position on territorial issues may vary according to their degree of mortgage debt, the local housing market investment potential, or their working-class identity (Ch. 13). Thus, status cleavages will form and become involved in turf politics in ways that are separable from conventional class alignments and which may be differentially expressed in various geographical contexts.

Economic and welfare state restructuring at the level of the locale has generated spatial changes which underpin many battles over territory. For example, deindustrialization and reindustrialization under post-Fordist production relations have led to a bifurcation of income distribution among urban workers. This is manifest in: (1) the creation of low-wage service job opportunities and high-wage professional and managerial jobs in central business districts; and (2) a decline in traditional blue-collar manufacturing positions paying middle-income wages. Both trends are contributing to the reorganization of inner-city neighborhoods. These are being gentrified by high-income households and converted to nonresidential uses as the demand for central land rises, and as capital becomes available for construction of high-rise offices and related shopping, hotel, and entertainment facilities. Such neighborhood upgrading has exacerbated a low-cost housing shortage, which was already severe as a result of declining real incomes for many economically dislocated households, as well as state retrenchment of human services and housing programs. For growing numbers of the most marginalized households, the result is homelessness.

Status groups impacted by social change have sought to capture new territory, defend their turf, and in some instances, to gain a greater degree of local autonomy. The strategies employed in these turf struggles include assistance from professional advocates, grassroots mobilization, and heightened demands on the local state. For instance, one impoverished black neighborhood in Houston, under intense pressure from central business district expansion, has relied on environmental design professionals committed to aid them. But they have steadily lost ground. A local capital/local state growth coalition, using central state transportation and urban renewal programs, has battered the community. Moreover, the economic dependency of design professions on state and corporate clients, plus their project-length time horizons and the locality-specific nature of their work tend to isolate them from the community residents. These factors pose fundamental barriers to the delivery of effective advocacy services for the community (Ch. 15).

In some urban locales, the housing squeeze has led to political demands for rent control and affordable housing. Failure by the local state to meet these demands has triggered new grassroots political movements geared toward increasing local self-determination. Using a "capture the state" strategy, gays and seniors in West Hollywood built a coalition around rent control and local autonomy. They succeeded in creating a local state apparatus designed to serve their needs (Ch. 16). In other instances, status groups have used economic advantage, collective effort, and pressure on the local state to develop new neighborhoods designed to serve their specific needs. In Montréal, economic restructuring has led to increasing employment of women professionals in downtown government and service industries. Many are financially-insecure, single heads of household with extensive child-care responsibilities. These professionals have occupied certain lower-cost inner-city residential districts. There they have been able to meet their needs for accessible housing and other services (child-care, for instance), and to develop alternatives to domestic labor in a traditional male-headed household (Ch. 6).

Some status groups are noticeably uninvolved in struggles over territory. Their lack of defined – and hence defensible – space for collective consumption mirrors their lack of cohesion as a distinct group. In prior years, gays fell into this category. For the physically disabled, the social reproduction process is anchored in a pervasive, culturally-defined conception of acceptable bodily images (Ch. 17). A profoundly negative image of the disabled is promoted by the prevailing norms of beauty, centered predominantly on a narcissistic athleticism emphasized in the media. For the disabled, opportunities for social change are likely to emerge only when they oppose such norms. Also needed are alternative perspectives on bodily image which facilitate an acceptance of disability, and a greater willingness to organize political movements for a greater share of social resources and for more accessible physical environments. In many ways, the continuing plight of the physically disabled provides an archetypal account of the problems facing any group wishing to break the mold of the socio-spatial reproduction process.

Notes

We are grateful for helpful comments by Robin Law, Allen Scott and Michael Storper on an earlier draft of this chapter.
1 Social relations in capitalism are, of course, much more complex than a simple binary structure of opposing classes. However, it is not our purpose to explain here the fundamentals of the capitalist system, although we shall take this opportunity to stress *territory* as one significant source of complexity in capitalist social relations.
2 Note that the category of material crisis subsumes the conventional notion of accumulation crisis, and that the legitimation crisis of the state is a special case of the broader legitimacy crisis.
3 In the concept of status crisis, we are building on traditional Weberian notions of social class, which emphasize the complexities of class structuration beyond economic categories. We incorporate such important categories as race, gender, and ethnicity in this sphere.

References

Althusser, L. & E. Balibar 1970. *Reading capital*. London: New Left Books.
Castells, M. 1976. Theory and ideology in urban sociology. In *Urban sociology: critical essays*, C. G. Pickvance (ed.). London: Tavistock.
Castells, M. 1977. *The urban question: a Marxist approach*. London: Edward Arnold.
Dear, M. 1988. The postmodern challenge: reconstructing human geography. *Transactions of the Institute of British Geographers* N.S. **13**, 1–13.
Giddens, A. 1973. *The class structure of the advanced societies*. London: Hutchinson University Library.
Lefebvre, H. 1976. Reflections on the politics of space. *Antipode* **8**, 30–7.
Marx, K. (1971). *Capital*. Moscow: Progress Publishers. vols 1 and 3 (originally published 1867).
Scott, A. J. & M. Storper 1986. Industrial change and territorial organization. In *Production, work and territory: the geographical anatomy of industrial capitalism*. A. J. Scott & M. Storper (eds). London: Allen & Unwin.

Part II

Industrialism, the state, and civil society

2

The geographical foundations and social regulation of flexible production complexes

MICHAEL STORPER & ALLEN J. SCOTT

The turning point

In all of the advanced capitalist countries, problems of deindustrialization have commanded considerable attention for some time now. Since the late 1960s, in many of the core industrial cities and regions of Britain, France, Italy, West Germany and the USA, there have been steep declines in output and employment in the mass production industries that served as the propulsive engines of economic growth in the post-war period. These declines accelerated in the early 1980s and have now become visible even in Japan. As production was restructured over this period (by plant closure, technical rationalization or decentralization) the resulting business cutbacks and job losses in core regions frequently engendered deep fiscal crises of local government and much community distress. In the face of these problems, a large literature on deindustrialization and its local effects in both the US and Western Europe has come into being (Bluestone & Harrison 1982, Massey & Meegan 1984, Markusen 1985, Stoffaës 1978).

At the very time when the deindustrialization of core regions was reaching its peak in the late 1970s and early 1980s, however, a number of other areas were experiencing rapid growth of industrial output and employment. This growth, moreover, has in many cases been posited on the development of alternative ways of organizing production systems and local labor markets. Indeed, it now seems that a new, hegemonic model of industrialization, urbanization, and regional development has been making its historical appearance in the US and Western Europe, based on a four-fold shift from the model that dominated over much of the post-war period. The shift may be characterized in the following terms:

(a) The *central sectors* of the production system are no longer focused to the same degree as previously on the production of consumer durables and associated capital goods. They consist, rather, of ensembles of flexible production sectors such as (i) selected high technology industries, (ii) revitalized craft specialty production, and (iii) producer and financial services. These ensembles are becoming increasingly central in the sense that they account for a steadily rising share of employment and output

21

growth in the North American and Western European economies.

(b) In all these sectors, *flexible production methods* constitute a basic principle of organization in contrast with the Fordist mass production methods of the preceding hegemonic model (cf. Piore & Sabel 1984). Note, however, that despite the recent tendency of the Fordist model to break down, it has far from disappeared altogether and it remains still a significant way of organizing production in many sectors.

(c) The *geographical foundations* of industrial growth have been shifting, in some cases quite radically. Many flexible production sectors have been locating in places that are often far removed from the old centers of Fordist mass production. At these places, moreover, there has been a definite re-creation of the process of spatial agglomeration. We might say, indeed, that a new set of core industrial regions is beginning to take shape on the landscape of contemporary capitalism.

(d) Every dominant model of industrialization is associated with a distinctive set of broader social institutions. In the post-war period, for example, mass production was coupled to Keynesian welfare-statist institutions designed to stimulate mass consumption and to maintain broad social order. As such, these institutions helped to regulate macroeconomic accumulation over the long-run. The transition to flexible production is beginning to bring with it wide ranging changes in social organization in North America and Western Europe. New forms of *collective social and institutional order* are now appearing in the specific locales of flexible production, as well as in the body politic at large.

Our purpose in this chapter is to explore in some detail these four dimensions of the turn from Fordism to flexible production organization, and to delineate some of the principal socio-spatial underpinnings of the latter system.

Two technological-institutional models of production

Capitalist production apparatuses may assume many alternative technological and institutional configurations, ranging, for example, from the early putting out system through classical factory methods to Fordist mass production, and so on. Each particular configuration consists in a historically determinate technological-institutional model of production comprising a web of production techniques, employment relations, methods of organizing the intra- and inter-firm division of labor, managerial and entrepreneurial relations, and so on. Each such model is also roughly equivalent to what theorists of the French Regulationist School call a *regime of accumulation* and a *mode of social regulation* (cf. Aglietta 1976, Boyer 1986, Lipietz 1986), i.e., a structure governing the production, appropriation and reinvestment of the economic surplus. Two contrasting technological-institutional models of production are of particular relevance in North America and Western Europe at the present time, namely, the aging model of Fordist mass production and an ascending model of flexible production organization. We shall briefly describe the principal characteristics of these two technological-institutional models before going on to a discussion of the dynamics of the transition that seems currently to be occurring from the one to the other.

Fordist mass production

The Fordist model of production was decisively installed in the major capitalist economies in the inter-war years. It flourished through a period of high Fordism corresponding more or less to the long post-war boom after which, in the 1970s and 1980s, it entered into an extended period of crisis, restructuring, and spatial reorganization that is still far from over. Fordism was based essentially on mass production, though we should keep in mind that other species of productive activity (such as small-scale batch production and skilled artisanal industry) continued to prosper throughout the Fordist period. In its classical guise, Fordism was underpinned by large and highly capitalized units of production consisting of either (a) continuous flow processes, as in the cases of petrochemicals or steel production, or (b) assembly line processes (and deep technical divisions of labor), as in the cases of cars, electrical appliances or machinery. Production was geared to an insistent search for internal economies of scale via increasing standardization of outputs, routinization of processes, and rigidly dedicated capital equipment. There was, as a consequence, a tendency for both physical output per plant and productivity per worker to rise steadily over time. Major plants in leading sectors were, and are, located at the center of networks of upstream producers providing necessary physical inputs and services, sometimes on the basis of non-Fordist artisanal labor processes. As selected production activities within the Fordist system reached technological maturity, they tended to become embodied in branch plants and then to decentralize to cheap labor sites in national and international peripheries.

These physical features of Fordism were matched in institutional terms by the emergence of oligopolistic multi-establishment corporations and big industry-wide labor unions. A basic division of labor also appeared between white-collar managerial workers who planned and directed production, and blue-collar manual workers caught up in a deepening dynamic of task fragmentation and deskilling. Through their union representatives, blue-collar workers secured contractual codification of job categories, seniority rules, and productivity-based wage-setting practices. The rigidities brought into being by these contractual arrangements were then typically matched by countervailing lay-off and recall processes as the mechanism by which producers attempted to adjust their demand for labor to changes in the economic cycle (rather than adjustment by means of wage rate variations or the re-allocation of labor on the shop floor).

The entire Fordist system tended generally to expand over time, despite cyclical downswings and periodic economic depressions. In the effort to regulate the system and to stabilize its overall growth path, a series of governmental measures inspired by Keynesian welfare-statist conceptions of economic and social control made their appearance after the 1930s in all

23

the major capitalist nations. These measures differed in detail from country to country, but they were all nonetheless generally (and at first successfully) directed to attempts (a) to mitigate recessionary conditions and rising unemployment by means of deficit spending and subsidized consumption as well as (b) to maintain a social and industrial peace by means of redistributive legislation in matters of unemployment insurance, housing, public welfare, and so on.

Flexible production and accumulation

In spite of the dominating position of Fordist mass production over much of the 20th century, more flexible patterns of industrial organization were – as we have already suggested – always present, and even preceded the historical emergence of Fordism (cf. Sabel & Zeitlin 1985). However, the specific forms of flexible production that are now shifting into an increasingly central position in the advanced capitalist economies have their origins in the period following World War 2, and have grown to major economic prominence only since about the end of the 1960s.

When we speak of flexible production systems we refer to forms of production characterized by a well developed ability both to shift promptly from one process and/or product configuration to another, and to adjust quantities of output rapidly up or down over the short run without any strongly deleterious effects on levels of efficiency. Both of these types of flexibility are achieved through a variety of intersecting strategies. Within the firm, flexibility may be attained through the use of general purpose, non-dedicated equipment and machinery (often programable) and/or craft labor processes. There is also a tendency within the flexible firm for job descriptions to break down into a restricted number of broadly-ranging categories, and for concomitant extensions to occur in the redeployability of labor on the shop floor. In the domain of inter-firm relations, flexibility is achieved by extensions of the social division of labor facilitating rapid changes in combinations of vertical and horizontal linkage between producers, thus leading to intensification of external economies of scale in the production system as a whole. In addition, the labor markets associated with flexible production systems tend to be typified by high rates of turnover, and by the proliferation of part-time and temporary work as well as homework.

As a consequence of these proclivities, individual units of production in flexible production systems are usually less specialized and smaller in size than mass production units. They are technologically capable of achieving great flexibility of production within their own spheres of operation and, at the same time, this flexibility is multiplied by the system effects of the social division of labor, which permits the formation and re-formation of interdependent combinations of producers. Product differentiation increases as a result, and markets become increasingly competitive. Flexibility is yet further enhanced by a certain tendency for a super-structural tier of firms to emerge involving specialists in system coordinating and marketing functions, like the celebrated *impannatori* of Prato in Italy (Becattini 1987).

The rise of the new flexible technological-institutional model of production has coincidentally been accompanied in the majority of the advanced capitalist countries by the electoral success of governments committed in varying degrees to attempts to dismantle the apparatus of Keynesian welfare-statism. Backed up by resurgent neoconservative ideologies, these governments are attempting to install new policies putatively designed to reinforce economic competition, entrepreneurial-ism, privatization, and self-reliance.

The historical geography of the transition from Fordism to flexibility

Industrial organization and location

Fordism emerged on the geographical foundations of a series of great manufacturing regions based on mass production, including the Manu-facturing Belt of the USA and the industrial region stretching from the English Midlands through Northern France and Belgium to the Ruhr, with outliers in such places as the Industrial Triangle of northern Italy and parts of Sweden and the Netherlands. In the two decades following World War 2, all these regions developed an economic geography which we now recognize as being typical of Fordism. The production units of propulsive lead firms were surrounded by complexes of input and service providers. There was a concomitant spatial massing of industries and of the workers (many of them unionized) employed in them. The net result of this massing was the formation of great industrial regions consisting of large metropolitan areas surrounded by networks of smaller industrial cities. A series of peripheral spaces could be identified around these core regions. In the early post-war period, there was large-scale labor migration from these peripheries to the core industrial regions of Fordist production, as in the northward movement of blacks from the American South to the cities of the Manufacturing Belt, or the migration of southern Italians to the Milan/Turin/Genoa region. In-migration of labor to core regions was complemented by a steadily accelerating and widening decentralization of capital – in the form of branch plants – from the core regions to peripheries.

By the late 1960s and eary 1970s, the hegemony of this model of (Fordist) industrialization was increasingly threatened. Internally, market saturation and thorough-going spatial decentralization in response to worker militancy were creating high levels of unemployment and productivity slowdowns in the core. Externally, competition from Japan and the newly industrializing countries was cutting dramatically into domestic markets in North America and Western Europe. The net result was intensified industrial restructuring and rationalization in the core, leading to more plant closures, more decentralization, and greater unemployment. In this context of economic stagnation, declining productivity and foreign competition, the fiscal bases of Keynesian welfare-statism began to crumble, as manifest in the prolonged stagflationary crisis of the 1970s. The oil shocks added to these woes. By the end of the 1970s, the Fordist model of industrialization in the core regions, and its associated macroeconomic and political arrangements, were permanently in disarray. The emerging structures of flexible production helped to intensify the crisis of Fordism by exerting strong competitive pressures on mass production industries. At the same time, the advent of flexible production organization was potentiated by the problems of Fordist industry, for significant labor resources were released into the economy, and the political power of unions to impose rigid work rules and high wage levels on new industries was weakened.

The recent development of flexible production is particularly evident in the three rapidly growing ensembles of sectors mentioned in our introductory section, to repeat: (a) high technology industry, (b) revitalized craft production, and (c) services. Even during the extended period of economic crisis in the 1970s and early 1980s, many of these sectors expanded rapidly. In addition, many sectors once dominated by Fordist mass production methods, e.g., cars and components, have now become, or are becoming, increasingly reorganized on the basis of flexible production methods and labor relations. This, of course, has long been the case in Japan, but it is also to an increasing degree characteristic of North American and Western European industry. All of these ensembles are marked by deepening social divisions of labor manifest in widening networks of specialized producers in which small firms are disproportionately represented. Two extreme forms of organization of these networks are: (a) the now classical Japanese *kanban* system with its many-tiered hierarchy of subcontractors, and (b) a diffuse labyrinth or complex of producers interconnected with one another through constantly changing linkages, as in much central business district office activity or specialty electronics production (cf. Scott & Angel 1987, Uekesa 1987). In both cases, what is observable is a progressive externalization of production, thus enhancing the potentiality of system readjustment and engendering in the process significant external economies of scale.

Because of this tendency to externalization of the transactional structures of production, selected sets of producers with especially dense interlinkages have a tendency to agglomerate locationally (Scott 1988, Storper & Christopherson 1987). This locational strategy enables them to reduce the spatially-dependent costs of external transactions. In flexible production systems, the tendency to agglomeration is reinforced not only by externalization but also by intensified re-transacting, just-in-time processing, idiosyncratic and variable forms of inter-unit transacting, and the proliferation of many small-scale linkages with high unit costs. Production flexibility and locational agglomeration are further consolidated by the higher levels of fluidity and readjustment made possible by the pooling of labor demands and supplies (Scott 1988). This encourages high rates of turnover, reduces the costs of labor market search, and streamlines the subtle but important tasks of matching specialized workers to specialized tasks. *Accordingly, the turn towards flexibility has been marked by a decisive re-agglomeration of production and the resurgence of the phenomenon of the industrial district* (cf. Becattini 1987, Bellandi 1986). These agglomerations, it must be pointed out, are by no means self-contained, but are also situated within wider, indeed expanding, regional, national, and international divisions of labor. Nonetheless, the emergence of flexible production has belied earlier predictions of the demise of locationally dense production systems due to the development of advanced communications technologies. Indeed, there is no end in principle to the division of labor in capitalism and hence no end to recentralization and agglomeration. Rather, as we have learned from recent history, advanced communications technologies encourage the appearance of new specialized production activities, which themselves then frequently cluster together in geographical space.

New industrial spaces

We have claimed that flexible production is, to a high degree, accompanied by agglomeration. This claim leaves open the question as to where precisely this agglomerated industrial growth is occurring. We may note at once, however, that it tends to come about in geographical contexts that are insulated from older foci of Fordist mass production. These foci constitute difficult environments for flexible production, on a number of counts. Above all, their extended historical experience of industrial production and work tends to have as its correlate the rigidification of labor relations and the unionization of large segments of the working class. Some of the consequences of this are, in the workplace, ossified work rules and job demarcation, limitations on managerial discretion, high wages and fringe benefits, and, in the community at large, the ability of workers to impose additional costs and restrictions on producers through local government planning and legislation (Clark 1986). Efficient flexible production, however, is dependent upon an ability rapidly to adjust process and product configurations and levels of output and hence upon an ability to redeploy workers within the firm, to adjust labor demands, and to experiment with new technologies and forms of organization.

Thus, as flexible production began its historical ascent in North America and Western Europe, it tended to flourish most actively at places where the social conditions built up in Fordist industrial regions either could be avoided or were not present. Moreover, the peculiar kinds of agglomeration economies available in old centers of Fordist industry – linkage into mass production complexes and access to their attendant labor skills – had little attraction to firms in what would ultimately become the dominant ensembles of flexible production. Indeed, in their early phases of development, these firms were as yet relatively free from dependence on external economies; in other words, they enjoyed a "window of locational opportunity," meaning that for a time at least they were comparatively free to locate in a diversity of geographical environments (Scott & Storper 1987). To illustrate, high technology industry tended to make its appearance at new suburban extensions of larger metropolitan regions, and in previously unindustrialized communities often far from former heartland regions; design-intensive industries grew in some cases on the basis of renaissant craft communities and in other cases were established at entirely new locations; and new service sectors initially gravitated to pre-existing central business district areas where their dynamic of development greatly intensified agglomeration economies over the 1960s, 1970s, and 1980s.

Let us enlarge upon these remarks. New high technology growth centers are to be found scattered across North America and Western Europe at such locations as Silicon Valley, Orange County, Route 128, Dallas–Fort Worth, the Cambridge/Reading/Bristol axis, the Scientific City of the southern Paris region, Grenoble, Toulouse, Sophia-Antipolis, southern Bavaria, and so on. Artisanal or design-intensive industry is currently highly developed in the Third Italy, where it occurs in specialized geographical clusters such as Prato (woolen textiles), Bologna (machinery), Carpi (knitwear), Sassuolo (ceramics), Arezzo (gold jewelry), and so on; further examples of this phenomenon can be found in France, Germany, Spain, Denmark, and even in parts of the USA (e.g., the clothing industry in New York and Los Angeles, motion pictures in Los Angeles). Lastly, major agglomerations of producer, commercial, and financial service activity are to be found in New York, London (especially after the so-called "Big Bang"), Tokyo and other global cities, together with dependent back offices and more recently-developed business agglomerations in parts of their suburban fringe areas. These outcomes have had two major geographical consequences. First, pre-existing clusters of flexible producers (in central business district areas or in traditional craft communities) have been revitalized and have grown very rapidly of late. Second, the geographical margins of industrialization have been pushed outwards, and a series of new industrial spaces has come into being in the various Sunbelts, "third development zones," and suburban peripheries of the advanced capitalist countries. In all of these production locales, the social and/or geographical distance from the old foci of mass production is great. Furthermore, as the various growth centers based on flexible production organization have acquired increasingly elaborate agglomeration economies, so they have become self-confirming foci of accumulation, and the window of locational opportunity mentioned earlier has tended to close around their powerful centripetal attractions.

The problem of social and institutional order in the new industrial spaces

Flexible production systems are, as indicated, often rooted in functionally-polarized agglomerations situated within a wider national and international system of markets and exchange. These agglomerations are composed of collectives of interdependent producers whose peculiar transactional interrelations induce them to converge locationally upon their own common center of gravity. Dense local labor markets form around these centers and boost their centripetal attractiveness. Workers in their turn are housed in adjacent residential communities.

All of these intricate events are mediated in part through market mechanisms and processes of individual locational decision-making and behavior. Markets, however, can only function effectively in the context of an institutionalized social order with definite regulatory capacities. In agglomerated flexible production complexes this problem of social regulation involves major questions about (a) the coordination of inter-firm transactions and the dynamics of entrepreneurial activity, (b) the organization of local labor markets and the social reproduction of workers, and (c) the dynamics of community formation and social reproduction. Each of these questions merits detailed attention in its own right.

The coordination of inter-firm transactions and the dynamics of entrepreneurial activity

We have learned that intensifying flexibility of production arrangements is commonly accompanied by and dependent upon vertical disintegration, leading as a consequence to spatial agglomeration. However, if agglomeration helps to resolve some of the predicaments that arise from increased externalization of production activities (above all by reducing the spatial costs of inter-firm transacting), there are additional problems that it leaves untouched and yet others that it actively creates. Many of these problems have their origins in the peculiar structure of institutional separation and interdependence that is engendered by an advanced social division of labor. Two major manifestations of these problems are of particular importance in flexible production complexes: one concerns the need to reconcile vertical disintegration with selective re-internalization (or quasi-internalization) of functions; the other concerns the problem of impacted information as described by Williamson (1975).

The first of these two cases occurs where production systems are evolving generally in the direction of increased vertical disintegration but where some producers nevertheless find it mutually beneficial to dedicate a portion of their production capability or input demands to one another. This phenomenon can take the form of production equipment that is specialized to a downstream firm's needs; or it can take the form of agreements about reliable precisely-timed deliveries without the need for constant re-contracting. Modern just-in-time systems may embrace elements of both of these kinds of relationships. Typically, such relationships can be seen as being part of a wider system of strategic alliances between producers. However, because these alliances tend frequently to involve producers in the hazards of monopolistic or monopsonistic relations with one another, they are often embedded in contractual governance structures that limit the ability of any member of the alliance unilaterally to impose additional costs on other members.

The second case arises where information is irremediably distributed in an unequal way so that one party to a given transaction is able systematically to gain at the expense of the other(s). This kind of information impactedness is evident, for example, where a subcontractor profits opportunistically by using inferior materials and where the putting out firm cannot easily control quality. Or it may occur where R.&D. services are contracted out, and the subcontract firm then makes additional discoveries that can be commercialized on its own account. In normal circumstances, problems of these sorts would tend to induce some vertical reintegration of production. However, in established place-bound business communities, entrepreneurs tend over time to learn collectively about one another's habits and capabilities, and this learning may substitute in part for hierarchical control of transactions through vertical reintegration. In brief, communities of trust and the social construction of unwritten business norms are important foundations for the maintenance of an effective social division of labor (cf. Brusco & Sabel 1981). Such communities are sustainable by reason of the costs that any defector would have to pay as a result of diminished reputation and consequent loss of business. Recent studies of traditional European craft communities have pointed to the particularly significant role that trust relations play in localized agglomerations of producers (cf. Ganne 1983, Raveyre & Saglio 1984). Trust and personal experience are also important pre-conditions of much re-contracting behavior in modern business complexes, as Storper & Christopherson (1987) have recently shown for the case of the motion picture industry of Los Angeles. In addition, these qualitative attributes of business complexes are often complemented by the formation of voluntary associations with powers of supervision and sanction (Brusco 1986).

If transactional communities thrive on informal and formal trust relations, they also frequently depend for their success on system-wide coordinating institutions. The latter may be profit-making or nonprofit-making agencies, though in either case they will tend to have wide responsibilities for the orchestration of inter-unit production activities and the distribution of outputs on national and international markets. Many examples of profit-making coordinating agencies are provided by the merchant organizations of the Third Italy, such as the Benetton Corporation or the *impannatori* of Prato, or by the new-style film production companies of Hollywood with their extensive putting out relations. Rather dramatic examples of nonprofit-making coordinating agencies are furnished by the Regional Centers for Innovation and Technology Transfer (CRITTs) recently put into place by the French government in the attempt to bring research workers and high technology entrepreneurs together in particular industrial localities for the purposes of information exchange and process/product development (OECD 1986a). Moreover, some industrial communities – for example Herning in Denmark and Prato in Italy – are now beginning to experiment with attempts to facilitate business transactions by means of telematic networks linking different producers together through a central computerized information system. All of these coordinating agencies tend further to encourage the social division of labor and spatial agglomeration by facilitating the flow of information through business complexes and out to wider market locations and back again. That said, it may well be that the advent of telematic information systems may loosen the bonds of mutual locational attraction between some kinds of producers while, through expansion of the division of labor, inducing intensified agglomeration between others.

The shifting networks of transactional interrelations in agglomerated business complexes are at once the medium and the outcome of entrepreneurial activity. These networks define a multiplicity of actual and potential market niches and commercial openings. At the same time, the exploitation of these structured opportunities by entrepreneurs and would-be entrepreneurs engenders significant change in the organizational structure of production, thus giving rise to new market niches and commercial openings and hence to new entrepreneurial possibilities. In this sense, entrepreneurship is not, as behavioristic theories suggest, a purely atomistic phenomenon, but is rather a collectively-defined activity, dependent in both substance and form upon the existence of a business milieu with its system of structured opportunities. Further, in the process of transacting and re-transacting within industrial networks, entrepreneurs invariably learn in practice about the manifold facets of production and work in the local complex. The end result is the constitution of a place-bound business culture in which practical forms of knowledge are socially reproduced and individual sensibilities about production processes, labor skills, materials, product design, markets, and so on are finely honed. In many cases, the socialization of useful local knowledge is carried yet further by the public provision of educational services tailored to the needs of the adjacent production system. These social mechanisms for the maintenance and transfer of useful knowledge, in conjunction with the circumstance that innumerable practical problems and questions are being constantly generated out of the transactional interaction between producers, constitute the basis of ever-active processes of informal technical change and innovation. Thus, Russo (1985) has shown that much technical innovation in specialized tile production equipment in the Sassuolo ceramics district of the Third Italy is generated out of the transactional interrelations between tile producers with their particular needs, and equipment producers with their specialized capabilities. These processes of innovation give rise to further qualitative changes in the structure of the division of labor and its associated transactional networks. Varieties of these phenomena of coordination and entrepreneurial activity can be found in virtually all localized production complexes no matter whether they are founded on high technology industry, craft-based or design-intensive production, or services.

31

The organization of labor markets in localized production complexes and the habituation of workers

Various forms of labor market flexibility, both internal and external to the individual unit of production, are possible in theory and observable in practice.

We have already pointed out some of the main sources of internal flexibility in the employment relation (e.g., neo-Fordist work teams, extended job categories, redeployability of individual workers from job to job, and so on). In some cases, flexible work rules within the individual firm may engender significant quantities of firm-specific human capital, as, for example, in the case of work teams where workers must learn to deal with one another's idiosyncrasies in environments that depend critically upon mutual cooperation, or in the case of research and development groups that are working collectively towards some long-term goal. The former case seems to be characteristic of many core Japanese firms; the latter of IBM and other large high technology corporations. In all such cases, internal flexibility is likely to be complemented by relative security of tenure of employment, and this will, in part at least, diminish any firm's margin of external labor market flexibility. That said, flexible production complexes are typically associated with high levels of external flexibility in the local labor market through such mechanisms as accelerated turnover, part-time work, short-term labor contracts, home-working, and so on. These latter forms of flexibility have evidently been on the increase in recent years throughout the advanced capitalist economies (OECD 1986b).

Flexible production systems in general employ two distinctive segments of workers: a highly remunerated segment consisting of professional, craft, and technical workers, and a poorly remunerated segment made up of politically marginalized social groups such as women, ethnic minorities, and rural-urban migrants. External flexibility is especially high in the case of unskilled workers, but it has also been increasing at the upper levels (cf. Christopherson & Storper 1988, Scott 1984). Many upper-tier workers are, of course, employed in stable jobs. Others are highly mobile and they often shift at quite short intervals from job to job. This mobility is expressed in both high lay-off rates and pervasive quit behavior in the upper segments of flexible labor markets. Here we may cite the examples of the Silicon Valley engineer, skilled crafts workers in the Los Angeles motion picture industry, or the increasingly mobile professional and managerial workers in metropolitan service industry complexes. So long as overall employment is stable or expanding, this job instability tends not to be translated into extended periods of unemployment. To the contrary, these workers can move with great rapidity through the job market because their skills, while at a high level, are attuned to the needs of a variety of firms in the production complex. Their skills are agglomeration- and sector-specific rather than firm-specific. Among lower-tier workers, by contrast, job instability in flexible production systems is usually reflected in longer periods of unemployment consequent upon the relative over-supply of unskilled labor. Both upper-tier and lower-tier workers deal with these forms of external instability through network relations providing information about labor market conditions, and these networks constitute important institutional bases of flexible production complexes. They may variously take the form of professional associations, kin and friendship ties, ethnic organizations, trade unions (where they exist), and so on. In addition, lower-tier workers survive extended periods of unemployment by reliance on such strategies as the pooling of family incomes, participation in the informal economy, or (as in the case of the Third Italy) occasional agricultural labor.

These different institutional responses to flexibility in the local labor market are steadily built up over time by means of both spontaneous action and planned intervention, and they contribute in significant ways to the total stock of agglomeration economies at particular places. Agglomeration economies are further enriched by communal processes of the habituation of workers into the local culture of production through associations, schools and universities, neighborhood contacts, local media, and so on. In these ways, among others, a Marshallian "industrial atmosphere" is created and sustained in the local area as part of its overall logic of territorial reproduction (cf. Bellandi 1986).

Community formation and social reproduction in flexible production agglomerations

By definition, industrial agglomerations of all kinds consist of an interdependent system of production spaces (the locus of work) and adjunct social spaces (the locus of the domestic life of workers). Where the technical, social, and occupational divisions of labor are far advanced, there will be a corresponding demand for many different skills and human attributes on the part of producers. Under these circumstances, the local population is likely to be heterogeneous, both in socio-economic terms and in cultural, ethnic, and racial terms too. In flexible production agglomerations, such heterogeneity is generally articulated with the primary dualization of local labor markets and local social structure as described above. At the same time, this dynamic of internal social differentiation in any given production agglomeration creates a fundamental and pervasive predicament. On the one hand, a socially-differentiated population comes together in one place in response to the labor demands of the production system; on the other hand, via individual residential behavior and choices within urban housing markets, the same population tends to become sorted out spatially according to the differentiated social norms, individual characteristics and economic capabilities of various occupational groups and social strata. The result is a definite though always incomplete and partial tendency for the social space of the agglomeration to disaggregate into a mosaic of distinctive neighborhoods and communities within which subtle and intricate processes of family life, childrearing, and social interaction take place. These social areas then function as sites of symbolic representation of

33

social distinction, identity, and status, as portrayed in general abstract terms by Baudrillard (1973) and Bourdieu (1979) with their theories of the political semiology of consumption and cultural activity. In short, the distinctive neighborhoods and communities that emerge within any agglomeration become integral to the legitimation and stabilization of socio-economic divisions in the local area.

Thus, residential segregation is brought about through the housing choices of individual families, and it functions as a vehicle of more effective domestic existence and social reproduction within the community. This suggests at once that the more diverse the bases of production and work are in any given center, the more varied its internal community structure is likely to be as a consequence. But a qualifying addendum to this speculative remark is necessary, viz., that there are additional principles of neighborhood formation that bear no direct relationship whatever to the division of labor and the structure of work within the agglomeration. Moreover, just as the local production system becomes an object of regulation and collective action, so also does the social space of the agglomeration. In these different ways, community life in any agglomeration takes on a significant logic of its own, and it in turn begins to feed back upon and to re-structure the development of the production system.

As the preceding remarks suggest, the communal dynamics of flexible production agglomerations pose many difficult and unresolved questions. For example, we know rather little analytically about such events in the new high technology growth centers of the US Sunbelt as the arrival of new migrant groups, the development of new politically marginal populations, the intensification of labor market segmentation, the relative absence of a traditional skilled blue-collar population, the rise of new scientific and technical cadres, the shift from redistributive to entrepreneurial forms of urban planning, and so on. These phenomena are evidently all important facets of the overall functioning of Sunbelt growth centers in the late 20th century, though their precise socio-spatial logic remains as yet to be deciphered. Most especially, if we are to make progress in the task of comprehending in its fullness the internal communal order of localized flexible production complexes, we need very much to investigate the interrelations between all such phenomena as these and the overall trajectories of local economic growth and development.

The politics of place in flexible production complexes

By the politics of place we mean to suggest processes of the formation and appropriation of systems of place-bound norms integral to the functioning of any locale as a center of economic and social life. These norms emerge in part from the historical geography of each given place and they are also an object of efforts on the part of various groups and factions to shape them actively to serve their interests. For example, in an earlier era of industrialization in the USA, attempts were made by elite groups to enhance the efficient operation of the dominant centers of growth through political movements such as Progressivism (and its manifestations in the City Beautiful and the City Efficient movements) and Boosterism. As the Fordist regime of accumulation deepened and as high Fordism came on to the scene in the post-war years, the politics of place came to center on a consumerist representation of urban life as manifest in ideals about the nuclear family, suburban residence, and private car ownership. This indeed is what in part kept Fordism going as a macroeconomic phenomenon (Walker 1981). This representation was accompanied by increasingly rigid practices, rules, and norms governing the interrelations between business and labor in the workplace, in the labor market, and in the municipal environment. One consequence, we have shown, was a dynamic of continually rising wage levels, and this in turn fed upon and helped to intensify the consumerist ideology of high Fordism. With the weakening of Fordism after the late 1960s, the accumulated culture and traditions of mass production centers reinforced the crisis and prompted producers in new production ensembles to seek out alternative locations where accumulation could be reconstructed unhindered by a Fordist politics of place.

Thus, as the window of locational opportunity (to which we alluded earlier) opened in the late 1950s and early 1960s, producers in many of the new flexible production ensembles actively took advantage of it in an attempt to maximize their geographical and social distance from established centers of Fordist production. High technology industry aggressively pioneered new production locations on the extensive margins of industrialization, as did some of the new artisanal industries. However, in many cases artisanal production together with burgeoning service industries tended to grow at sites where there was already some pre-existing activity of the same type, for these antecedent activities were already well insulated from Fordist labor markets. In all of these places, a new post-Fordist politics of place has been constructed, though with considerable variation from industry to industry and from location to location. Three examples may be invoked in illustration of these remarks.

First, new high technology growth centers in the US Sunbelt have developed a politics of place that is, in a sense, an attempt to perfect the model of suburban life that came to be strongly associated with urbanization under Fordism. This phenomenon may be identified partially in terms of low densities of urban development, highly privatized forms of domestic life based on individual homeownership, and abundant recreational opportunities. As we have shown, growth was largely initiated by producers in the new high technology industrial ensemble seeking out places without a prior history of Fordist industrialization, where the relations of production and work could be reconstructed anew. These experiments have turned out so far to be extremely successful, for a politics of place has been instituted and maintained in which neo-conservative attitudes about work and life have become remarkably pervasive (cf. DiLellio 1987). At the same time, these attitudes conform well to the need for high levels of flexibility among producers, for they

35

inhibit precisely the kinds of organizational responses on the part of workers that create dysfunctional forms of inertia and rigidity in the production system. The more experimental and changeable the technology, the more these qualities are important to producers. In addition, local governments have reinforced the process of growth by taking an entrepreneurial stance in which they compete among one another in providing the tangible bases of a good business climate.

Second, in the craft communities of the Third Italy, despite a long tradition of active labor union organization and leftist municipal government, an effective politics of place has been constructed around a social agreement about growth among labor unions, employers, and local governments. Labour unions have helped to mobilize and reproduce craft skills and consciousness among workers (Scarpitti & Trigilia 1987). Employers have combined in order to coordinate local production activities through institutional arrangements. In addition, local governments have taken an active role in the promotion of the small firm sector, in aiding the coordinating activities of firms, and in enhancing local labor market flexibility by providing special unemployment payments and retraining programs (Hatch 1987).

Third, and finally, over the last couple of decades, dense service production complexes have grown in and around the central business districts of major world cities such as New York, London, Paris, and Los Angeles. These developments have been posited upon both massive downtown redevelopment projects and extensive gentrification of inner-city neighborhoods to house the burgeoning white-collar workforce of the central service complex. Downtown office redevelopment has been facilitated by the emergence of growth coalitions which have been able significantly to shape the direction of urban planning and renewal in central cities (cf. Davis 1987). Gentrification has, in its turn, been associated with the rise of active and influential neighborhood and community groups that lobby on behalf of their own interests (neighborhood preservation, continued upward momentum of property values, improved services, and so on). As all of this has come about, the centers of major world cities have been remade, both physically and symbolically, as domains of leisure and self-identification for the new service and managerial fractions, as centers of international taste and culture, and as loci of mass spectacle of the sort described by Debord (1967).

These examples demonstrate the great variety of the politics of place in and around contemporary flexible production agglomerations. The cases we have examined have all thus far developed without encountering major social and political breakdowns. This condition is underpinned by the recency of their emergence as modern flexible production agglomerations, for oppositional forms of consciousness and organization – if they are to appear – need time in order to be built. If the history of previous industrial localities is any indication, we may well expect one form or another of organized opposition to materialize, whether it be based on class or community relations. We may also expect that if such opposition should appear, complementary political responses by producers and local government would also come into being, though in some cases (in the USA above all) producers' own neoconservatism and the historically low level of coordination of small business may turn out to be barriers to the formation of such responses. In view of the wayward character of history, we take the precaution of refraining from any attempt to imbue these expectations with further substance.

One final important point needs to be made in this context. In some accounts of new industrial development, and above all in many analyses of high technology industrial location, the observed communal characteristics of Sunbelt growth centers are described unproblematically as signifying a "high quality of life." This, in and of itself, is then said to attract large numbers of technical and scientific workers from other places to these centers. The same accounts frequently go on to the conclusion that the location of high technology industrial enterprises can accordingly be largely accounted for in terms of the prior residential preferences and migratory patterns of these workers. In view of what we have written above about the logic and dynamics of the formation of industrial agglomerations, this conclusion is surely incorrect; indeed, it represents a highly ideological view of things, for it reduces the complex totality of growth center development to a simple set of consumer preferences and behaviors. As we have shown both here and elsewhere (Scott & Storper 1987), we can surely more effectively approach this problem in terms of the intricate interrelations between the technological-institutional bases of production, the dynamics of industrial organization and local labor market development, and processes of local social regulation.

Summary and analytical prospect

Territory and location have always played a major role in the shaping of capitalist production systems. In the era of Fordist accumulation an initial phase of industrial agglomeration around the lead plants of mass production industries was followed by active decentralization from core areas to peripheral areas. Now, as the transition to flexible accumulation gathers momentum, strong agglomerative tendencies in flexible production sectors have appeared once more as an important territorial expression of capitalist industrialization. Concomitantly, the significance of place as the foundation for efficient and effective production apparatuses has been re-affirmed.

In this chapter we have sought to highlight some of the detailed elements of this theme. We have examined, in particular, the changing logic of production systems, the emergence of new industrial spaces, and the problems of social regulation and reproduction within these spaces. We began by contrasting two technological-institutional models of industrialization, namely, Fordist and flexible production organization. We then dealt with the transition that is currently proceeding from Fordism to flexible accumulation, and we traced out some of the principal geographical changes that are associated with this transition – above all, the resurgence of new growth centers based on flexible industrial ensembles. We developed a detailed argument about social and institutional order in the new industrial spaces of capitalism with particular reference to the coordination of inter-firm transactions, the social organization of local labor markets, and processes of community development. Lastly, we focused on what we have termed the politics of place, looking at both its ideological and institutional forms and the role it plays in the rise, reproduction, and deliquescence of industrial localities.

All such localities are embedded in national and international markets and caught up in a wider global division of labor. As such, the problem of the social regulation of flexible production systems is not just confined to particular places and regions (our emphasis in this investigation), but extends through many different levels of socio-economic reality to the national and international spheres. We may thus expect the further unfolding of the emerging technological-institutional model of flexible production to be accompanied by the rise of new kinds of collectivized order and regulatory practices at these different levels. This is all the more probable in view of the circumstance that the old forms of system governance built up around Fordism are increasingly unable to deal with the looming problems of macrosocial and macroeconomic coordination at the inter-regional and international levels. Moreover, as new patterns of inter-regional and international social regulation make their historical appearance, a process of associated restructuring and readjustment at the purely local level is likely to begin to operate.

A next and imperative phase in the attempt to comprehend the geography of flexible production is to initiate the difficult task of identifying and anticipating these wider mechanisms of social regulation, and of deciphering their detailed spatial meaning. It seems to us that a progressive politics in the post-Reagan era must grapple resolutely with these issues.

References

Aglietta, M. 1976. *A theory of capitalist regulation*. London: New Left Books.
Baudrillard, J. 1973. *Le miroir de la production*. Tournai: Casterman.
Becattini, G. (ed.) 1987. *Mercato e forze locali: il distretto industriale*. Bologna: Il Mulino.
Bellandi, M. 1986. *The Marshallian industrial district*, Dipertimento di Scienze Economiche, Università degli Studi di Firenze, Studie Discussioni, No. 42.
Bluestone, B. & B. Harrison 1982. *The deindustrialization of America*. New York: Basic Books.
Bourdieu, P. 1979. *La distinction: critique sociale du jugement*. Paris: Editions de Minuit.
Boyer, R. 1986. *La théorie de la régulation: une analyse critique*. Paris: Editions La Découverte.
Brusco, S. 1986. Small firms and industrial districts: the experience of Italy. In *New firms and regional development in Europe*, D. Keeble & E. Wever (eds). London: Croom Helm.

Brusco, S. & C. Sabel 1981. Artisanal production and economic growth. In *The dynamics of labor market segmentation*, F. Wilkinson (ed.), 99–113. New York: Academic Press.

Christopherson, S. & M. Storper 1988. Flexible specialization and new forms of labor segmentation. *Industrial and Labor Relations Review* (forthcoming).

Clark, G. L. 1986. The crisis of the Midwest auto industry. In *Production, work, territory: the geographical anatomy of industrial capitalism*. A. J. Scott & M. Storper (eds), 127–48. London and Boston: Allen & Unwin.

Davis, M. 1987. Chinatown, part two: the internationalization of downtown Los Angeles. *New Left Review* no. 164, 65–86.

Debord, G. 1967. *La société du spectacle*. Paris: Buchet/Chastel.

DiLellio, A. 1987. Changing citizenship in "High Tech" communities: the case of Dallas (US) and Grenoble (France). Paper presented at the International Conference on Technology, Restructuring and Urban Development, Dubrovnik, Yugoslavia, June 25–30.

Ganne, B. 1983. *Gens du cuir, gens du papier*. Paris: Editions du Centre National de la Recherche Scientifique.

Hatch, R. 1987. Manufacturing networks and reindustrialization: strategies for state and local economic development. Newark, N.J.: New Jersey Institute of Technology.

Lipietz, A. 1986. New tendencies in the international division of labor: regimes of accumulation and modes of social regulation. In *Production, work, territory: the geographical anatomy of industrial capitalism*, A. J. Scott & M. Storper (eds), 16–40. Boston: Allen & Unwin.

Markusen, A. 1985. *Profit cycles, oligopoly, and regional development*. Cambridge, Mass.: MIT Press.

Massey, D. & R. Meegan 1984. *Spatial divisions of labour; social structures and the geography of production*. London: Macmillan.

OECD 1986a. *La politique d'innovation en France*. Paris: Organisation for Economic Cooperation and Development.

OECD 1986b. *Flexibility in the labor market: the current debate*. Paris: Organization for Economic Cooperation and Development.

Piore, M. & C. Sabel 1984. *The second industrial divide*. New York: Basic Books.

Raveyre, M. F. & J. Saglio 1984. Les systèmes industriels localisés: éléments pour une analyse sociologique des ensembles de P.M.E. industriels. *Sociologie du Travail* 2, 157–76.

Russo, M. 1985. Technical change and the industrial district: the role of interfirm relations in the growth and transformation of ceramic tile production in Italy. *Research Policy 14*, **14**, 329–43.

Sabel, C. & J. Zeitlin 1985. Historical alternatives to mass production: politics, markets and technology in nineteenth century industrialization. *Past and Present* no. 108, 133–76.

Scarpitti, L. & C. Trigilia 1987. Strategies of flexibility: firms, unions and local governments – the case of Prato. Paper presented to the Conference on New Technologies and Industrial Relations: Adjustment to a Changing Competitive Environment, Endicott House, Dedham, Mass., February.

Scott, A. J. 1984. Territorial reproduction and transformation in a local labor market: the animated film workers of Los Angeles. *Environment and Planning D: Society and Space* 2, 277–307.

Scott, A. J. 1988. *Metropolis: From the Division of Labor to Urban Form*. Berkeley and Los Angeles: University of California Press.

Scott, A. J. & D. P. Angel 1987. The U.S. semiconductor industry: a locational analysis. *Environment and Planning A* **19**, 875–912.

Scott, A. J. & M. Storper 1987. High technology industry and regional development: a theoretical critique and reconstruction. *International Social Science Review* **112**, 215–32.

Stoffaës, C. 1978. *La grande menace industrielle*. Paris: Calmann-Lévy.

Storper, M. & S. Christopherson 1987. Flexible specialization and regional industrial agglomerations. *Annals of the Association of American Geographers* **77**, 103–17.

Uekesa, M. 1987. Industrial organization: the 1970s to the present. In vol. 1: *The political economy of Japan*, K. Yamamura and Y. Yasuba, 469–515. Stanford, CA: Stanford University Press.

Walker, R. 1981. A theory of suburbanization. In *Urbanization and urban planning in capitalist society*, M. Dear & A. J. Scott (eds). London: Methuen.

Williamson, O. 1975. *Markets and hierarchies: analysis and antitrust implications*. New York: The Free Press.

3

Collective consumption

STEVEN PINCH

In the past decade debate within urban studies about the role of the state in capitalist societies has been dominated by the concept of collective consumption. This term was coined by Castells (1977) to indicate the increasing tendency throughout most of the 20th century for governments in advanced industrial societies to intervene in the provision of goods and services. Unfortunately, this term collective consumption has become such common currency in urban studies in recent years that, rather like the terms "community," "restructuring," and "privatization," it can be said to have attained "a high level of use and a low level of meaning." Indeed, Pahl (1977) argues that the term is so ambiguous and confusing that it is best abandoned. Dunleavy (1986) has preferred to replace the term with the concept of "socialized consumption" which now incorporates "collective consumption," "quasi-collective consumption," and "quasi-individualized consumption"!

Part of the reason for this confusion over the meaning of the term collective consumption must stem from the fact that many of these public goods and services could easily be provided by private markets. Indeed, state intervention takes many forms. For example, the state can provide the good or service directly through public sector institutions, it can subsidize the provision of a good or service through the use of public funds to lower the price below that which private markets would otherwise dictate, it can regulate the price, quantity, and quality of products and services provided through private markets, or it can adopt some hybrid of these arrangements. In most societies the balance between different forms of provision is in a continual state of change over time. Theorizing about these complex arrangements is made even more complex by territorial diversity. The quantity and quality of goods and services and the ways in which they are provided display enormous geographical variations at all scales – between different nations, between political and administrative jurisdictions within nations, between towns and cities, and between neighborhoods and communities within urban areas. It is, therefore, hardly surprising that most attempts to float some theory to explain collective consumption have foundered when confronted with the enormous complexity of the subject matter. At the risk of extending the metaphor, this chapter attempts to sift through the debris of these wrecked theories to see what can be salvaged to analyze the ever changing nature of collective consumption.

Definitions of collective consumption

Private consumption goods versus public consumption goods

Most efforts to explain the intervention of states in the provision of goods and services have been based upon the creation of polar opposites. This is evident in the highly influential Theory of Public Goods expounded by economists after World War 2 (Samuelson 1954, Musgrave 1958). Whereas private consumption goods were defined as those which could only be consumed by one individual or at most a household – such as clothing – public consumption goods were defined as those goods with properties which made it impossible for provision through private markets. These properties were defined as: first, *joint supply*, which means that if a good can be supplied to one person it can be supplied to all other persons at no extra cost; second, *non-excludability*, meaning that it is impossible to withhold a good from those who do not wish to pay for it; and third, *non-rejectability* which means that once a good is supplied it must be equally consumed by all, even those who do not wish to do so. The Theory of Public Goods is therefore based upon the characteristics of the goods and services which, it is maintained, have properties which make it impossible, or extremely difficult, for allocation via private markets. The classic example which is always quoted to support the theory is defense. A little thought, however, reveals that the defense of a nation will vary in character across space. Some areas may be more vulnerable to attack than others, while some territory might be sacrificed in a time of war.

In reality then, there are some basic "facts of geography" which mean that public goods and services are not equally available at equal cost to either consumers or producers, thus undermining the purity of public goods theory (Tiebout 1956, Teitz 1968). These geographical factors which serve to undermine the notions of joint supply, non-excludability and non-rejectability can be grouped into two basic types. First, there is the phenomenon of jurisdictional partitioning. Most countries find it expedient to devolve the administration of many services to units that are smaller than the nation state. Sometimes these units have elected representatives with varying degrees of political autonomy, and in other cases they are special administrative districts with varying degrees of political and consumer representation. One of the main consequences of this devolution is that these political and administrative units vary enormously in the quantity and quality of the goods and services that they provide. A second basic factor which serves to undermine the theoretical purity of public goods is the existence of externalities. These are in essence unpriced effects (Margolis 1968). They can be benefits received by those who do not pay for them or costs incurred by those who are not directly compensated. Pure public goods are in theory extreme cases of externalities since, if available to one person, they should be available to all others at no extra cost. In reality, these externality effects exhibit decreasing effects with increased distance from the source of the activity.

There are, in fact, few technical reasons why, defense apart, goods and services should be provided by any one particular mode rather than another. It is certainly true that certain services such as electricity, water supply, and mass transport systems are difficult to supply via unregulated markets with numerous competitive entrepreneurs, but they can be provided through private companies if they are regulated monopolies. The decision to allocate certain goods and services via the public sector or to interfere with private markets has therefore come about, not so much for technical reasons and the characteristics of the goods and services, but because of political pressures translated into policy (Pickvance 1982, Pinch 1985). Often this pressure has risen because of a dissatisfaction with the inequalities generated by private sector mechanisms. Public goods theory therefore provides us with relatively little explanation of the imperative to provide goods and services by some public sector intervention.

Collective versus individual consumption

Another polarization that has been constructed to explain public service provision is that between services that need to be consumed individually and those services that, literally, require a group of persons to come together to consume them. Lojkine (1976) uses this distinction as one of his criteria to define collective consumption. According to this approach, schools, hospitals, and buses are "collective" services since they obviously require groups of individuals to come together to use them. In the language of Hägerstrand's (1978) pioneering work on time-space budgets, there are important "coupling constraints" which require that individuals combine to participate in many forms of activity. However, it is difficult to see the explanatory payoff to be derived from maintaining Lojkine's distinction, since it bears no relationship to whether the service has any degree of state intervention or is provided by the state or private sectors. Thus, "collective" hospital services are predominantly private in the United States and mostly public in Great Britain. As Pahl (1977) notes, the net result of this perspective is almost absurd. If being in school is a collective service, is taking homework an individual service? Similarly, if being in hospital is a collective service, is consulting a single doctor an individual service?

It is certainly true that services vary in the extent to which they are allocated on an individual versus collective basis, and this is related in some degree to technological developments and consumer preferences in particular spheres. For example, it has been argued by Gershuny & Miles (1983) that increasing productivity in manufacturing has meant that households are able to replace labor intensive services with consumer products that undertake the same tasks within the home. The classic examples are the substitution of microwave ovens for eating out in restaurants, washing machines for laundries, private cars for public transport and video machines for live entertainment. Whatever the truth of these assertions in the private sphere, the extent to which services are consumed on an individual or collective basis in the public sphere would seem to bear little relationship to the characteristics of the service, but is often related to political issues. In many advanced industrial nations there are at present powerful forces attempting to put the responsibility for caring for dependent groups back upon families (which usually means putting responsibility back upon women). These issues are discussed in greater detail later.

Goods versus services

In recent years a number of theories have been put forward to explain the rapid growth of services in advanced industrial societies. Services now comprise the majority of jobs in most advanced industrial economics and are the sectors of the economy that are growing most rapidly. However, compared with manufacturing activity, little is known about service growth. One of the reasons for this state of affairs is that, until recently, researchers have adhered to a crude export base approach which envisages services as essentially dependent upon manufacturing activity, either providing inputs as producer services or servicing final demand through consumer services (Daniels & Thrift 1985). In recent years there has been a realization that services no longer merely supply inputs to the production process, but have their own internal dynamics.

A major limitation of theories attempting to explain such developments is that they are based upon somewhat arbitrary definitions of goods and services (Walker 1985). Services are usually categorized as those activities that do not result in the physical transformation of a product. When considered in more detail, it is extremely difficult to maintain the difference between goods and services. For example, computer software involves mental effort but the actual output is a physical product (as indeed is the case with this book)! Yet in most analyses of service growth, sectors are usually allocated to manufacturing, producer services or consumer services sectors in a somewhat arbitrary fashion. From the perspective of collective consumption the goods and services division makes comparatively little analytical sense, since the public sector is involved in the provision of both tangible physical products, such as public housing, and intangible services such as information, help, and care. It is, of course, true that much of what can be termed collective consumption is not concerned with commodities that can be sold on the market but with services that involve the expenditure of labor, but this fact alone provides little insight into the reasons for the enormous diversity of ways in which services are provided in different capitalist societies.

Market versus non-market

As stressed above, many public services have been introduced because of a basic dissatisfaction with the inequalities generated by private market mechanisms. Another division that can be erected, therefore, is that between goods and services allocated on the basis of need, or some other non-market criteria, and those allocated on the basis of the ability of consumers to pay. It is often assumed that collective goods and services satisfy fundamental human needs such as shelter, health, and education. These may be contrasted with the vast array of commodities such as blue jeans, audio equipment, and private cars that make western capitalism such an alluring prospect for many peoples around the world. It is sometimes argued (by the political left but also ironically by many on the political right) that if left to their own devices many people – but perhaps especially the poorer and less educated – would spend their income on satisfying short-term desires for these consumer goods and services and forgo spending upon long-term insurance against disability, poor health, and old age. The state therefore adopts a paternal role through insurance and social provisions.

In practice it is enormously difficult to measure needs, as demonstrated by the enormous amount of academic literature on this topic. It is especially difficult to separate out fundamental human "needs" from less important "wants." Although humans require food, water, warmth, and sexual activity for survival, there are a vast range of culturally determined practices that serve to satisfy these needs. Needs and wants are therefore social constructions and this makes it difficult, if not impossible, to draw a firm distinction between fundamental needs and wants (Eyles 1987).

This distinction between market and non-market criteria is, however, useful for it serves to undermine the public versus private divide which has dominated much debate, and draws attention to the multiplicity of ways in which goods and services can be provided. Thus, many publicly-owned services are traded on a commercial basis while many privately allocated goods and services are bolstered by public subsidy. If one is looking at housing for example, one needs to consider subsidies to the private sector as well as direct state provision. Sometimes these non-market elements are designed to meet the needs of deprived or dependent groups, but in other cases they are introduced by powerful groups who wish to derive concessions from the state. Indeed, Harrison (1984) argues that we are in a period of "welfare corporatism" in which powerful interests are able to reap considerable tax and other concessions from the state.

Figure 3.1 is an attempt to incorporate some of the polar dichotomies described above to outline the structure of consumption. Given the difficulties of maintaining firm distinctions between many of the defining concepts, it should be apparent that the allocation of activities to particular categories is to some degree arbitrary. For example, the extent of governmental regulation of the provision of goods and services is very largely a matter of degree, since taxes and regulations affect most aspects of commodity exchange in advanced industrial societies. It should also be noted that Figure 3.1 is based upon those sectors responsible for the allocation of services and says nothing about the sector responsible for the production of the commodity. In some societies the state controls the manufacture of commodities that are allocated by private markets. Most of what is generally referred to as "collective consumption" may be found within the category of public sector services allocated on a non-marketed basis. However, it should also be noted that some of the services fit more

Figure 3.1 A typology of consumption.

Defining characteristics   **Examples**

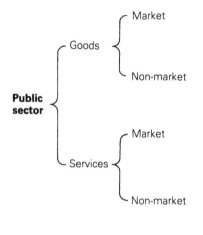

Quasi-individualized consumption
State coordinated consumption
of goods, e.g., cars in Soviet bloc

Collective consumption
Public housing, state subsidized
health aids

Quasi-collective consumption
Utilities, gas, water, electricity,
public commercial services

Collective consumption
Social services, education
hospitals in UK, fire protection,
police, roads

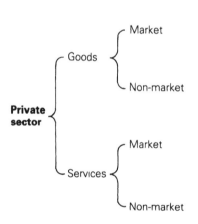

Individualized consumption
Consumer durables, private cars,
stereos, clothing

Quasi-individualized consumption
State subsidized goods, owner-
occupied housing, company cars

Individualized consumption
Commercial services, hospitals
in US, private nurseries

Quasi-collective consumption
State regulated monopolies, British
telecom, "privatized" services,
e.g. cleaning, catering

than one category, while different nations provide the same type of service in different ways. What also needs to be recognized is that in many countries, and especially the United Kingdom, many of the organizations are changing their location from the public to the private sectors. Figure 3.1 does, however, draw attention to the diversity of organizational forms that are to be found in advanced societies. The major limitation of such a static typology is that it tells us nothing about the reasons for activities and organizations being in particular categories.

Production versus social reproduction

The final division to be considered here is that between production and social reproduction. These notions are essentially Marxist in origin and as such are based upon a number of different assumptions to the majority of the ideas quoted above. The intention is to produce "committed explanation" (Gregory 1978) which represents the interests of the common people and not the interests of "generals, politicians and corporate chiefs" (Harvey 1984,7). These approaches conceptualize the sphere of production as that in which the class of laborers in a capitalist society is exploited by the interests of capital for the purpose of producing profit. According to the early work of Castells (1977), the collective consumption process is that which ensures that the labor force is renewed both on a daily and intergenerational basis for the function of engaging in productive work. The reproduction of labor power involves not only biological reproduction but the continual reproduction of all the skills and attitudes necessary for the maintenance of an adequate workforce in a capitalist society.

Castells's ideas spawned an enormous debate, but have also received extensive mauling at the hands of his critics (see Lowe 1986). For example, one of the earliest criticisms was of Castells's definition of cities as centers for the reproduction of labor power. Writers such as Lojkine (1976) pointed to the importance of collective services in cities in affecting productive infrastructure. Other critics focused upon Castells's notions concerning the potential of collective services to generate so-called "urban social movements" that would threaten the status quo in society (Pickvance 1976). Yet another source of criticism has been Castells's definition of collective consumption in terms of state provision (Saunders 1979). Pahl (1977) goes so far as to describe the concept of social reproduction as a "functionalist banality."

If, however, we adopt a more extensive view of social reproduction than Castells's early focus upon state provision, then one of the advantages of this approach is that it diverts attention from the simple public versus non-public distinction and draws attention to the variety of ways in which the workforce can be socially reproduced. For example, in order to reproduce themselves all societies have to construct arrangements to care for the elderly and young children. This can be undertaken within households (usually by the labor of women); by private markets, through the informal neighborhood economy (either directly for money or indirectly for barter); by charitable and voluntary organizations; and finally, by the state either at local or national level. In addition, there are numerous hybrid arrangements whereby central and local governments support provision by the voluntary and private sectors through tax incentives and subsidies. Not only do these arrangements obviously vary enormously between nations, but they also vary within nations depending upon variations in local class structures, political parties, work arrangements, and gender roles within households.

Unlike much previous geographical analysis which takes the capitalist mode of production for granted, Marxist theories attempt to relate changes in services to the crises and contradictions of international capitalism. However, relating the sphere of production to that of social reproduction raises many problems. Whilst many of these theories based upon the notion of social reproduction have avoided the absolute autonomy approach – in which the spheres of production and reproduction are seen as evolving independently of one another – many have fallen into the traps of either reductionism or functionalism. Reductionist approaches conceptualize the sphere of social reproduction as being shaped entirely by the sphere of production, whereas functionalist approaches envisage the sphere of social reproduction as not entirely shaped by the sphere of production, but playing a vital role in its operation and survival. Humphries & Rubery (1984) argue that the systems of production and social reproduction are relatively autonomous and are in a process of continual interaction. Many would no doubt agree with this point, but this still leaves many problems for future studies. Dunleavy (1986) argues that existing theories derived from political economy tend to be partial, in that they are either economic explanations related to the imperatives of capitalist accumulation or else are ideological explanations which stress the legitimation crises facing capitalism or the role of welfare expenditures in undermining the work ethic of capitalism. These theories tend to produce generalized descriptive accounts which ignore the specificity of political processes and the role of institutional arrangements in determining the structure of welfare systems.

Realism, space, and collective consumption

Currently the most popular solution to the problems that deterministic and functionalist Marxist approaches experience in relating small-scale empirical developments to wider societal forces lies in the philosophy of realism. According to this approach the task of realist science is conjecturally to probe beneath the "realm of appearances" to uncover the generative mechanisms that give rise to empirical outcomes (Bhaskar 1975, Sayer 1982). Central to realist approaches is a distinction between external contingent relations and internal necessary relations. The distinction between these two types of relations is seen as essential for the process of abstraction – isolating one-sided aspects of objects. It is argued that any empirical situation consists of a variety of necessary and contingent factors which may or may not have been successfully theorized in a "rational abstraction." In the physical sciences it is possible to discover the necessary tendencies in things by setting up a closed system and testing if the hypothesized causal mechanisms result in empirical outcomes. In marked contrast, social systems are open systems with highly complex sets of both necessary and contingent factors, and one cannot assume any direct relation between structures and outcomes. This means that empirical outcomes in social systems are always contingent. All one can do is to develop an adequate conceptualization of necessary relations and discover the coexistence of necessary and contingent relations in any given empirical context.

Given the overwhelming complexity of collective consumption, it is not difficult to see the attraction of this approach. The relations between capital and labor can be envisaged as the necessary ones in a capitalist society, and the diverse paraphernalia of state structures and subsidies may be seen as the contingent outcomes in different societies and areas. Many of the concepts that have previously been used to describe structures in the sphere of social reproduction, such as are incorporated into the theory of public goods, have, according to the realist approach, been concerned with highly contingent factors observed in particular societies that have been extended into general theories.

Ironically, the criticism made above also applies to numerous Marxist theories. Indeed, despite the attractions of realism, it still poses major problems for those concerned to undertake empirical analysis. A key difficulty is that, when reduced to essentials, very little in the world is, in effect, truly essential. It is certainly highly desirable that capitalist societies have welfare systems of various types, and that women undertake a great many tasks caring for dependent groups that would otherwise be a burden on the state, but neither of these types of arrangement is absolutely essential for capitalism to survive. For example, it is possible for capitalist societies to have minimal welfare states and instead be bolstered by repression. It is also possible to envisage a capitalist system reproducing the workforce through a system of state childrearing agencies that would benefit from economies of scale, and which would provide a workforce with the necessary skills and attitudes for the capitalist system to survive (Humphries 1977). The researcher is therefore confronted with the overwhelming diversity of contingent factors in the world around us, and there is a danger that one will end up with only extensive descriptions of contingencies. This is a particular problem for those researchers concerned to incorporate a geographical dimension into their analysis. There is now widespread acceptance that, by itself, space can have no independent effect, but assumes importance in the context of particular social processes. The so-called "facts of geography" or the "logic of space" (as incorporated into concepts such as jurisdictional partitioning, tapering effects, and externalities) are related to the level of technology and social and economic development of the societies concerned. Yet such a view would imply that space is always contingent, whereas many have argued that the operation of processes over space can change their character (Massey 1985, Urry 1985).

The crucial issue for further studies is thus how to avoid, on the one hand, endless descriptive catalogues of contingencies with crude links with underlying structural mechanisms, and, on the other hand, generalized theories evolved from limited observations with no reference to wider determinants. In addition, geographers and other social scientists concerned to understand territorial organization must avoid conceptualizing space in absolute terms as simply a reflection of broader social processes or as merely a contingent element in causal processes. Obviously, there is no one simple solution to these formidable problems. Nevertheless, some of the recent work on the restructuring of production provides us with some indications as to how these problems might be avoided. The so-called "Restructuring Thesis" has drawn attention to the ways in which businesses have been responding to diverse pressures such as changing capital markets, changing technology, and the need to control the workforce (Massey 1984). The important point is that the outcomes of these changes are not a simple response to "top down" pressures, but interact with, and in turn are affected by, the enormous diversity that is to be found in different localities at different times.

Two important points should be made at this stage. First, it is important to adopt a perspective which makes allowance for "human agency" and the variety of solutions that can be adopted by individuals and groups within organizations to apparently identical circumstances. Second, space and territorial organization should be envisaged as something more important than merely the contingent element in the manifestation of social processes. At any given level of economic, social, and technological development "spatial factors" (for want of a better term) provide both constraints and opportunities for actors in the sphere of collective consumption. For example, without imputing spatial determinism, it is possible to draw attention to links between spatial form and modes of service provision. Thus, dispersed suburban settlements are much more amenable to ideologies of individualism and privatized forms of consumption than high-density communal living environments. Where the latter comprise concentrations of deprived minorities, they can form bastions of resistance to pressures to undermine collective welfare provision.

The rest of this chapter considers some of these contemporary pressures in capitalist societies and the types of conceptualizations that are necessary to understand their manifestation in the sphere of collective consumption in particular localities.

Restructuring and collective consumption

Increasing international competition in the last decade has forced most capitalist nations to put issues of profitability at the top of the political agenda. This imperative has had a number of effects upon the structure of production, and upon the sphere of social reproduction. In the United States the Reagan Administration has attempted to increase the average rate of profit throughout the economy. This broad goal has involved numerous other objectives: lessening the resistance of workers to having their wages reduced; persuading previously unemployed high-wage manufacturing workers to accept less well-paid jobs; increasing the mobility of workers between sectors; and ensuring that there is an adequate supply of labor for the low-paid jobs in the newly emerging private service sectors (Rosenberg 1983). An important element of these policies has been reductions in public service employment and welfare assistance. Reagan's policies when Governor of California were based upon voluntarism – a combination of self-help and mutual aid as an alternative to state welfare policies. Upon reaching the White House, Reagan appointed a domestic policy advisor who argued that welfare programs

> have created a new caste of Americans – perhaps as much as a tenth of this nation . . . almost totally dependent on the state, with little hope . . . of breaking free. . . . Practical welfare reform requires that we reaffirm our commitment to the philosophical approach of giving aid only to those who cannot help themselves (quoted in Goodin 1986,27).

In Britain there has been a similar overall objective, but the details of policy have obviously been different. There has also been a concern to reduce the scale of public services and the value of the "social wage," and to create conditions suitable for private investment. This has led to efforts to decrease the power of trade unions through weakening employment protection legislation, reductions in welfare benefits, and the toleration of high levels of unemployment. It has been suggested that these policies have had a stronger ideological role in Britain, since trade union strength and middle-class fear and resentment of organized labor are stronger elements there than in the United States (Adams & Freeman 1982).

At the same time that welfare expenditures in many sectors have plummeted, expenditure related to defense has increased dramatically. Indeed, Castells (1985) argues that these changes in state spending represent a fundamental shift in the function of the state as it shifts its role from profit accumulation and redistribution towards military reinforcement – a trend he dubs the growth of the "warfare state." Miller (1978) terms this the "re-commodification" of capitalism. He argues that many states have been unable to bargain successfully with international capital because excessive welfare expenditures and taxes have created environments that are not conducive to investment. The response has been a reduction in corporate taxes, reductions in the scale of the public sector, and expansion of the private sector.

The manifestation of these processes at the level of local governments and administrative organizations charged with providing services can be envisaged as a response to a combination of three different pressures. First are "top-down" pressures from various types of central and federal government, and second, "bottom-up" pressures resulting from the structure of the localities. Both are affected by a third set of "mediating" factors which result from characteristics of the local governmental organizations concerned. Table 3.1 is an attempt to list just a few of these various types of factors.

The outcomes of these pressures can be envisaged as analogous to the complex outcomes observed through the restructuring of private companies in different areas. There are, of course, numerous differences, not only in terms of the types of pressures, but also in terms of the different objectives of private and public sectors. Thus, many public agencies are

51

Table 3.1 Pressures for restructuring upon local governments and local administrative bodies.

External pressures from "above"
- cuts in central fiscal support
- central controls over standards of provision
- pressures for privatization and contracting out
- changing technology
- legislative changes and legal judgments

Internal pressures from "below"

- changing demographic structures leading to changing demands and needs for services (e.g., increasing numbers of the elderly, young, unemployed)
- changing local tax base resulting from population and employment change
- changing client pressures for increased quality of service (e.g., local education pressure groups, minority and feminist lobbies)
- growth of anti-bureaucratic and dencentralist tendencies, tax revolts

Mediating factors
- entrenched professional interests
- attitudes of local trade unions
- local political parties

charged with providing roughly equal levels of services in different parts of a nation. They do not usually reveal the complex hierarchies of specialization in different areas that characterize modern corporations, and, by and large, are not able to play off workers in one area against those in another (Massey 1984). Nevertheless, many public and private organizations display similar processes transforming the character of working practices and relationships with client and consumer groups. Furthermore, just as with the restructuring of production, space is an integral part of the process of change. There is no one single role of space here, but it appears in many guises depending upon our conceptualizations and the empirical questions being asked. At times it will represent variations in the local tax base of political jurisdictions which in turn reflect the uneven development of capitalist relations across space. At other scales it will reflect inequalities in the capacities of various communities to petition for their demands. In yet other contexts it will reflect variations in social wellbeing resulting from the differential abilities of various groups to travel towards fixed point service facilities.

The final section of this chapter considers processes of change in the sphere of collective consumption and their manifestation in different organizations. These concepts are an amalgam of many of the ideas put forward by Massey (1984), Urry (1986), and Gershuny & Miles (1983).

Changes and manifestations of collective consumption

Rationalization

One of the most widely discussed processes in the field of industrial restructuring in the last decade has been the process of rationalization – the closure of plant capacity with little or no new investment. Most research in this field has been undertaken in the sphere of manufacturing, although many services such as laundries and cinemas have undergone similar declines. There are, however, many public and welfare-oriented services that have also undergone processes of rationalization in recent years, as represented by the closure of schools, hospitals, and welfare centers. Of course, the criteria for closure are very different from the calculations of profitability that lie behind the private sector. As Honey & Sorenson (1984) point out, however, even where the criteria for school closures are based upon explicit rational criteria (such as the age of buildings), they can often have detrimental effects in the poorest communities. Rationalization in the public sector is therefore inherently a political process and displays wide variations from one area to another, both within and between local jurisdictions.

In Britain the extent of rationalization in particular areas is often related to the political character of the local council. Labour-controlled authorities have attempted to resist central pressures for cutbacks in service infrastructure, although it should be noted that many Conservative-controlled authorities have been less than enthusiastic about the increasing financial straitjacket imposed upon them. The class composition of local areas also affects the outcomes of rationalization, since vocal middle-class groups may be able to defend services better than less organized communities. The issue of gender is also relevant here. As Webster (1985) notes, not only are the services most vulnerable to cuts dominated by women providers, but they are also the services where the main beneficiaries are women. Women workers dominate many of the services that have been cut – residential accommodation for the elderly, day-care services and schoolmeals – and are left to shoulder the main responsibility for caring for dependent groups following the withdrawal of services that act as family-care substitutes.

In general, it would appear that there are a number of poorer, often discriminated-against and politically marginalized groups that can suffer from policies of fiscal retrenchment. Thus, following the debt crisis of New York in the mid-1970s, many of the blacks who had been recruited into the public sector under equal opportunity programs in the 1960s were laid off (Sheftner 1980).

Privatization and subcontracting

Hand in hand with the rationalization has been the so-called "privatization" of many public services. As Le Grand & Robinson (1984) point out, privatization is not simply a shift from state to market provision, but involves many alterations to state intervention. Privatization can involve a direct reduction in state intervention such as with the sale of public housing, the closure of local authority residential homes for the elderly, the expansion of privately provided medical care, increased reliance upon private insurance companies to provide sickness benefits, and greater dependence upon charities. A second possibility is the reduction of state subsidies, as with the reduction in subsidies to remaining council tenants. A third strategy is the reduction of state regulation, as with the easing of rent controls and the deregulation of transport services. In addition, there are wide variations in the types of institutions that are used to take over direct provision by the state. Some are profit-maximizing entrepreneurs operating in an unregulated environment, some are highly regulated private monopolies, and others are nonprofit-making organizations such as charities, community associations, and consumer cooperatives.

Much of the so-called privatization in Britain might best be described as "contracting out," since it has involved the use of regulated private companies to undertake services such as cleaning, catering, and maintenance that were previously undertaken by the internal labor force of the public institutions concerned. This subcontracting may be seen as part of an attempt by managements to reassert their authority in situations where unions have often had considerable power to dictate the nature of their working practices. Thus, contracting out has usually involved a decline in the size of the workforce, inferior fringe benefits, and the absence of trade unions. Once again, the extent of privatization has varied enormously between different areas. Some right-wing Conservative-controlled authorities have been extremely keen to contract out services such as refuse disposal and school meals, while other left-wing dominated councils have put up staunch resistance to central pressures for privatization. It has even been argued that some district health authorities have resisted putting services out to contract in some hospitals where trade union strength was particularly strong (Paul 1984).

Intensification

Another widely discussed concept in industrial geography in the last decade is intensification – increases in labor productivity via managerial and organizational changes with little or no new investment or major loss of capacity. There are also many examples of intensification in the state sector, since efforts have been made to increase the efficiency of public services defined in terms of the ratio of outputs to given inputs. In the British National Health Service, for example, the number of patients treated per member of staff has been increasing in recent years. Intensification has also been brought about in the public sector by recent policies of redundancies, the non-replacement of retiring staff, and a greater reliance in many services upon part-time work. All these processes put increased strain upon a diminished number of workers. The pressures have been particularly acute in the social services where increased numbers of "at risk" groups such as the elderly, the unemployed, and the poor have increased workloads. The subcontracting mentioned in the previous section has, of course, also frequently led to an intensification of the work process.

Investment and technical change

In the study of industrial restructuring, investment and technical change refers to situations in which heavy capital investment in new forms of production results in job losses. Far less is known about the impact of such changes in the context of services, although many forms of investment are underway in the sphere of non-marketed services. Recent years have seen the computerization of health and welfare records, the introduction of electronic diagnostic equipment in health care, and the use of distance learning systems through computers and videos in education (Gershuny & Miles 1983). In many cases these innovations have increased numbers of workers through the need for more specialist staff. However, there does seem to be widespread agreement that those jobs most vulnerable to technological innovation in the service sector are secretarial jobs, which can be reduced with the introduction of word processing systems.

Increasing self-reliance and partial self-provisioning

As mentioned previously, the work of Gershuny & Miles (1983) has drawn attention to the ways in which technological innovations can also affect households. Changes in the relative costs of goods and services have brought about shifts in the extent to which households buy services in the formal economy or undertake tasks in the home. These ideas are highly controversial and, clearly, technological innovation can also transform the nature of services purchased in the market outside the home. Moreover, these ideas have also been criticized for technological determinism. However, if one also considers social and political struggles, then it is possible to envisage a variety of trends currently leading to increased self-reliance and self-provisioning of all types of work. The fiscal retrenchment in many western economies has been accompanied by moral exhortations for families to become increasingly self-reliant and less dependent upon state benefits. There is, in fact, nothing new in this trend – in Britain the Elizabethan Poor Law stipulated that the poor, old, lame, and blind should be looked after by parents, grandparents, or children (Goldin 1986). The latest version of this idea is the notion of "community care." In theory, decentralized community-based welfare facilities can overcome the disadvantages of large institutions, but in Britain the notion of community care has often been used as a smokescreen to close facilities with little or no compensating support in the community. In North America deinstitutionalization has created large service-dependent populations in inner city areas (Dear 1980, Wolch 1979).

A research agenda

These changes in the structure of collective consumption raise many questions for future research, some of which are listed below:

1 *What are the relationships between the types of pressure, the nature of the organization, and the effects upon workers and client groups?* Much more evidence is needed before we can make any firm conclusions, but there are some indications that tempt speculation. For example, Mohan (1986) notes that, given the political obstacles to wholesale privatization, intensification and subcontracting have been the most attainable objectives in the British National Health Service. In a similar vein, Le Grand (1984) speculates that the ability of governments to restructure services depends upon the relationships between the social class of the producer groups and the social class of the beneficiaries of the services. Where, as in the case of the National Health Service, the middle classes are an important part of the workforce, and also benefit from treatment, it is difficult for a right-wing government to dismantle the structure in any radical fashion without alienating many of its supporters. Where, in contrast, the producers and principal clients are low-class groups, such as with refuse collection or state housing, then privatization is possible. Dunleavy's (1986) work on the political implications of sectoral consumption cleavages is the most sophisticated articulation of these ideas to date. Clearly, much more work is needed to explore these issues for, as indicated above, race and gender also have an impact upon who gains and loses from rationalization.

2 *To what extent have recent changes led to a deskilling and polarization of the workforce in the public and nonprofit-making sectors?* There is now a growing amount of literature relating to manufacturing and, to a lesser extent, to private services about the impact of recent employment changes upon the composition of the workforce. It is also important to examine the extent to which the public sector is reinforcing or ameliorating the effects of private sector changes.

3 *What effect has recent restructuring had upon the composition of the workforce in terms of skill, gender, age, and race?* Again, there is a growing body of literature relating to private manufacturing and services. It is becoming well established that many employers are especially reluctant to take on young people. In some sectors women have suffered as a result of redundancies and non-replacement of jobs, while in other sectors they have gained from the growth of part-time employment. It is clearly of considerable interest to know to what extent the public sector is able to ameliorate these tendencies, and to what extent it reinforces them.

4 *What has been the effect of these changes upon service provision in terms of criteria of equity, quality, and efficiency?* These issues are enormously complex for each of these criteria and can in turn be measured in many different ways. Efficiency can be regarded as the ratio of

resource inputs to a given level of output by an organization (Spencer 1984). Calculations of efficiency will, therefore, concentrate upon the internal costs of the organization compared with contracting out costs. The quality and equity of service provision, on the other hand, relate to the outputs of the service and their impact upon the local community.

5 *What has been the effect of restructuring in the public sector upon self-employment?* Recent years have seen a considerable increase in self-employed persons, many of them women in the service sectors of education, health catering, and cleaning (Daniels & Thrift 1985).

6 *To what extent are there geographical variations in the extent of service restructuring in different parts of Britain?* An important related issue here is the extent to which the political complexion of the political unit, the nature of the service and the character of local industrial relations, affects the type of restructuring that is to be found.

7 *To what extent do public sector services and industries have an effect upon, and to what extent are they affected by, local work cultures in different localities?* Are working relationships consensual, antagonistic, or paternalistic? To what extent are these relationships the result of occupational, sectoral, or locality effects?

8 *How important are variations within the public sector?* Most of the examples discussed above have been taken from local government and the British Health Service. There are, however, many other aspects of the public sector, together with recently privatized services in Great Britain, that require research: post and telecommunications, transport services, utilities, public research institutions, and miscellaneous bodies such as the Ordnance Survey.

9 *To what extent are families (or more particularly women within families) able to undertake self-provisioning?* The answer to this question is again likely to display wide differences between different areas. Whilst modern technology has produced labor-saving machines that ease the burden of domesticity, the decline of the extended family and the breakup of older community networks following World War 2 has lessened the ability of families to cope with the burden of caring for dependent groups. It may be that poorer families in areas surrounded by extensive informal self-help networks can cope with pressures towards self-reliance better than more affluent mobile families.

Concluding discussion

The basic thrust of this chapter has been that the notion of collective consumption, together with the related concept of social reproduction, are both too complex to be encapsulated by any simple watertight definition. Rather than focus upon some essentially static definition – however complex and sophisticated, it is preferable to adopt a dynamic approach which is capable of analyzing the continual transformations being made to the nature of work, both in public and private institutions, in the formal and informal economy, and within and outside households. Underlying many of these changes are attempts to increase the rate of profitability in western economies, but the effects upon service structure cannot be "read off" in any simple manner from the "needs of capitalism." The important task is to keep in perspective the instrumentality underlying many of the recent changes without falling into the various pitfalls of reductionism, functionalism, and the absolute autonomy approach, or ending up with mere catalogues of contingencies. A variety of concepts from the study of industrial restructuring has been suggested to illuminate these processes of transformation, and to indicate the factors that lead to divergent outcomes in different areas.

References

Adams, P. & G. Freeman 1982. Social services under Reagan and Thatcher. In *Urban policy under capitalism*, N. I. Fainstein & S. S. Fainstein (eds.), 65–82. Beverly Hills: Sage Publications.

Bhaskar, R. 1975. *A realist theory of science*. Brighton: Harvester Press.

Castells, M. 1977. *The urban question*. London: Edward Arnold.

Castells, M. 1985. High technology, economic restructuring and the urban-regional process in the United States. In *High technology, space and society*, M. Castells (ed.), 11–40. Beverly Hills: Sage Publications.

Daniels, P. & N. Thrift 1985. *The geographies of the U.K. service sector: a survey*. Mimeo.

Dear, M. J. 1980. The public city. In *Residential mobility and public policy*, W. A. V. Clark & E. G. Moore (eds.), 219–41. Beverly Hills: Sage Publications.

Dunleavy, P. 1986. The growth of sectoral cleavages and the stabilisation of state expenditures. *Environmental and Planning D: Society and Space* **4**, 129–44.

Eyles, J. 1987. *The geography of the national health: an essay in welfare geography*. London: Croom Helm.

Gershuny, J. I. & I. D. Miles 1983. *The new services economy: the transformation of employment in industrial societies*. London: Frances Pinter.

Goodin, R. E. 1986. Self reliance versus the welfare state. *Journal of Social Policy*, 25–47.

Gregory, D. 1978. *Ideology, science and human geography*. London: Hutchinson.

Hägerstrand, T. 1978. Survival and arena. In *Timing space and spacing time. Vol. 2 Human activity and time geography*, T. Carlstein et al. (eds.). London: Edward Arnold.

Harrison, M. 1984. The coming welfare corporatism. *New Society* **67**, 321–3.

Harvey, D. 1984. On the history and present condition of geography – an historical materialist manifesto. *Professional Geographer* **36**, 1–10.

Honey, R. & D. Sorenson 1984. Jurisdictional benefits and local costs: the politics of school closings. In *Public service provision and urban development*, P. Knox, A. Kirby & S. Pinch (eds.), 114–30. Beckenham: Croom Helm.

Humphries, J. 1977. The class struggle and the persistence of the working class family. *Cambridge Journal of Economics* **1**, 241–58.

Humphries, J. & J. Rubery. 1984. The reconstitution of the supply side of the labour market: the relative autonomy of social reproduction. *Cambridge Journal of Economics* **8**, 331–46.

Le Grand, J. 1984. The future of the welfare state. *New Society*, June 7, 385–6.

Le Grand, J. & R. Robinson (eds.) 1984. *Privatisation and the welfare state*. London: Allen & Unwin.

Lojkine, J. 1976. Contribution to a Marxist theory of capitalist urbanisation. In *Urban sociology: critical essays*, C. G. Pickvance (ed.), 119–41. London: Methuen.

Lowe, S. 1986. *Urban social movements: the city after Castells*. London: Macmillan.

Margolis, J. 1968. The demand for urban public services. In *Issues in urban economics*, H. S. Perloff & L. S. Wingo (eds.), 536–66. Baltimore: Johns Hopkins University Press.

Massey, D. 1984. *Spatial divisions of labour*. London: Macmillan.

Massey, D. 1985. New directions in space. In *Social relations and spatial structures*, D. Gregory & J. Urry (eds.), 9–19. London: Macmillan.

Miller, S. M. 1978. The recapitalisation of capitalism. *International Journal of Urban Regional Research* **2**, 202–12.

Mohan, J. 1986. *Spatial aspects of health and employment: historical perspectives and contemporary issues*. London: Queen Mary College, London University, Department of Geography, mimeo.

Mohan, J. 1988. Spatial aspects of health-care employment in Britain, 2: current policy initiatives. *Environment and Planning A* **20**, 203–17.

Musgrave, R. A. 1958. *The theory of public finance*. New York: McGraw-Hill.

Pahl, R. E. 1977. Collective consumption and the state in capitalist and state socialist societies. In *Industrial society: class cleavage and control*, R. Scase (ed.), 153–71. London: Tavistock.

Paul, J. 1984. Contracting out in the NHS: can we afford to take the risk? *Critical Social Policy* **10**, 87–94.

Pickvance, C. G. 1976. On the study of urban social movements. In *Urban sociology: critical essays*, C. G. Pickvance (ed.), 198–218. London: Methuen.

Pickvance, C. G. 1982. *The state and collective consumption*. The Open University, Course D202. Urban change and conflict. Milton Keynes: The Open University Press.

Pinch, S. P. 1985. *Cities and services: the geography of collective consumption*. London: Routledge & Kegan Paul.

Rosenberg, S. 1983. Reagan social policy and labour force restructuring. *Cambridge Journal of Economics* **7**, 179–96.

Samuelson, P. A. 1954. The pure theory of public expenditures. *Review of Economics and Statistics* **36**, 387–9.

Saunders, P. 1979. *Urban politics: a sociological approach*. London: Hutchinson.

Sayer, A. 1982. Explanation in human geography. *Progress in Human Geography* **6**, 68–88.

Sheftner, M. 1980. New York fiscal crisis: the politics of inflation and retrenchment. In *Managing fiscal stress*, C. Levine (ed.), 251–72. Chatham N.J.: Chatham House.

Spencer, K. 1984. Assessing alternative forms of service provision. *Local Government Studies*. March/April, 14–20.

Teitz, M. B. 1968. Towards a theory of urban public facility location. *Papers and Proceedings, Regional Science Association* **31**, 150–7.

Tiebout, C. M. 1956. A pure theory of local expenditures. *Journal of Political Economy* **64**, 416–24.

Urry, J. 1985. Social relations, space and time. In *Social relations and spatial structures*, D. Gregory & J. Urry (eds.), 20–48. London: Macmillan.

Urry, J. 1986. Services: some issues of analysis. *Lancaster Regionalism Working Paper* **17**.

Walker, R. 1985. Is there a services economy? the changing capitalist division of labour. *Science and Society* **49**, 42–83.

Webster, B. 1985. A woman's issue: the impact of local authority cuts. *Local Government Studies* March/April, 19–46.

Wolch, J. 1979. Residential location and the provision of human services. *Professional Geographer* **31**, 271–6.

4

The politics of turf and the question of class

KEVIN R. COX

Introduction

The divorce between workplace and living place concomitant with industrial capitalism is now widely recognized as of immense significance. One result is that the living place is physically separated from the site of working-class exploitation. And following on from this, conflicts in the two spheres have often developed in very different ways. The standard image is that workplace conflicts have been fought more frequently along self-consciously class lines, but this has been much less common for conflicts in the living place.

For Marxists this observation clearly has to be squared with theory regarding the centrality of class struggle in an understanding of society. Accordingly, it is an issue that has concerned all Marxist work on living place politics. This has been apparent in the literature on urban social movements which has been dominated by Marxist writings. Work on neighborhood activism and the related politics of turf has tended to receive less attention from Marxists, but, where it has, the relation to class has also been a prime consideration.

It is the relation between turf politics and class that I want to take up in this chapter. I do so with particular reference to the post-1950 surge in controversies around urban renewal, highway construction, re-zonings, and suburban growth in North American metropolitan areas. This is not to say that I think the arguments have no applicability beyond North America, but evaluating that applicability would require another chapter.

I start out by defining the politics of turf as a politics of collective consumption. After identifying some pertinent issues that arise from a reading of the literature on the topic, I identify some critical social changes with respect to which I believe the emergence of turf politics should be situated. It is within the context of those social changes – Fordism, the socialization of consumption, etc. – and subsequent struggles around them, that one can understand the emergence of the separate sphere: a set of practices and associated meanings, with the living place as their locale, constructed around consumption, status, and family. In turn, the separate sphere has generated its own distinctive politics. This is built around a competition over distribution between different interest groups, a competition that is aided and abetted by state and capital anxious to preserve the separateness of the separate sphere in the division of the product between capital's accumulation fund and labor's consumption fund.

In the second part of the chapter I approach turf politics more directly, by situating these interests and relationships with respect to location. I place considerable emphasis upon the concept of local dependence as the root of a necessary politics of location: this refers to the problem of spatial fixity, and hence to the interests which both capital and labor acquire in location.

Two types of location politics around living place issues are considered: a class politics corresponding to a class politics of collective consumption; and a territorial politics, which is a projection on to space of competing distributional groupings in the consumption sphere. Given the fact that the politics of turf is a territorialized politics of location, the question of relating it to class then becomes one of understanding why it should exist when there is an alternative class politics of location. Of crucial significance here are the activities of growth coalitions. Businesses, locally dependent at metropolitan, urban, even neighborhood levels, come together with branches of the state – city or county, for example – around development programs whose goal is channeling value through the locality on which they are dependent. These policies, however, threaten labor's collective consumption, and this in turn poses a threat to capital's accumulation fund. The problem is one of channeling subsequent tensions away from a class axis and along one of competing interest groups: an important form of this has been the territorial.

This channeling of tensions has been achieved through a variety of political structures which tend to incorporate/exclude, not necessarily intentionally, along territorial lines; and also through a discourse of territoriality. The political structures in question provide opportunities for cooptation across class lines and for dividing territorially. They form a necessary precondition for a concomitant discourse at the center of which are localities, organized horizontally and hierarchically, competing with one another by structuring flows of capital and labor to what they see as their respective community advantages. This obviously makes a particular sense of growth coalition politics. Within the context of this discourse, it is then possible to re-define growth coalition threats to labor's use values in the living place, which could quite conceivably become class issues, into the innocuous terms of the territorial – of city versus neighborhood, perhaps, or city versus suburbs – that are so characteristic of the politics of turf.

This is not to underestimate the receptiveness of labor to this type of ideological blandishment. Privatization around family, status, and consumption produces its own distinctive politics: one which is antithetical to the solidarities of class and long-term commitments; and more amenable to the opportunism and search for the immediate "fix" which capitlist interests have always found so congenial and easy to exploit.

The politics of turf defined

The concept of turf politics is deceptively straightforward. Although a number of core meanings attach to it, their resituation with respect to different contexts adds significant complexity. A central focus is that of collective consumption: the collective consumption (or, more accurately, experience) in the living place of such values as education and physical amenity; and of such dis-values as congestion and local taxation. Collective consumption has important spatial aspects. This accounts for the pervasive territorial imagery that goes along with the concept: notions of exclusion or inclusion by territorially-based coalitions of residents, and of the spatially competitive character of those processes (Williams 1971, Wingo 1973). Of primary significance in achieving inclusion and exclusion is the power of the state.

Interest in the concept has closely followed the appearance of something approximating it in the real world. The urban historian Zane Miller tried to capture some of the flavor of this in his study (1981) of the development of a Cincinnati suburb:

> After the mid-1960s, then, the metropolitan area seemed to be composed of a melange of service and economic areas, varied in size, which existed to foster pursuit by individuals of personal goals and objectives. Each of these areas competed for economic resources, power and 'top notch' citizens, and the competition not only pitted Cincinnati against its suburbs but also big city neighborhood against big city neighborhood, suburb against suburb, and neighborhoods within a particular suburb against one another. Each of these communities, in other words, comprised a community of advocacy. And the larger units, such as Forest Park or Cincinnati, constituted a community of advocacy made up of smaller communities of advocacy (Miller 1981, 239).

Other less casual observations have documented the idea that something new happened in the sphere of urban politics after about the mid-20th century. A study of neighborhood organizations in Seattle over the period 1929 to 1979 (Lee *et al.* 1984) documents a contrast between an earlier emphasis on social functions (street carnivals, pageants, flower shows, construction of community clubhouses), with a more contemporary and political orientation. More specifically, this is concerned with warding off threats to interests deriving from homeownership and stage in the life cycle. To the extent that neighborhood organizations in the earlier part of the century had political activities, it was largely a matter of pushing for local improvements and was, therefore, non-exclusionary in form: acquiring street lighting and street paving, for instance.

Studies of neighborhood activism in Columbus, Ohio, over the period 1900 to 1980 are broadly supportive of this change (Cox 1984, 288–92, Sutcliffe 1984). Using content analysis of issues reported in a local newspaper, a contrast is drawn between a later exclusionary type of activism which emerges with vigor during the 1950s, primarily around re-zoning issues; and an earlier activism which largely assumed the form of petitioning for improvements – better roads, mass transit, street cleaning, sewer provision, fire stations, street lights, and so on. The other major conclusion from this work is the immense surge in activism occurring after 1950.

However, although the imagery attaching to these practices has been primarily a (crudely) spatial one, there is also a minority view which makes the concept a little richer. This sees the politics of turf as more fundamentally a politics of status and of relative, as opposed to absolute, consumption advantage. This is the stand taken by Danielson (1976) in his study of exclusionary zoning. Studies of housing classes in spatial context (Rex & Moore, 1967 Saunders 1978) have adopted a similar view.

Given this, it is not surprising that alongside these positive approaches there have emerged standpoints of a more normative nature. There are, for example, concepts of turf politics which we may dub "liberal reform," and which emphasize its perversely redistributive character (Downs 1973, Newton 1975, Danielson 1972). Partly in rebuttal of this is a public choice literature of a more conservative cast emphasizing the efficiency properties of resultant residential segregation; and also of those variations in public provision which liberal reformists see as the product of residential exclusion (Ostrom *et al.* 1961, Bish & Ostrom 1980). One of the interesting things about this more normative literature, and something which adds significant complexity to the concept, is that it makes what is ostensibly a *local* issue into a national issue. Concern over the redistributive effects of so-called exclusionary zoning, for instance, has been the focus of national policy debate. This is one of the central paradoxes that must be confronted in any attempt to link turf politics to questions of class.

Having said this, there are certain points of a more critical nature which should be made in any attempt to decipher the politics of turf. In particular, there are some major aspects to the politics of turf about which the literature is silent. Four issues seem to me to be crucial:

(a) *The role of class.* Although status plays a role in a number of these accounts, class, as a relation of production, and the point of origin for the central tensions of capitalist society, is conspicuously absent. This is despite Harvey's (1978a) suggestion of these conflicts as somehow "displaced" forms of class conflict. Empirically also, in a number of conflicts the issue seems to have been less one of competition between fractions of labor for a fixed consumption fund; and more one of conflict with capital over the magnitude of that fund. The writings of Mollenkopf (1976) and Clavel (1986) on growth coalition politics are apposite, as is the work of Fincher (1982).

(b) *The constitution of interests.* It is clear where class comes from. More enigmatic are interests in neighborhood schools, or in neighborhood as a status symbol. Even more problematic, because it is cut loose from that of "interest," is the notion of "preference" apparent in some (unselfconscious) treatments of the subject. For what is being consumed, and why, or why a neighborhood externality is important, is not always self-evident; use values are bound up in webs of meanings and associated structures of relations including, somewhere along the way, that of class.

(c) *The role of the state.* The politics of turf is one in which the state is crucially involved. Although Castells (1977) recognized this in his early discussion of social movements, this is not an insight that has penetrated the – quite separate – turf politics literature. For the most part the role of the state is taken for granted. This is with respect to the state as precondition: in defining expectations regarding consumption, in providing a source of accessible power, in constituting coalitions through its interventions, as in zoning, and in giving a material basis to the concept of collective consumption. Also taken for granted is the historical emergence of the state: the state is not simply a precondition, it is a precondition with a history.

(d) *The role of space.* Whatever nuanced meanings we wish to attribute to it, turf politics is undeniably a politics of location. The politics of location, however, is not an unproblematic category. It touches on questions of the relation between mobility and immobility which by and large have not been addressed either in the turf politics literature or in the broader literature on the politics of location. There is some limited recognition of the importance of this in discussions of the tradeoff between collective action and relocation, but for the most part this is considered only in the most idealistic of terms (Orbell & Uno 1972, Wingo 1973, Fainstein & Fainstein 1980).

Yet the question of the relation between mobility and immobility returns us once again to the question of class. For we are talking about mobility and immobility as refracted by a particular form of class society. Likewise, class and class conflict form necessary preconditions for the role of the state in the creation of a politics of turf; and also for understanding the production of those interests in particular use values with which that politics of turf is characteristically associated. The adequacy of my interpretation of turf politics, therefore, rests on the degree to which I can convincingly link class to these other issues about which the literature has been so silent. How successful I am is for the reader to judge.

Fordist social relations

In the first place I want to situate class struggle with respect to that distinctive ensemble of capitalist social relations known as Fordism. This concept captures many of the central themes described elsewhere under such rubrics as "late capitalism," "monopoly capitalism," and "an intensive regime of accumulation," and designed to encapsulate the distinctive character of post-war capitalism. Since 1970 there has been a sense among academics, at least, of its disintegration. In its place is seen the possibility of a new accumulation regime: so-called "flexible accumulation." I address the issues raised by this social restructuring in the conclusions to this chapter.

First and foremost, Fordism is a term used by the French Regulation School to indicate a regime of accumulation underpinned by a distinctive relation between production and consumption: a regime, that is, in which the departments producing means of consumption and production respectively are brought into close relationship one with another through a revolution in consumption (Aglietta 1979). Through enhanced wages underpinned by changes in the relation of capital to labor and tied to steadily increasing productivity levels, capital is able to solve a problem of effective demand. Other concepts such as that of the Keynesian welfare state (Offe 1984, Ch. 8) capture aspects of this, while the monopoly-competitive capital dualism underlines the uneven establishment of Fordism within the American economy.

A crucial precondition for this articulation has been substantial state intervention into the reproduction of labor power. This has been crucial in two distinct ways. On the one hand there has been intervention into the capital-labor relation. This has included the legalization of labor unions and the granting to them of powers of negotiation. In addition, welfare state legislation – covering minimum wages, unemployment compensation, social security – has served to provide organized labor with the bargaining power it needed to exploit the formal powers granted it by the state. Without these state mediated powers it seems unlikely that substantially improved terms of employment could have been imposed on capital.

An important consequence of this state intervention is that consumption comes to be mediated more and more by commodity exchange. This is in contrast to a previously more substantial use of mutual aid within the extended family or the ethnic neighborhood; and a heightened use of inputs of domestic labor within the household (Aglietta 1979). What has been more obvious, however, has been the mediation of consumption by the state. By a mixture of regulation, redistribution through taxation, and actual provision, the state has sought to substitute both for mutual aid and for commodity exchanges.

A second crucial area of state intervention has been in formal education. Fordism is distinguished by a massive expansion of secondary, college, and university education, and for good reason. Fordism as an economic system rests firmly on what Marx described as the real subsumption of labor to capital (1976, 1019–25). Only through that intervention into the labor process, which Marx intended to signal by this concept, can capital hope to revolutionize it and so lay the basis for those increased levels of productivity on which increased consumption depends. Expertise – managerial, technological, scientific – therefore, acquires a premium; and the labor market comes to be mediated to a very significant degree by the possession of qualifications acquired in institutions of education.

66

The establishment of the separate sphere

Obviously, Fordism has a history. Familiarity with that history affirms the role played in the creation of its preconditions and in its instantiation by class struggle. On the one hand there is the real subsumption of labor to capital which created the technical basis for mass consumption. On the other hand, the welfare state is incomprehensible outside of the sustained pressure of the working class.

In turn Fordism has provided a context, a set of conditions, for subsequent struggles. Broadly speaking, these struggles have not been to labor's advantage. Rather there has been a progressive displacement of the (related) prisms through which political issues are judged: from class to competition; from production relations to consumption; and from equality of outcomes to equality of opportunity. Hirsch has written of a "thorough capitalization of the whole society (commodity form of social relations, individualization, and social disintegration)" (1983, 76).

In brief, the net effect of capitalist activities has been to construct a separate sphere for labor around living: a realm of experience, that is, which is in sharp contrast to the experience of working. In order to make the constraint, alienation, and meaninglessness of the workplace acceptable, labor is invited to build new worlds of freedom, humanity, and significance beyond the factory gate.

Already in the late 1920s, Lynd & Lynd (1929) were struck by the orientation of workers to consumption: something which they regarded as quite novel, and which they linked to a problem of meaning subsequent to deskilling. Thus, "for the working class both any satisfaction inherent in the actual daily doing of the job and the prestige of kudos of the able worker among his associates would appear to be declining" (72). So

> [T]his whole complex of doing day after day fortuitously assigned things, chiefly at the behest of other people, has in the main to be strained through a pecuniary sieve before it assumes vital meaning. This helps to account for the importance of money in Middletown, and, as an outcome of this dislocation of energy and expenditure from so many of the dynamic aspects of living, we are likely to find some compensatory adjustments in other regions of the city's life (Lynd & Lynd 1929, 52).

Consumption has become a primary means of expressing social significance. As such it has displaced more multiplex status systems deriving partly from skill in the workplace and also from close personal relations with others. The other marker of social status is occupation. The mental-manual divide has assumed enhanced importance. While parents want their children to obtain the educational certifications that will facilitate upward mobility, this is not simply because of the access it gives to increased consumption. As Sennett & Cobb have pointed out (1973), interpretive functions in the division of labor are seen as bestowing not only a freedom and control absent among manual workers; they are also seen to endow, for the same reason, respect. Accordingly "the great pressure toward education on the part of the working class is . . . another phase of this desire to escape to better things" (Lynd & Lynd 1929, 80).

Aside from consumption and related interests in social status, the third pillar on which the separate sphere has been constructed is that of the nuclear family. It is the family that has provided the basis for a privatized existence built around consumption; the family has also functioned as the principal mediator of upward mobility. Building on the opportunities implicit in the welfare state for ridding the family of its external appendages, it has been reconstructed as a "haven in a heartless world" (Lasch 1977), a center of calm in a world in flux. It has become a social relation in which people can be valued for their own sakes and where human agency can be reaffirmed, if only in the form of sacrifice (Sennett & Cobb 1973, Ch. 2): a child-centered family, therefore. Child-centeredness and interests in upward mobility combine to give point to childrearing. The education of the children comes to be a touchstone of family life, affecting all social relations.

Consumption, status, and family have thus become the key prisms of people's lives. Class plays a quite subordinate role. In implementing this vision capital, as a condensation of class power, has had immense advantages. Most importantly, through the long boom capital has been able to convince that Fordism and the welfare state deliver: the idea of enhanced consumption has, by and large, been realized. Further, there has been its ability to control discourse through control of the media and through advertizing (Ewen 1976). Part of the success of the advertizing industry and of the mass media in this regard is due to the severe diminution of extended family and neighborhood as sites of discourse. This in turn is related to the socially atomizing impacts of the welfare state, and to the emergence of the privatized nuclear family as a preferred site for consumption.

The politics of the separate sphere

The effect of state intervention in consumption processes has been to spawn a massive politicization. Through its interventions the state, in effect, legitimates certain standards of, for example, health, nutrition, housing, education, and income. Given its dependence on mechanisms of capital accumulation over which it has no direct control, however, it cannot hope to maintain those standards for everybody at all times. The subsequent danger is that under electoral pressure the state will upset the balance between consumption and accumulation – and to capital's detriment. A problem from capital's standpoint, then, is whether or not the separate sphere can indeed be kept separate. This also becomes a problem for the state, given its dependence on the accumulation process. To the extent that the separate sphere *can* be kept separate, a politics of consumption organized around class can be rebuffed; tensions can then be channeled into a politics in which interest groups form in the consumption sphere and have to compete against each other.

According to Hirsch, an important way in which the state has moved to limit consumption demands has been through mechanisms of what he calls mass integration. He describes a "system of modern 'mass-integrative' apparatuses (large bureaucratic parties, mass alliances, labor unions and especially those organizations which relate to the laboring class)" whose function is to "control . . . social conflicts and labor struggles; and . . . at the same time, to allow for material demands, while also filtering and reducing them to a size which conforms to capital's requirements" (1981, 597). Elsewhere Hirsch has commented on the specific mechanisms through which this filtering is achieved:

> internal compromises, mobilizing counter-interests, playing them off against each other, diversion, delay, postponement and reduction to the possible. The functional level of this process is the self-interest of the ruling groups (i.e. professional politicians and civil servants) in the preservation of the prevailing social structure, to which they owe both their material existence and their capacity to exercise power (1978, 224–5).

Although in making these arguments Hirsch is basing his case primarily on West German materials, he also makes it clear that his arguments have a broad applicability to all advanced capitalist countries (1983, 81). In the American case there can be little doubt that the cooptation – not to say, formation – of interest groups by the political parties works substantially in the direction of limiting consumption demand and in many of the ways anticipated by Hirsch. Incorporation imposes a bargaining mode of operation on organized interests. The playing off of one interest against another within, or between, parties, serves to structure that bargaining and to convert the politics of consumption into a politics of competing interest groups: the demands of environmental groups are counterposed to those of consumers and employees; demands for enhanced welfare spending to those of taxpayers.

A necessary precondition for this politics has been the hegemonic role of an ideology of growth. Appealing to widely held interests in enhanced consumption, this belief made its historic entry during the post-war recovery (Mishan 1967, 27). According to Mishan "economic growth has become an official feature of the Establishment" (1967, 27). Increased consumption, it is argued, is dependent on continuing economic growth. Growth depends on profit. If wages, private or social, increase relative to profit, then growth will be undermined and so will increased consumption. Appetites for increased consumption need, therefore, to be controlled and a correct order of national priorities maintained.

Given dependence on the rate of accumulation, the state has a clear stake in this ideology. Accordingly it provides the basic assumption for bargaining within the state and within and between its apparatuses of mass integration. It is the terrain on which the technocracy makes its appearance testifying to the detrimental effect of this or that measure on the rate of economic growth, national competitiveness and the like. And it is the ideological fulcrum around which capital constructs its own public pronouncements on policy issues.

This does not mean that class is, or can be, entirely abolished from popular politics as a realm of experience. This is especially so for those for whom economic salvation through the rules of the system appears remote. This will apply to those left out of dominant Fordist structures: to those locked for whatever reason into the relatively unprotected, low pay, competitive sector of the economy. And there are those for whom, by dint of hard experience, the possibility of climbing, individualistically, some ladder of opportunity seems an absurdity. To these groups the appeal of salvation – through collective action aimed at the enlargement of labor's consumption fund at the expense of capital's fund for accumulation – may have real appeal.

The politics of turf: an interpretation

Politics, however, is also spatial. Conflicting interests are in particular localities and, as will be pointed out shortly, necessarily so. So a politics of location is inevitable. Yet a critical question is: what sort of politics of location will merge? For while a politics of location is necessary, its specific forms are contingent matters.

We have seen that to the extent that popular politics comes to be regarded as revolving around questions of collective consumption, it can assume two broad forms. One of these facilitates class interpretations, while the other is quite antithetical. A distinct possibility is a politics organized around an opposition between capital and labor: between interests in maximizing the magnitude of capital's accumulation fund on the one hand, and interests in maximizing labor's consumption fund on the other. However, it is also possible to see the politics of collective consumption organized around the antagonistic relations of different consumer groups.

Aligned with this distinction are two distinct types of politics of location. Alongside a class politics of collective consumption it is possible to define a *class politics of location*. Just as capital's accumulation fund interacts with labor's consumption fund, the relation has a spatial expression. For the geography of capital accumulation is from one standpoint a geography with respect to worker's living places. There is a pattern of physical facilities, of factories, highways, housing, of rates of change in physical facilities as in urbanization, which impinges, through diverse externalities, upon workers in their places of residence. There is also, though harder to identify, a fiscal geography indicating, across territorial units, the fiscal burden carried by labor in support of accumulation.

Expressing the intersections of space and social relations in this way, of course, implies a distinctive antagonism of interests and, therefore, a distinctive politics of location: one in which people in their living places interpret conditions there as also conditions for capital accumulation; and who therefore come to see the drive for accumulation, however inadequately they conceptualize it, as the fundamental parameter of their collective consumption. According to this view, actual outcomes represent the changing balance of power between capital and labor.

Alternatively, it is possible to define a *territorial politics of location*. Again, there is a particular view of spatial patterns: but here, interpretations of the geography of collective consumption are less in terms of class, and more in terms of the relations of localized consumers one with another. In this case, however, it is a spatiality – of externality fields, tax bases, public provision, children's peer groups, schools – mediated less by the capital-labor relation and more by the spatial relations of localized consumption groupings. Assuming priority are the relations of neighborhoods, suburbs, with each other and with respect to a spatial flux of people and dollars. Spatially defined coalitions now displace social classes as the antagonistic, politically organized entities. Outcomes are defined not by class struggle, but by a politically mediated competition of territorial coalitions. This competition is aimed at a selective filtering of the spatially mobile, attracting some elements in and keeping others out: Zane Miller's competition for "economic resources, power, and 'top-notch' citizens" (see Miller 1981, 5).

The necessity of a politics of location

Distinctive class and territorial versions of the politics of location are but contingent outcomes of social relations rooted in the spatiality of capitalism. It is, of course, a truism that social relations are necessarily spatial: in order to exist a social structure must occupy space, and the relations through which it is defined must transcend it. However, Sayer (1985) has argued that this tells us nothing about *specific* locations: they are a contingent matter. While this is true, quite how helpful or unhelpful this necessary relation is in understanding location depends upon the level of abstraction at which a specific piece of research is pitched. If the necessary is defined at a high level of abstraction, as is implied by Sayer's example of "capital," then it is, of course, extraordinarily *un*helpful, and Sayer's pessimism is justified. On the other hand, at lower levels of abstraction, defined perhaps by the needs of firms for particular use values, it seems that we can say much more about specific locations than Sayer's statement would admit. This is certainly implicit, for example, in Allen Scott's work on the locational implications of vertically disintegrated labor processes (see, for example, Scott 1985).

This necessity of location within particular structures is what Andrew Mair and I have tried to capture elsewhere through our concept of *local dependence* (Cox & Mair 1988). People depend for their reproduction on social structures existing at various geographic scales. For diverse reasons, language perhaps, or complementarity of social roles, people are locked into, immobilized within, these social structures. Given limits to the stretchability of social relations, reproduction depends not only on reproducing a social structure, but also on reproducing its distinctive spatial patterning. This local dependence thus exercises constraints on individual relocation. It also gives the individual an interest in the location of others.

Under capitalism social structures are defined by commodity exchange. What tends to bind people into particular structures of exchange is, firstly, the fixity of the physical facilities: fixed capital, that is, which must be amortized, often over a protracted period of time. And secondly, there is the fixity of commodity exchanges themselves. Commodity exchange can be usefully viewed as a learning process in which not all buyers or sellers of the same use value are viewed equally. Rather, through trial and error they come to be distinguished according to their trustworthiness, responsiveness, and predictability.

In this context of spatial fixity of structures of commodity exchange, and in the language of mainstream economic analysis, the opportunity costs of relocation can be considerable. Avoiding these opportunity costs depends upon the degree to which value continues to flow through social relations, i.e., whether or not commodity exchanges continue to be realized.

Given the dynamic character of capitalism, and not least its *spatially* dynamic character, local dependence will be experienced as contradictory: as a contradiction, that is, which must be suspended if the economic agent is to reproduce him/herself. We encounter here the fundamental reason why the antagonistic relations of capitalism inevitably have a spatial moment: interests, be they of capital or of wage labor, are attached to particular locations. It is not just a wage or a profit that is at stake, but a wage or a profit in a particular locality. The exclusionism in which both capital and labor have periodically indulged is only the most obvious symptom of this elementary relation.

To some degree it may be possible to socialize these risks. For firms this may be through the spreading of risks across diverse locations, as in a multi-locational, branching structure. Socialization may also occur through the state. The state has assumed responsibility for many of industrial capital's fixed investments. State sponsored programs of unemployment compensation allow a pooling of risks for locally dependent labor. The result of these ameliorative strategies, however, is not to abolish the tensions that result from local dependence: rather it is to shift them to another geographic scale. Instead of being dependent on a context of commodity exchanges defined by a particular metropolitan area, the necessary context is defined by the region within which the firm has acquired branches. Or dependence on local philanthropic structures for unemployment relief is traded for reliance on a state-sponsored welfare system. A politics specifically of location is, in consequence, unavoidable under capitalism and cannot be abolished, though how state-mediated it is, and what forms it takes, are contingent matters.

With this as background, we turn and consider in more detail the respective local dependencies of capital and of labor. Interest in particular attaches to local dependence in late 20th-century North America and how it affects conflicts within urban areas. In broad principle, however, the arguments will apply to other advanced capitalist societies.

THE LOCAL DEPENDENCE OF CAPITAL

The local dependence of firms is rooted in a combination of the literal fixity of fixed investments of long life; and the particularization of commodity exchanges. Within characteristic contexts of distance-constrained commodity exchanges this means that one can define, at least theoretically, some locality at some scale, upon which the firm is dependent. Thus, and to exemplify: a public utility, due to its heavy investments in distribution networks, would be a classic case of local dependence due to the immobility of fixed facilities; the independent insurance agency or law firm with its heavy reliance on personalized relations with customers and "brand loyalty" nicely exemplifies that local dependence which owes more to the fixity of the commodity exchanges themselves.

Yet while local dependence is necessary, it is not easy to predict which types of firm will be locally dependent at which scale. One generalization that can be offered with some confidence is the tendency for most industrial capital to have become dependent on localities at substantially larger scales than those defined by the metropolitan. There has been a marked delocalization of industry resulting from large-scale branching networks or the establishment of national markets which has been associated by many observers with a withdrawal from local politics.

In other cases delocalization has been blocked by legislative fiat. In the US, anti-trust legislation has tended to confine public utilities and banking to highly localized markets. Not surprisingly, they have been vigorous supporters of attempts to channel investment into the metropolitan areas forming their major markets. In this they are usually joined by locally owned media, particularly newspapers, those wholesaling and retailing firms which confine their operations to particular local markets, and smaller builders and developers. At lower levels within the metropolitan area other structures of local dependence can also be observed. Typically, independent suburbs have clusters of interests in local suburban growth: the suburban newspaper, small retailers, landowners, and construction and builders' supply firms in particular.

THE LOCAL DEPENDENCE OF LABOR

A fundamental spatial effect of capitalism has been the divorce of workplace from living place (Harvey 1978b). This has had important ideological implications, not the least of which are the ensuing problems of drawing the links between conflicts experienced in the two locales: conflicts which are experienced as distinct and unrelated to any subsuming social mechanism. The subsequent mystifications have been strengthened by disparate patterns of local dependence.

It is possible to argue that in terms of access to the means of subsistence there has been a considerable weakening of dependence at very local, submetropolitan levels and a growth in dependence at broader scales. There has been some socialization of risk to state or national level or to both through programs of unemployment compensation and the more general "safety net" of the welfare state. At smaller scales, enhanced personal mobility, either through public transport or widespread automobile ownership, has allowed some socialization of risk to the metropolitan level. The effectiveness of this has been increased by the diversity of employment available in metropolitan areas and the resultant possibilities of risk spreading within households through the employment of spouses.

In terms of many aspects of collective consumption, however, local dependence has tended to remain – may even have become more firmly rooted – at very local, neighborhood, definitely submetropolitan, levels. Principal, though not exclusive interest attaches here to homeownership. For diverse reasons, ranging from the home as an expression of one's individuality, to the transaction costs of moving, homeownership is an extremely immobilizing relationship. At the same time, everyday life is situated. Eventually this gives practices a habitual nature creating a small world of familiarity and confidence: the source of Marris's (1975) conservative impulse.

That the resultant local dependence is experienced as a problem is a result of metropolitan development and housing market processes. These connect one neighborhood to another and to the city as a whole, exposing the individual household to influences emanating from the furthermost corners of the metropolitan area. To a degree, concerns are expressed in terms of use values. The anarchic character of the land development process places heavy and localized burdens on the collectively-consumed generating concern about congestion. The intrusion of the "facts of production" into the living place produces similar concerns.

Other issues involve exchange values. In the context of the fragmented local government of metropolitan areas, one of these is fiscal. New development may boost local tax bases but can also, and ultimately, impose new tax burdens in order to mitigate subsequent congestion. A more pervasive concern tends to be that of property values. The critical issue here is the flow of value through the housing stock. To the extent that the flow of value is not maintained within the context of average turnover time for the housing stock, values will fall; ultimately this may be intensified by the refusal of home finance agencies to grant mortgage loans.

The significance of this is only partly related to the fact that the house represents a substantial part of the average household's savings. It is also related to maintaining current patterns of externalities through which the collective consumption aspects of neighborhood are defined. What is at issue here is the role of property values in mediating neighborhood change: for example, change in social composition, change in land use patterns, change in local tax base and public provision.

Homeownership and the (spatially) habitual nature of everyday life tend to interact. For reasons suggested earlier, homeowners tend to be relatively immobile. On the other hand, homeownership may be entered into precisely because of feelings in favor of stability in day-to-day routines. This type of argument is particularly likely in the context of school age children and their socialization/schooling needs.

There is, however, more to the local dependence of labor in the living place than this might indicate. It is certainly not a relation confined to more affluent homeowners. The spatial entrapment of low income segments of the population has often been remarked on. An effect of the housing market, in combination with the exclusionary zoning of more affluent neighborhoods, is to confine the poor to the – residentially – least desirable sectors of the city. This may be intensified by racially discriminatory housing effects. Relocation for most, therefore, is not a serious option. In this context the battle to maintain consumption standards can only be fought through some sort of collective action: rent strikes, opposition to gentrification, demands for enhanced city services, for police protection, and so forth. Remnants of non-capitalist community and their immobilizing effects, particularly among recent immigrant groups, may intensify the likelihood of these responses.

All these interests, however, only make sense in the context of the widespread alienation of social relations implied by collective consumption and subsequent commodification of neighborhood. Collective consumption, especially in the form of the welfare state, made redundant those social structures which were formerly the basis of survival in an uncertain world. This was most clear in the case of the family, and the kin-based neighborhood. Ethnicity and church provide other examples. What had been necessary relations between people – ones of mutually defined responsibilities and freedoms – were replaced by much more contingent ones: whom one related to in the living place was no longer necessary, merely possible. Likewise reproduction, both material and ideological, was no longer assured. What had been internal to the individual became external.

On the side of the individual the disintegration of old unities has implied a considerable strengthening of conceptions of individual preference and choice. On the other side, that of society, there are collective consumption goods (and bads): public provision, tax rates, externality effects. As a result neighborhoods are not evaluated as, in Wingo's parlance (1973), experiential environments: bundles of externalities and publicly provided goods and services. Housing values are bid up or down accordingly so that "neighborhood" becomes commodified in a much more extensive form than hitherto. Instead of appearing as a relation between people, which it really is, rent now appears as a relationship between neighborhoods. As such "neighborhood" becomes available for ideological mobilization into a location politics of a territorially fetishized sort: one grounded in concepts of neighborhoods, localities, competing and doing battle one with another.

Harnessing the state

Consider now, once more, the problems of local dependence of capitals in metropolitan areas. Typically, strategies aimed at ameliorating the problem have focused on some sort of socialization of risk, as in the shift to a multi-locational structure, or through the aegis of finance capital. Not all strategies have been a matter of individual initiative. In the 19th century industrial capitals often came together behind programs, usually sponsored by the local chamber of commerce, to attract new and complementary industries into the area. More recently growth coalitions of this nature have tried to harness the powers of the state to the overall end of socializing risk.

In some cases the state itself has assumed some of the risks of fixed investment of long life. Local government investments in industrial estates would be one instance, and the ownership of downtown shopping center developments another. There is also a state sponsorship of convention centers, sports stadia, which could quite realistically be developed privately and which have substantial spinoffs for local hotels and retailing.

More common, though not necessarily unrelated, is enlisting the powers of the state to make the locality attractive for new investment: in short, to expand the demand for the products and services of locally dependent firms. At the level of city government resultant programs have ranged from the granting of tax subsidies to incoming businesses, to the spatial restructuring of the city through urban renewal and freeway construction programs. The state, moreover, is not an unwilling partner in these ventures. Collectivization of consumption implies dependence on a localized tax base. This has affected local government as much as it has branches of government at other geographical scales.

The upshot of all this is the formation of coalitions between (locally dependent) state and capital around programs of development, or even around specific projects. These coalitions vary in their consistency over time and in the degrees to which they are able to develop cohesive programs. Cities may have identifiable growth strategies as in the post-industrial city strategy of many of the larger cities in the US. In other cases internal conflicts of interest may prevent this. Membership, therefore, also fluctuates. What is for the most part a collaborative exercise between local government and a business-organized growth committee may at times and for specific projects have as its only active protagonists a local government along with a particular firm or landowner, or with a sector-specific business group: say, the local tourism council or a retail developers group.

To these features should be added the scale independence of the phenomenon. Alongside some sort of metropolitan growth committee struggling to attract investment into the area as a whole, there will be more localized organizations aiming to attract resultant investment and its spinoffs into the particular parts of the metropolitan area on which they are locally dependent. Independent suburbs typically have their own growth coalitions. Further, with the transformation of the economic geography of the area subsequent to development new tensions are set in motion between the old and the new, stimulating further struggles over the flow of value. Accordingly, a common effect of the arrival of suburban shopping centers in suburbs has been the organization of retailers and property owners in older shopping areas around programs of offstreet parking, some modest redevelopment, tree planting and the like.

Relations between growth coalitions extend vertically as well as horizontally. For growth coalitions at lower levels attracting in the "right sort" of investment may hang on the securing of subsidies by local government from state or federal sources: this may have to be traded off against limits to local discretion. The locational programs of those governments themselves become an issue: the location of state and federal offices, or of a county landfill. Alternatively, the withdrawal of subsidies or the closure of some public facility can present obstacles to the realization of growth coalition aims.

This suggests that the geography of accumulation is mediated by growth coalitions at diverse geographic scales. Locations are fought for and fought over in order to structure the flow of value to the advantage of locally dependent firms. In this struggle, growth coalitions inevitably threaten labor's use values in the living place. Programs of spatial restructuring, new growth, tax concessions for business, undermine labor's (collective) consumption fund. This may not happen in all places at all times. Costs may be externalized in various ways, possibly on to higher levels of government, mitigating the likelihood of conflict. Given the structural characteristics of a capitalist economy, however, there are serious limitations to this process. In some places and at some times, therefore, tensions will accumulate to the point at which some sort of opposition is likely to be forthcoming, particularly from those fractions of labor which are more locally dependent. Furthermore, a second aspect of local government's own local dependence – its dependence on a local electorate – means that these protests cannot be ignored. There will in consequence be a politics of location. The important question is, however: what sort of politics of location will it be?

The dominance of territorial politics

In contemporary North America, the dominant impression is one of territorial responses: a rhetoric and practice, that is, of neighborhood "clout," of "not-in-my-backyard," of "beggar-thy-neighbor," and of "exclusion" and "inclusion" processes; in brief, what might be called a spatially fetishistic politics.

The territorialized form of location politics is dominant, but it is not exclusive. There is some evidence in American cities of a class politics of location. This has been apparent in battles over urban renewal (Hartman 1974, Mollenkopf 1976, Fincher 1982). Sometimes these class-based initiatives have self-consciously opposed the overall strategies of growth coalitions on an inter-neighborhood basis. Such was the case with DARE (Detroit Alliance for a Rational Economy), a largely black organization in

Detroit opposed to the spatial restructurings and tax policies employed to convert the city into a post-industrial form (Hill 1983).

Usually this class politics of location is associated with lower income populations whose dependence on a locality has very little to do with homeownership. In part it is often a matter of the remnants of local community: of the social structures developed by poor Appalachian whites to mediate the rural-urban transition, for instance. Partly, local dependence may be a product of externally imposed barriers to movement elsewhere: housing market, zoning, and housing discrimination constraints.

Clearly, to the extent that a class politics does assume predominance it becomes a threat to capital in general and to locally dependent capitals in particular. Given the electoral dependence of local government, popular approval for growth coalition initiatives is critical. In consequence, dissent must be contained and not allowed to congeal along class lines. Rather, opposition needs to be channeled into much more innocuous forms, an important instance of which has proven to be the territorial: in this way the electorate can be organized into supporting rather than opposing growth coalition projects, and policies. Major and complementary roles in this process are played by spatially-based structures designed to coopt the opposed; and a spatial discourse focusing on territories and their interrelations.

STRUCTURES OF INCORPORATION

Significant in the first place, then, are institutional structures whose effect, intentional or otherwise, is to structure the access to policy making of the different groups having a material interest in it: to selectively limit, therefore, rights to influence local policy and to the means, particularly fiscal means, for implementing it. There is, in other words, a selective incorporation of interest groups into the local branches of the state, and an exclusion of others. In some respects the structures developed are similar to those one recognizes at the national level as "corporatist"; in other cases they lack the purposefulness inherent in that concept.

Specific forms of incorporation include the appointment of interest group leaders to these *ad hoc* committees of local government often set up as a response to growth coalition initiatives. Local labour leaders may find themselves being appointed to committees exploring the possibility of downtown renewal or of a revived mass transit system. In the aftermath of failed bond levies, opposition leaders may be invited on to committees set up to review and revise the projects which the bonds were intended to fund. Alternatively, the municipality and its various agencies often find it useful to develop more or less institutionalized practices of consultation with local groups that have in the past shown an ability to resist.

In this context, and in a paper on state responses to grassroots discontent, Clarke & Meyer (1986) have referred to the development of "new interest representation modes." These include:

the opportunity to participate in bureaucratic allocation processes which affect the group and the opportunity to authoritatively participate in broader policy formation and implementation processes. The former, described as "bureaucratic enfranchisement" by Norman and Susan Fainstein, is associated with American citizen participation demands of the 1970s and social policy arenas. The latter linkage is more recent; we describe this linkage as a "fused policy" mode and associate it with joint public/private partnerships and policy forums, particularly economic development institutions. In contrast to bureaucratic enfranchisement processes which bring citizen groups into bureaucratic decision processes by retaining the distinction between state authority and social groups pressuring for public benefits, the fused policy links blur the distinction between public authority and private interests. Decision authority is shared among those brought into quasi-public organizations in order to reach consensual decisions on complex interdependent issues – such as transformation of land uses and siting issues – that cannot be resolved through independent bargaining among sectoral, functional or territorial interests alone.

Some of these fused links are project-specific. Often, they are institutionalized elements of city policy-making processes. The key distinction between these linkage modes and electoral ties is the designation by the state of affected interests in particular policy areas to share state policy authority. This top-down structuring of interest representation contrasts with the pluralist bargaining processes by which most groups seek to influence policy-making processes (Clarke & Meyer 1986, 409–10).

From the standpoint of conflicts over local collective consumption the "opportunity to participate in bureaucratic allocation processes" is perhaps best exemplified by recent developments in the practices of zoning authorities. More and more, it would seem, they are reluctant to consider controversial re-zonings unless there is evidence that the party requesting the re-zoning has already tried to iron out differences with affected residents; and subsequently incorporated compromises, agreements, into revised land use plans.

In other instances city governments have set up their own network of "representative" neighborhood institutions one of whose tasks is to make (non-binding) recommendations on re-zoning decisions. Such neighborhood bodies also serve as vehicles for representing, selectively, neighborhood interests in Clarke and Meyer's so-called "fused policy" mode. Localized congestion may be the occasion for setting up a committee comprising representatives of city government and of the neighborhood affected whose task is to come up with recommendations.

These collaborative arrangements have several effects, none of which serve to strengthen grassroots and neighborhood groups. Rather they tend to smooth the way for realizing the broad objectives and specific projects of growth coalitions. The shift from adversarial tactics to ones of consensus through consultation is especially significant. Among other things it encourages bureaucratization of opposition groups, so that bargaining can take place, and deactivation of rank and file. At the same time "rights" of consultation impose obligations on opposition groups to behave "responsibly": not all land use changes can be opposed for fear that "rights" will be withdrawn and, subsequently, the possibility of concession elsewhere voided. Moreover, acceptance of the bargaining mode inevitably puts opposition groups at a disadvantage due to the fact that they have little idea of what the desired end-states of growth coalitions/developers are. For one effect of increased citizen participation in urban development policy has surely been a studied adaptation on the part of state and capital: a tendency, that is, to ask for "too much," recognizing the political virtues of an apparent "concession" as they strategically withdraw to the position they *really* want to be at.

A less malleable and therefore less predictable structure of incorporation/exclusion is represented by the jurisdictional fragmentation characteristic of metropolitan areas in the US. Not only municipalities, but school districts, park districts, water and sewer districts, tend to partition electorates and tax bases, severing groups from the exercise of powers in which they surely have an interest; and conversely providing political monopolies for others. Ken Newton has stated it well:

> The creation of a large number of political arenas at the local level has a tendency to reduce political conflict and competition. Social groups can confront each other when they are in the same political arena, but this possibility is reduced when they are separated into different arenas. Political differences are easier to express when groups occupy the same political system and share the same political institution, but this is more difficult when the groups are divided by political boundaries and do not contest the same elections, do not fight for control of the same elected offices, do not contest public policies for the same political units, or do not argue about the same municipal budgets. If different groups in the same municipality disagree about how public money should be spent, then they can, in principle, compete for control of the budget. If they are divided into distinct jurisdictions, they have no legal or political claim on one another. Norton Long has written that "American government has been largely based on placing its fundamental politics out of reach of its formal politics," and the multiplicity of local governments is a major cause of this disassociation. (1978, 84)

One consequence is that cooptation through intentionally corporatist structures (within municipalities) is often that much easier; and this because the material effects on a local electorate are diminished. In the language of externalities, effects on local collective consumption that would otherwise be internal can be externalized, perhaps through the careful siting of facilities, or through rights of extra-territorial taxation. Although there are strong tendencies towards the socialization of consumption through the state, the fragmentation of the state in its spatial aspects tends to run counter to this and to produce inequalities and distortions which can be mobilized by growth coalitions in all manner of ways.

THE DISCOURSE OF TERRITORY

According to Middlemas (1983) corporate bias in state institutions is "necessarily sustained by continuous opinion management." (33) A discussion of discourse, therefore, necessarily complements the identification of incorporating institutions as critical to those projects aimed at rebuffing class politics.

A crucial role here is played by a discourse of territoriality, orchestrated by the growth coalition, and facilitated by the spatial structure of the state we have just been discussing. This discourse is pursued over the long term and in the debates around specific controversies. Through the media, through the public statements of local politicians, bureaucrats, businessmen, people are addressed as residents or as citizens of that particular locality. Events – local, national, global – are interpreted in terms of their relevance to people as locals, rather than in terms, say, of their social class or race.

Typically, the interpretations emanating from this particular prism are couched in terms of a balance sheet of advantage or disadvantage for locals. What is projected, in other words, is a conception of a common or community interest which cuts across class, gender, race, and other cleavages, and opposes those of people in localities elsewhere.

These localities exist in some sort of spatial hierarchy which is grasped through a rhetoric of "cities," "communities," "suburbs," "states," "federal government," etc. The interests people have in events touching their lives, therefore, exist both in horizontal and vertical planes. In the horizontal plane it is what is happening in other localities; while in the vertical dimension it is largely a question of what is being handed down from above through the different layers of a spatially structured state.

These interests refer primarily to implications for capital mobility: the implications of national tariff decisions for the viability of local industry, perhaps; or the effects of taxation levels in adjacent localities on location decisions. These implications are important, it is argued, because of what they mean for local growth. This in turn brings, for example, jobs, enhanced tax base, an improved quality of life, social justice for the deprived. If these common interests are to be realized, however, people must stand together against other localities and must be willing to make sacrifices for the future.

In other cases interests are in residential mobility. Here the emphasis shifts from the dangers posed by the *in*clusionary policies of other places to the threats implied by their *ex*clusionary actions: the impact of the exclusionary policies of cities or neighborhood groups on supply and demand in the housing market and hence on rents and housing prices; the impact of those same policies on tax levels and public services.

Now these ideologies of locality are potent tools in the hands of capital. In particular they are powerful aids in the context of struggles between capital seeking to enlarge its accumulation fund and labor struggling to maintain its level of collective consumption. Within the framework that these ideologies provide, opposition – to a freeway perhaps, or to tax concessions to business – can be set against the claims of territorial entities; and what might conceivably have developed as a class issue is reduced to a matter of the contestation of different territorial coalitions. In some instances opposition may be localized so that the issue is interpreted as the respective claims of city and neighborhood; in other cases opposition may be spatially much broader so that the issue is reduced to one of a life-and-death struggle between one city and another.

A good example of this type of local ideology in action is provided by a recent editorial in the *Columbus Dispatch*, the major daily newspaper in the Central Ohio area (see Fig. 4.1). The issue concerned a proposal to impose impact fees on commercial and residential developers in the City of Columbus. This proposal had originally come from a fact-finding committee set up by the city to identify ways of relieving congestion in a rapidly developing part of the city: members of the committee included both representatives of a large umbrella organization for different neighborhood groups in the area and mayoral appointees.

The issue is evaluated here precisely in terms of the overarching territorial interests defined by the localist ideologies of growth coalitions. The editorial is clearly opposed to impact fees. This is on grounds of what they will do to the attractiveness of Columbus for developers and therefore to local growth. To some degree growth is presented as something about which there can be no argument ("without developers there would be no city"); there are also, however, references to the implications for property values and hence tax base – something to appeal to everybody and to balance against failure to relieve congestion in the area affected. Particularly impressive is the conception of spatial competition implicit in the editorial: the conception, that is, of a geography of business climates of which Columbus is a part and to which it must adapt if developers are not to be driven elsewhere.

The image then, is of an objective spatial process distributing life chances across communities and structured by competitive community policies. The reason we must sacrifice our parochial interests, it is argued, lies in the competitive, inclusionary policies of other places. In other cases, however, the territorialization of class tensions may be achieved by pointing to their *exclusionary* policies. Rising rents may be presented as due less to speculation and landlord greed, and more a result of suburban zoning policies. Blame for high taxes and poor city services can similarly be projected outward and the heat taken away from fiscal policies aimed at stimulating growth.

As a result of transformations such as these popular politics in the city is, more often than not, experienced as a multiplicity of battles between territorial coalitions, and at different scales. Consider the issue of "opening up the suburbs," typically portrayed as a confrontation between affluent suburbanites and the deprived of the central city anxious to obtain better housing, schools, lower taxes, etc. (Downs 1973). Closer examination shows that considerable initiative for policies of dismantling zoning

Figure 4.1 "Impact of impact fees." Source: *Columbus Dispatch*, June 25, 1987, 14A. Reprinted, with permission, from *The Columbus Dispatch*.

Impact of impact fees

City government officials should think carefully before imposing "impact fees" on developers, fees that would require them to offset the costs of street improvements and other services needed to accommodate commercial and residential growth.

Columbus and Westerville are among a number of cities across the country studying the potential consequences of such fees. Columbus City Development Director G. Raymond Lorello is attending a conference on the topic in San Francisco. He says he's not convinced impact fees are needed.

Nor are we.

Last week, the Columbus Retail Developers Group voiced opposition to the impact fees. The group represents the city's leading shopping-center developers and is understandably opposed to fees that would drive up costs. The developers cited areas where they had helped the city relieve traffic congestion.

The developers want the mayor to appoint a committee to study impact fees, traffic congestion and possible zoning changes. They asked that the committee include broad representation not only from civic associations and government, but from the banking, real estate, planning and development professions, the Columbus Area Chamber Commerce and the Board of Education.

Several important factors must be considered in the impact-fee debate. Columbus' growth is an important part of its past success and its future potential. In the strictest sense, without developers there would be no city. There is a limit to what developers can be asked to do and on what they will tolerate in a competitive market.

At a time when other communities are offering tax incentives to new development, impact fees might be seen as a penalty to a developer for developing his land – a punishment that will encourage him to invest elsewhere.

Also, development often leads to other forms of growth, which raises property values and generates added tax revenues.

Columbus developers should not be asked to commit substantially to street construction and utility extension in one community when those things are provided at no cost in another community nearby. Columbus City Center probably would not be under development today if impact fees had been in effect. Areas hungry for development today would be punished even more by impact fees.

The burden of impact fees ultimately is borne by shopping-center customers and home buyers. One national study estimates, for instance, that impact fees on residential developments could add $7.88 to the monthly mortgage payments on a $75,000 home. Such hidden costs could eventually squeeze first time home buyers out of the market.

These issues should be studied and all consequences considered. Thus far, no persuasive arguments have been offered for impact fees.

barriers, whether national or local in origin, has come from local growth coalitions, both within the central city and operating at the level of the metropolitan area as a whole. The interest of the central city growth coalition has been in decanting low income populations seen as detracting from the ability of the city to lower tax rates (Danielson 1976). As far as metropolitan-level initiatives are concerned, Trounstine & Christensen (1982) report on an interesting case from Silicon Valley where the Santa Clara County Manufacturing Group, representing in particular the major electronics firms, has attempted to enhance the land use planning capabilities of the county. The goal, apparently, has been to counter those policies of the suburbs seen as producing land use patterns judged irrational from the standpoint of housing and journey-to-work costs.

Within the independent suburbs, on the other hand, residents have not been without significant allies. In his discussion of class monopoly rent, Harvey (1974) has written that "the man–made resource system created by urbanization is, in effect, a series of man–made islands on which class monopolies produce absolute scarcities" (24). Among these man–made islands Harvey includes political jurisdictions. Given the high status, highly favored educational systems associated with many suburbs and their implications for rents, landowners, and small builders specializing in upmarket houses clearly have a stake in exclusionary policies. In his discussion of the attempt to "open up the suburbs" Danielson (1976) pinpoints this group as a significant source of opposition and in coalition with suburban residential interests.

The gentrification case provides another instance. Less a conflict between developers and the displaced, it frequently appears as a set of battles between cross–class alliances in different neighborhoods: alliances, that is, between developers and the incoming middle class who, through their neighborhood organizations push for that further gentrification which they see as enhancing to local public safety, to their property values, and possibly to local schools. Each alliance vies for the attention of a city government dispensing various infrastructural expenditures and land use decisions which can give an alliance a critical advantage in the battle for neighborhood change. To the extent that serious opposition does develop on the grounds of residential dislocation, the argument, possibly propounded by the municipality itself, is likely to be in terms of creating a post–industrial city which will be to the benefit of everyone. Alternatively, and as we have already seen, opposition to gentrification may be defined by laying the blame for housing scarcity at the door of exclusionary suburbs.

The ideological blandishments of local growth coalitions are far from the whole story, however. City planners speak a language which, in its reliance on a fetishized concept of space, has close affinities to them. The advantage that the planners enjoy relative to the growth coalition is their seeming autonomy of material interest; and the legitimization of their views through appeals to expertise. According to Roweis (1983) "the main social role of urban planning in contemporary North America is to mediate ongoing territorial politics, mainly, but not exclusively, by providing professional interpretations of relevant territorial realities" (159). Suggestive of the sorts of interpretations they provide is Salisbury's (1964) alignment of them with "locally oriented economic interests" in his "new convergence of power" in cities. Piven (1975) and Bordessa & Cameron (1982) have underlined the subordination of planning to growth coalition interests.

Ultimately, though, one cannot ignore the fact that labor itself is peculiarly susceptible to the idea of a territorialized politics of location. What combines with local dependence in the living place to produce this susceptibility are the meanings which, through Fordism, the welfare state, labor has embraced as central to life itself: family, consumption, and status. These are meanings which privatize and make labor partial to the parcelizations, the shifting coalitions, and opportunism implicit in an interest group politics of consumption sectors. Moreover, through discourse and through timely promptings, state and capital can work to ensure that the tensions implicit in consumption in a capitalist society are indeed directed into internecine fights among labor rather than against itself.

Given the dominance of interest groups in the sphere of collective consumption, local dependence injects a uniqueness of circumstance which, in combination with a highly individualistic, competitive orientation to consumption, inevitably territorializes the resultant politics of location. The ways in which this is expressed may vary. The scales at which people are locally dependent can change. The local dependence associated with renting does not generate the same type of interest in locality as does homeownership. But local dependence is a necessary aspect of social existence.

This is not to imply that localized coalitions organized around issues of collective consumption have nothing in common, and that they are not prepared to come together for common purposes. There has been some aggregation of interests to the national level around issues of busing for racial balance and exclusionary zoning respectively. These interests, however, have to do with the institutional arrangements providing some localized consumption advantage; in the two cases mentioned here, the neighborhood school concept and local responsibility for land use regulation respectively. Once the institutional arrangements have been secured, successfully defended, these interests provide no further basis for solidarity across neighborhoods. Nor can we be optimistic about the outcome if they are overturned. The challengers, after all, subscribe to the same central meanings enshrined in the separate sphere: and will come under the same pressures to continue to realize those meanings in particular places as were those they have successfully usurped. One can anticipate with confidence that new institutional arrangements designed to protect localized advantage anew will come into existence, or old ones will assume new purpose. The issue of busing for racial balance is a case in point. For alongside the widely publicized flight to the suburbs, there has been a plethora of attempts to reintroduce differentiation into central city schools: magnet schools, alternate schools, and the freezing of school assignments of pupils so that specific neighborhoods are permanently linked to particular schools.

Concluding comments

The essence of capitalism is a class relation. To the extent that labor becomes aware of this and struggles against exploitation, then that class relation and ultimately, therefore, capitalism is threatened.

With the achievement of limited democratic rights, both inside and outside the workplace, capital's options for coping with this threat have narrowed. No longer having the same freedom of outright repression and violence, it has sought to control through cooptation. Historically this has assumed such forms as Fordism and the welfare state. The difficulty here is that increasing labor's consumption fund threatens capital's fund for accumulation; depending on the forms assumed by state provision, it may also reduce labor's dependence on capital for means of subsistence.

The historical resolution of this has been state intervention, seeking alliances with fractions of labor so as to control the nature and level of demands from other fractions. In this way conflict has seemed to be less about class and more about competition, less a confrontation between capital and labor and more one between different fractions of labor identifiable by different interests in collective consumption.

These relations have a spatial counterpart, and necessarily so. The essential spatial character of social structures gives economic agents interests in production relations which are tied to particular places. An important manifestation of this in North American metropolitan areas is the emergence of growth coalitions whose goals are to channel value through their respective (localized) social relations. Coalescing around policies of spatial restructuring, property tax incentives, and a liberal regulatory environment for land development, they inevitably threaten somewhere, sometime, labor's use values in the living place. The opposition between capital's fund for accumulation and labor's consumption fund is therefore posed anew.

Democracy, therefore, becomes a problem. To some extent this has been suspended through the development of political structures which selectively incorporate and exclude so as to achieve the necessary degree of consent. Playing an important role in this process is the multiplicity of local governments into which North American metropolitan areas are typically divided.

This spatially fragmented pattern of local authorities in turn provides a condition for a discourse which likewise plays a significant role in overcoming opposition. This discourse, heavily laced with spatial imagery, relates labor's life chances in a particular place to movements of capital and labor through a spatial hierarchy of localities, and to the structuring of those movements by place competition. Within this context it is then possible to convert class issues into distributional struggles among fractions of labor organized territorially.

Yet the power of this discourse to convince has as its essential condition the social atomization of labor which has proceeded at much broader scales through Fordist structures. In combination with the uniqueness of spatially defined interests that comes from their own forms of immobility, concepts and practices of privatization and competitive consumption make labor receptive to the territorial ideologies of local growth coalitions and hence to the attractions of turf politics.

Now, in this argument considerable stress has been placed upon the social relations of Fordism. There is, however, significant and ongoing debate about their durability. A common view is that Fordism is being superseded by a different regime of accumulation: what is being called "flexible accumulation." Something, therefore, needs to be said about the impact of these changes on the more long-term validity of my arguments. Given the slow disintegration of Fordism after 1970, to what extent does the politics of turf experienced over the last ten or 15 years have different roots from that which came before?

Briefly, flexible accumulation is defined as a new malleability of labor processes, markets, products, and patterns of consumption. It is counterposed to the greater degree of regulation of markets by state and oligopoly characteristic of Fordism. Along with this new flexibility is said to go an increasingly rapid sectoral and spatial switching of capital. The latter has been of particular interest to geographers who have linked it to a supposed intensification of growth coalition activity.

Rather than an intensification, however, what seems to be happening is a shift in the territorial context of growth coalition activity. In the 1950s, and carrying on into the 1960s, this was primarily metropolitan. Major conflicts were between central cities and suburbs. Central city revitalization initiatives were a response to the devaluations implied by massive suburbanization. This is very clear from writings on that period (Mollenkopf 1976, Stone 1976). Since then "rural turnaround," "the Sunbelt shift," and "new international divisions of labor" have sharpened apprehensions at the metropolitan level with appropriate shifts in growth coalition organization and, indeed, in discourse.

With respect to labor's responses on the other hand, nothing much seems to have changed. Our argument has been that it is the separate sphere constructed around family, consumption, and status that has made labor peculiarly vulnerable to growth coalition ideologies, and this is true of the whole post-war period. Flexible accumulation may have intensified the pressures on locally dependent businesses and, ultimately, through local growth coalition programs, on labor in the living place. It has not, however, promoted any qualitative change in labor's responses. These have remained quite emphatically territorial in character.

The durability of the separate sphere, moreover, suggests that although my major concern in this chapter has been the politics of turf in North American cities, the arguments deployed have a good deal more generality. To the extent that the same processes of social atomization are replicated elsewhere, then we can realistically expect some form of the politics of turf to emerge. For local dependence is a universal of social existence. Only its forms vary, producing different configurations of interests at different geographical scales, vacating some and colonizing others. Comparative study is clearly called for, therefore, to evaluate further the relations between turf politics and the question of class.

Note

I am grateful to Michael Dear, Andrew Mair, and Jennifer Wolch for comments on earlier drafts of this chapter. The chapter is based upon work supported by the National Science Foundation under grant nos. SES–8112324 and SES–8520094. This support is gratefully acknowledged.

References

Aglietta, M. 1979. *A theory of capitalist regulation.* London: New Left Books.

Bish, R. L. & V. Ostrom 1980. Understanding urban government. In *The urban economy,* H. Hochman (ed.), 95–117. New York: W. W. Norton.

Bordessa, R. & J. Cameron 1982. Growth-management conflicts in the Toronto-centred region. In *Conflicts, politics and the urban scene,* K. R. Cox & R. J. Johnson (eds.). Harlow, Essex: Longman.

Castells, M. 1977. *The urban question.* London: Edward Arnold.

Clarke, S. E. & M. Meyer 1986. Responding to grassroots discontent: Germany and the United States. *International Journal of Urban and Regional Research* **10**, 401–17.

Clavel, P. 1986. *The progressive city.* New Brunswick: Rutgers University Press.

Cox, K. R. 1984. Social change, turf politics and concepts of turf politics. In *Public Service Provision and Urban Development,* A. Kirby, P. Knox & S. Pinch (eds.). London: Croom Helm.

Cox, K. R. & A. Mair 1988. Locality and community in the politics of local economic development, *Annals of the Association of American Geographers* **78**, (forthcoming).

Danielson, M. N. 1972. Differentiation, segregation and political fragmentation in the American metropolis. In *Governance and population: governmental implications of population change,* A. E. K. Nash (ed.), 143–76. Washington, DC: US Government Printing Office.

Danielson, M. N. 1976. *The politics of exclusion.* New York: Columbia University Press.

Downs, A. 1973. *Opening up the suburbs.* New Haven and London: Yale University Press.

Ewen, S. 1976. *Captains of consciousness.* New York: McGraw-Hill.

Fainstein, N. I. & S. S. Fainstein 1980. Mobility, community and participation: the American way out. In *Residential mobility and public policy,* W. A. V. Clark & E. G. Moore (eds.), Ch. 13. Beverly Hills: Sage Publications.

Fincher, R. 1982. Urban redevelopment in Boston: rhetoric and reality. In *Conflict, politics and the urban scene,* K. R. Cox & R. J. Johnston (eds.). Harlow, Essex: Longman.

Hartman, C. 1974. *Yerba Buena: land grab and community resistance.* San Francisco: Glide Publications.

Harvey, D. 1974. Class monopoly rent, finance capital and the urban revolution. *Regional Studies* **8**, 239–55.

Harvey, D. 1978a. The urban process under capitalism: a framework for analysis. *International Journal of Urban and Regional Research* **2**, 101–31.

Harvey, D. 1978b. Labor, capital and class struggle around the built environment in advanced capitalist societies. In *Urbanization and conflict in market societies,* K. R. Cox (ed.). Chicago: Maaroufa.

Hill, R. C. 1983. Crisis in the motor city: the politics of economic development in Detroit. In *Restructuring the city,* S. S. Fainstein et al. (eds.), Ch. 3. New York: Longman.

Hirsch, J. 1978. The crisis of mass integration: on the development of political repression in Federal Germany. *International Journal of Urban and Regional Research* **2**, 222–32.

Hirsch, J. 1981. The apparatus of the state: The reproduction of capital and urban conflicts. In *Urbanization and urban planning in capitalist society,* M. Dear & A. J. Scott (eds.), Ch. 22. New York: Methuen.

Hirsch, J. 1983. The Fordist security state and new social movements. *Kapitalistate* **10/11**, 75–87.

Lasch, C. 1977. *Haven in a heartless world*. New York: Basic Books.

Lee, Barrett A., R. S. Oropesa, Barbara J. Metch & Avery M. Guest 1984. Testing the decline-of-community thesis: neighborhood organizations in Seattle, 1929 and 1979. *American Journal of Sociology* **89**, 1161–88.

Lynd, R. S. & H. M. Lynd 1929. *Middletown*. New York: Harcourt, Brace.

Marris, P. 1975. *Loss and change*. New York: Anchor Books.

Marx, K. 1976. *Capital Vol. 1*. Harmondsworth, Middlesex: Penguin.

Middlemas, K. 1983. Corporate bias. In *States and societies*, D. Held et al. (eds.), 330–7. New York: New York University Press.

Miller, Z. L. 1981. *Suburb: neighborhood and community in Forest Park, Ohio, 1935–1976*. Nashville: University of Tennessee.

Mishan, E. J. 1967. *The costs of economic growth*. Harmondsworth, Middlesex: Penguin.

Mollenkopf, J. H. 1976. The post-war politics of urban development. *Politics and Society* **5**, Winter.

Newton, K. 1975. American urban politics: social class, political structure and public goods. *Urban Affairs Quarterly* **11**, 241–64.

Newton, K. 1978. Conflict avoidance and conflict suppression: the case of urban politics in the United States. In *Urbanization and conflict in market societies*, K. R. Cox (ed.). Chicago: Maaroufa.

Offe, C. 1984. *Contradictions of the welfare state*. Cambridge, Mass.: MIT Press.

Orbell, J. M. & T. Uno 1972. A theory of neighborhood problem solving: political action vs. residential mobility. *American Political Science Review* **61**, 471–89.

Ostrom, V., C. Tiebout & R. Warren 1961. The organization of government in metropolitan areas. *American Political Science Review* **55**, 831–42.

Piven, F. F. 1975. Planning and class interests. *Journal of the American Institute of Planners*, 308–10.

Piven, F. F. & R. Friedland 1984. Public choice and private power. In *Public service provision and urban development*, A. Kirby, P. Knox & S. Pinch (eds.), Ch. 15. London: Croom Helm.

Rex, J. & R. Moore 1967. *Race, community and conflict*. London: Oxford University Press.

Roweis, S. T. 1983. Urban planning as professional mediation of territorial politics. *Society and Space* **1**, 139–62.

Salisbury, R. H. 1964. Urban politics: the new convergence of power. *Journal of Politics* **26**, 775–97.

Saunders, P. 1978. Domestic property and social class. *International Journal of Urban and Regional Research* **2**, 233–51.

Saunders, P. 1981. *Social theory and the urban question*. New York: Holmes & Meier Publishers, Inc.

Sayer, R. A. 1985. The difference that space makes. In *Social relations and spatial structures*, D. Gregory & J. Urry (eds.). Basingstoke, England: Macmillan.

Scott, A. J. 1985. Location processes, urbanization, and territorial development: an exploratory essay. *Environment and Planning A* **17**, 479–501.

Sennett, R. & J. Cobb 1973. *The hidden injuries of class*. New York: Vintage Books.

Stone, C. 1976. *Economic growth and neighborhood discontent*. Chapel Hill: University of North Carolina Press.

Sutcliffe, M. O. 1984. *Neighborhood activism in socio-historical perspective: Columbus, Ohio, 1900–1980*. Unpublished PhD dissertation, Department of Geography, The Ohio State University, Columbus, Ohio.

89

Trounstine, P. J. & J. Christensen 1982. *Movers and shakers: the study of community power*. New York: St. Martin's Press.

Williams, O. P. 1971. *Metropolitan political analysis*. New York: The Free Press.

Wingo, L. 1973. The quality of life: toward a microeconomic definition. *Urban Studies* **10**, 3–18.

Part III

Industrial society

5

Class and gender relations in the local labor market and the local state

RUTH FINCHER

The local labor market and the local state contribute significantly to the way in which class and gender are experienced by residents and workers in urban communities. They are not independent of each other. The provisions of governments in local areas (part of the local state), in public transport and child day-care, for example, permit people to undertake paid work outside the home. In this sense the spatial scope of local labor markets is defined by virtue of their accessibility to people who must be sustained in paid work by community-based services. As well, the hours of operation of paid workplaces and their requirements of workers influence the extent to which existing public services are useful for those workers. It is important to emphasize the obvious point that in people's daily lives the paid work available and the community services that can be used are utterly interrelated. The many analyses that separate production in local labor markets and social reproduction in the use of community services cannot do this.

This chapter aims to indicate some of the characteristics emerging in the local labor markets and local states of Melbourne, and their implications for the formation of class and gender relations. It has the following structure. First, it discusses how the practices of the local state and the nature of the local labor market may structure people's experience of class and gender. In the second section, some of the characteristics of Melbourne's local labor markets are presented, with suggestions for the different and similar class and gender experiences associated with them in different parts of the metropolitan area. Section three documents some of the priorities of local states in Melbourne, and how these priorities are being influenced by changes in federal social policy. It suggests directions being taken in making services available to support people in paid work, and the implications of these in the formation of class and gender relations.

Class and gender in the local state and the local labor market

Before discussion of the presence of class and gender relations in local state and local labor market, a brief statement of the interpretation given here to class and gender relations is in order.

93

Class and gender relations

Theoretical statements (e.g., Wright 1985) usually differentiate class formation – a conscious process of grouping into class collectivities, from class structure – a formal map of class positions in a society, and how many people could be said to hold positions in the capitalist class, the working class and the middle class. Class formation is based in varying degrees on class structure, but is never entirely independent of it. Embodied in both the process of class formation and the map of class structure are class relations; these are the social relationships between people in the paid workplace, the community, and the home, through which they experience class (the unequal relationship between capital and labor in a capitalist society), even if they do not consciously give their experiences that label. Accordingly, Walker (1985, 187) has written:

> People experience class as it is built into their lives in particular ways and come to realise the force of class in and through the immediate circumstances they can experience and understand directly.

And Katznelson (1981, 209) notes that different "amounts" of consciousness of class (and therefore of conscious class formation) are present in people's daily lives in capitalism, this amount depending on a wide range of circumstances.

Many of the circumstances affecting the experience of class are institutional features particular to a time and place, that differentiate the population along lines not directly identifiable as class-based. Harvey (1975) long ago termed these features of class structuration, and listed them as the division of labor, the consumption habits and authority relations of a time and place, and the institutional and ideological barriers to mobility there (Harvey 1975, 362). This chapter is concerned with the class relations embodied in two institutional – though certainly not static – features of Melbourne's localities. Thus it is seeking to identify emerging gendered divisions or groupings within Melbourne's working class, that are being produced through class structuration (via the local labor market, and the local state with its authority over forms of community social reproduction), and the sorts of class relations they seem to embody.

The changing class relations of the paid workforce in contemporary capitalist cities have been noted by many authors (Massey 1984, Massey & Meegan 1982). They include the casualization, marginalization, and peripheralization of large groups within the traditional working class, e.g., male manufacturing workers. New class relations are appearing as well within groups like part-time women workers, and between them and other segments of the paid labor force. It is now quite widely accepted that the reproduction of labor power that occurs in the home and the community, and the fact that household members are often workers in the paid labor force as well as workers in the household and community, draw class relations into the household and community as well. Class relations exist in a wide range of daily activities, then, and through a number of strands of class structuration. Class relations, class formation and the means of class structuration that affect them both, are profoundly spatial phenomena; they are restricted or facilitated through use and reproduction of spatial structures.

Now, gender relations are the socially constructed relationships between men and women. I suppose analysts committed to the theoretical primacy of class categories over gender categories would term gender relations one form of class structuration; others would champion the primacy of socially constructed gender roles over class as the basis of many everyday experiences. Suffice it to say here that the comparative theoretical status of class and gender continues to provoke controversy (Foord & Gregson 1986, McDowell 1986). Research on changes in the nature of gender relations documents the different or similar tasks, responsibilities, and experiences of men and women in their relationships with each other, but it identifies as well the social processes giving rise to these. Often it is in the empirical identification of the causes of particular gender relations that the intertwining of the class and gender characteristics of times and places is best demonstrated: Nelson's (1986) skilful study of the suburbanization of low-wage office work in San Francisco is a case in point. Recent research on the class and gender relations of outwork or homework, where paid work and domestic production and reproduction all take place on site in the household, demonstrates as well the lived complexity of class and gender interrelationships (Walker 1987).

Underlying claims that class analysis must be accompanied by analysis of gender relations is not just a wish to "add on" gender to class, to adhere feminist concerns to historical materialist ones, but rather to accept the need for

> an expanded and reworked concept of a mode of production, wherein the production of labour power stands on an equally basic footing with the production of the means of production and subsistence (Livingstone *et al.* 1982, 9).

This conceptual intermeshing, as well, emphasizes that men's and women's activities in the paid workforce, in the household or family, and in the community are not separate spheres but are mutually interrelated. At the simplest level, the time-space requirements of getting to and being at work restrict people's opportunities to act elsewhere. But less tangible aspects of the social relations of one "sphere" affect the social relations of another "sphere": for example, a rigid hierarchy in the paid workplace, and strict sex segregation in occupations there, may have implications for the assumption or allocation of tasks and responsibilities in the household of a worker undertaking paid work in that workplace. The class groupings being sought in analysis of men's and women's experiences of local labor market and local state are gendered divisions.

Consider now the views of the literature on class and gender relations in the local state and the local labor market of contemporary capitalism.

Class and gender in the local labor market

Recent research has termed local labor markets or local economies "localities." They are made up of spatially and sectorally grouped establishments providing paid employment in a town or region, and the area from which their workers are drawn (see Cooke 1986a, Murgatroyd *et al* 1985). Some discussions have also emphasized the social relations of localities that are not found primarily in the workplace or the journey to work. So, Urry (1981) and Cooke (1986a) have mentioned traditions of community and kinship links, and a more or less significant gendered division of labor in production and domestic reproduction, as important features of localities. Savage (1987) has endeavored to link the national political alignments expressed in British election results with the changing characteristics of British localities. Massey's (1984) stories of the declining mining towns of south Wales and the changing economies of the villages of Cornwall remain some of the most fascinating for their illustration of the social traditions built up through lifetimes of involvement in certain sorts of labor process and occupation, and the new relationships being constructed between men and women in paid work as traditional occupational characteristics change. But the focus of much of this work has remained on the economic transformation of localities as local labor markets (see Cooke 1986b).

What, then, can be said about the structuring of class and gender relations that is affected by the characteristics of local labor markets and changes in them? First, consider how local labor markets have different characteristics; second, how these may encourage the formation of particular class and gender relations in certain contexts; and third, how changes in economic and other relations may alter all this, and what form some such recently noted changes have taken.

Local labor markets will often have different characteristics, related to the particular industrial and occupational sectors represented in them. Associated with a predominantly manufacturing local economy may be a sexual division of labor and a set of labor processes unlike those of a local economy dominated by personal service provision or tourism. The particular economic base of a local labor market will be important in determining its fate in a time of economic restructuring: traditional manufacturing centers have been hard hit since the 1970s in a number of capitalist countries, as the literature on the deindustrialization of the American north-east and the English north demonstrates. Particular variations are noted in the ongoing British localities research (see Cooke 1986b). Single sector communities, reliant on one or a small number of industrial sectors, often suffer most heavily (see Bradbury 1984, Sandercock 1986). At the same time as traditional manufacturing areas and their predominantly male full-time workforces are experiencing dislocation in the present economic circumstances, other forms of production are emerging and other social groups are finding work, albeit work of different characteristics. Particularly, female participation in the paid labor force is increasing, and is often associated with increasing numbers of part-time jobs.

The dominance of certain employment sectors and forms of work does not determine the class and gender relations of communities, regions, or nations. If this were the case, we would find identical social relations in places with similar economic bases. Rather, continuing economic changes combine with traditions forged out of histories of economic, political, and social interaction in a place, on the part of its long-term residents and indeed on the part of immigrants too. Massey (1978) conceived of layers of past economic and social relationships underlying and influencing the present circumstances of a region, and she has presented this metaphor again more recently (1984). Of importance, too, is the degree to which collective action by local residents and workers can create local circumstances, despite or in sympathy with exogenous change, that they prefer. Massey (1984) has briefly suggested links between the class and gender relations of south Wales and Cornwall. But Bleitrach & Chenu (1977) have documented and classified the class and gender relations of Marseilles in considerable detail, looking closely at the influence of the practices of the workplace on ways of living outside it.

The research has suggested the challenges to existing class and gender relations in manufacturing communities that recent economic upheavals have begun: (a) fewer jobs exist in manufacturing of the traditional type. Youth unemployment in places dependent on such jobs is now accompanied increasingly by the unemployment of middle-aged men who have been full-time factory workers. Though the decline of manufacturing in a community does not imply the appearance of a service sector employment base (Murgatroyd *et al.* 1985, Ch. 1), the part-time employment of women, especially in jobs classified as "services," has increased over this period. In some areas, then, one may expect to find a change in the gender characteristics of the workforce in paid employment. (b) This has implications for the characteristics of the unionized labor force, which most literature indicates to have been male-dominated (and often male-centered), and unfavorably disposed towards part-time work. A change in the gender and nature of the workforce and paid work done, then, will see changes in the proportion of the paid labor force that is organized in the traditional manner. (c) Outwork, or homework, is on the increase. As Walker (1987) indicates, this is not just to say that women clothing and textile workers who are based in their homes are increasing in numbers, though there is evidence of this from Britain and Australia. Rather, a large number of jobs are being performed from the home and there is every indication that this will increase (Huws 1985, Walker 1987). The implications for the organization of the labor force, of reduced spatial proximity between paid workers in their daily work, are obvious.

Across metropolises, where communities traditionally dependent on manufacturing are juxtaposed with residential suburbs that are also the sites of retail and office development, it appears that women in part-time work in those businesses are also becoming more numerous. And as flexible accumulation grows, a mode of profit-making in which production is disaggregated and spatially decentralized to different communities and households, so suburban residential sites that previously were locations of consumption are the new production sites as well (Nelson 1986, Harvey 1987). New social relations will emerge from the placement of these new groups of workers and new forms of employment into any local circumstances, but changes will be especially dramatic where they accompany decline in traditional ways of making a living. Women and men are doing different forms of paid work: their relationships with each other in the workplace and in the home and community will change because of this. New alliances within the working class will emerge from this, and old ones perhaps disappear.

Class and gender in the local state

The local state is interpreted here as the sphere of political relations associated with the local practice of government (see Fincher 1987). Studies of the local state therefore include studies of those community conflicts that are associated with the state apparatus and its policies as well as studies of the state apparatus that acts with respect to localities. Clearly, the local state is a sphere of relationships that contains class and gender relations, and can influence their formation or change elsewhere. Consider the following ways the local state (the apparatus and associated political conflicts) influences class and gender relations in paid work and its community support.

First, the local state, through its direct provision (or contract provision to for-profit firms or charities) of public services to people in paid work who have dependents (children, the elderly), affects the degree to which some people can participate in the paid labor force, quite apart from whether or not employment opportunities are available locally. If suitable child day-care and after-school programs, or domiciliary care for elderly dependents of potential labor force participants, are not available, then those individuals will be unable to take up waged work. Second, the local state may, through its direct involvement in employment creation schemes, cause scheme participants to experience work in new ways, and may decrease the incidence of unemployment amongst certain local groups (e.g., women, youth). Third, through its facilitation of different forms of built environment change, the local state can change the nature of work in communities and the demand for public services there, through a change in the nature of the local resident and workforce population. The classic case of this is gentrification, for example the situation in which an alliance of local state and major developer locates a luxury water front development of offices and expensive high density housing in an inner city working-class neighborhood. Also, if it fails to regulate evenly the distribution of different sorts and sizes of housing accommodation through a city, the local state has an effect on structuring accessibility to the housing market for different social and income groups, and therefore their access to work opportunities and community support services. (The class and gender structuration that is associated with changes in the housing market and built environment, though an important issue, is not the subject of this chapter.)

A large international literature on the restructuring of the welfare state in advanced capitalist countries since the mid-1970s indicates a ubiquitous reduction in central government financing of, and responsibility for, human services (see Mishrah 1984, Dear & Wolch 1987). To the extent that these reductions have locational effects, and are influential in causing local service providers to reduce or change human service provisions, then the local state is involved. Generous provision of child day-care and domiciliary care for elderly dependents would allow more women (especially, as they are the traditional "minders") to participate in paid work, or at least to have more flexibility in doing so. It would allow men more flexibility too, in households where they have a role in caring for dependents. Reduced provisions would be expected to be associated with the opposite effect, unless the household were able to draw upon informal help. The suggestion has been made many times that reduction of human service provisions supporting people in paid work, and the increasing cost of services like child day-care when it is available, are most likely to return women to the home by reducing their options to work outside it.

Though the local state is often seen largely as a service provider, as a set of institutions involved in the reproduction of labor power and the provision of collective consumption goods (see Fincher 1987), local state institutions are becoming involved in employment creation schemes, usually on a fairly small scale. The socialist municipalities of Britain have made their mark here (Boddy & Fudge 1984). Not only did scheme participants benefit from employment with these British schemes, but there was also an increase in public awareness that collectively controlled work is feasible. Gow (1986) lists Australian local employment initiatives, some of which give priority to socially useful production and employment. But the prospects for widespread expansion of such workplaces seem limited in the current political and economic climate of advanced capitalist countries: Lowe (1987) makes this clear in a recent review.

It is important to note from this discussion that the activities of the local state have implications for the nature of class and gender relations, as do the activities of employers in the local labor market. Furthermore, the local state influences accessibility to the labor market, and the local labor market creates needs for certain types of public service provision. The local state can be a player in the local labor market through its development of local employment schemes; local businesses can provide community services to support people in paid work, either through employee benefits like work-based day-care, or through being in the business of for-profit human service provision.

Class and gender relations in Melbourne's local labor markets

This section describes Australian, and then Melbourne's, contemporary labor market characteristics. It goes on to hypothesize the labor market sources of local class and gender relations found in the eastern and western suburbs of Melbourne. Melbourne's local labor markets embody many national trends. These national trends show Australia to be quite like other advanced capitalist countries: its labor market characteristics indicate increasing marginalization in paid work for some employees. The following are Australia's recent labor market features.

(1) Unemployment has risen drastically in Australia since the mid-1970s (Bureau of Industry Economics (BIE) 1979, 1). This increase has not been shared equally across different social and economic groups, according to their share of labor force participation. Youth, non-English speaking migrants and women have suffered high relative unemployment (BIE 1979, 17), this of course relating to their concentration in particular industries whose labor requirements have decreased (BIE, 1979, ch. 3, Blakers 1986, 142). The relative unemployment of married women began to decline in the early 1980s.

(2) There have been different trends since 1966 in full-time and part-time paid labor force participation (Bureau of Labour Market Research (BLMR) 1985, 41–3). Male full-time participation rates have decreased sharply since 1976 and male part-time rates have risen but not offset the full-time rate decreases; female paid labor force participation has increased since 1966, 90 per cent of it in part-time work. Married females accounted for all this increase.

(3) Over half the new jobs created since 1973 have been part-time, many of them in community services and retailing (Wajcman & Rosewarne 1986, 15). That many of these jobs have gone to women reflects the significant sex segregation of the Australian labor force: in September 1984 approximately 64 per cent of female employees were in clerical, sales, and service occupational categories (Blakers 1986, 142). Government claims to have improved employment prospects for women must be examined very critically: that there are now more jobs for women does not mean they are well-paid or secure jobs.

(4) The labor market is further segmented, particularly in Melbourne and Sydney, along another axis relating to the recency of migration of workers, and whether they have non-English speaking backgrounds. One analyst has suggested a four-way segmented labor force classification, where the first category is the most privileged and the last the most marginal:

First are the Australian-born and Anglophone male migrants who earn the highest pay and work in the best conditions primarily in the tertiary or white collar sector or in skilled jobs in the manufacturing sector. They have opportunities for promotion and well-established career structures in most cases. . . . The second segment, non-Anglophone migrant males, are overwhelmingly concentrated in the semi-skilled and unskilled jobs in the manufacturing, building and construction sectors. Their jobs are often hard and dirty, their average pay much lower than those in the first segment, and upward job mobility is minimal. . . . The third segment is made up of women who are born in Australia or who are Anglophone migrants. Their pay is less than that in both male segments on average, and they tend to work in the "tertiary" sector in "women's jobs". . . . Women migrants from non-Anglophone countries occupy the fourth segment. These women tend to be concentrated in the shrinking jobs in the manufacturing sector, often in the clothing, footwear and textiles industries. They are at the "bottom of the heap": their pay is lowest, their working tasks and conditions often the most repetitive, with few promotion possibilities (Collins 1984, 12).

(5) The number of people working at home now forms a significant portion of Australia's paid labor force – estimates from 1981 data (the latest available census figures) are that 7.2% of Australians work at home (Walker 1987). Walker's work shows the wide range of industrial sectors in which homework or outwork is done, and the different jobs done by men and women in them. (In 1981, almost 10 per cent of the women recorded as members of the paid labor force recorded their home as their workplace.) In these situations, the class and gender relations of home and workplace do not merely influence each other, but are utterly, spatially intertwined.

Different parts of Melbourne exhibit these national trends in different degrees: jobs have been growing and declining in number, and changing in type, in different locations within the metropolis. The early 1960s saw "suburbanization" of the considerable job growth in Melbourne. Manufacturing employment was expanding in the east, away from the traditional manufacturing base of the western suburbs. This suburban movement (except in the west) increased with commercial job growth between 1966 and 1971, though the inner city also regained some jobs in the early 1980s (Fothergill 1984).

The spread of job growth in Melbourne in the last 25 years has been uneven: for every job added to the western sector between 1961–66 there would have been an additional two jobs gained in the east. By 1976–81 this ratio had been extended to 1:40 (Fothergill 1984, i).

The eastern sector's job growth rates were the highest in the period for each industry category. There was some disparity, too, between the location of jobs and the location of workers' residences. Workers have been moving to the outer suburbs faster than jobs, which are still tending to concentrate in the middle ring suburbs – the extent of this disparity varies between regions of the metropolitan area, however (Fothergill 1984, ii). The state government's new metropolitan strategy indicates concern with the serious spatial discrepancy between the location of manufacturing jobs and manufacturing workers: the workers, for the most part, remain in the western region, but the jobs are increasingly in the east (Government of Victoria, 1987). (Figure 5.1 shows the regions of the metropolitan area; Figure 5.2 the distribution of employment and economic document.) Growth in the suburban share of office development has been marked since 1980 – it has increased from 9.8 per cent to 27.1 per cent (Edgington 1987, 22). New offices are locating in the middle eastern

Figure 5.1 Regions of the metropolitan area, Melbourne.

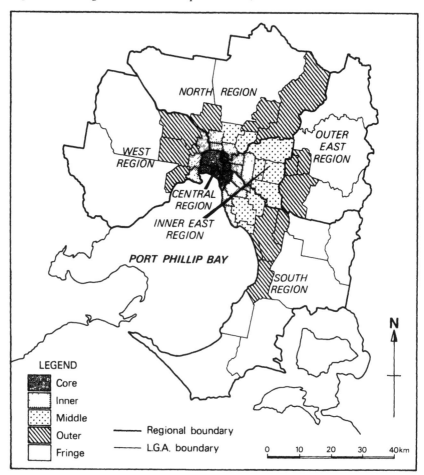

suburbs, and more recently in the outer east of the metropolitan area. The share of suburban offices built in the north-west region of the city has been far less than that built in the south-east region, and this imbalance appears to be growing into the 1980s (Fothergill 1987). This disproportionate growth of suburban jobs in the east, compared to the west, is a major difficulty in a large urban area in which regional laborsheds are relatively independent (O'Connor 1981, 9). It is certainly a problem for those without access to a car, and for those whose preferences have been to work locally (traditionally women and blue-collar workers) (Cass & Garde 1984, Howe & O'Connor 1982).

What changes in class and gender relations are indicated in the Australian labor market characteristic that are differentially exhibited or bunched in the different localities of Melbourne? Tables 5.1, 5.2, 5.3 and 5.4 are illustrative. First, manufacturing workers have in the past been

Figure 5.2 Distribution of employment and economic development across the metropolitan area, Melbourne.

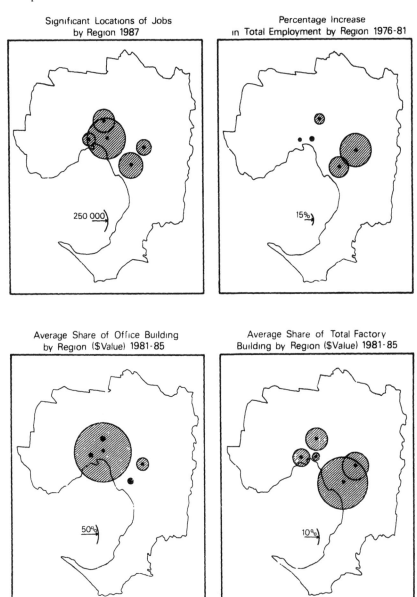

Significant Locations of Jobs
by Region 1987

250 000

Percentage Increase
in Total Employment by Region 1976-81

15%

Average Share of Office Building
by Region ($Value) 1981-85

50%

Average Share of Total Factory
Building by Region ($Value) 1981-85

10%

Table 5.1 Percentage distribution of all persons in paid employment, by occupation.

Occupation	Region								
	West			Inner east			Outer east		
	1971	1981	1985	1971	1981	1985	1971	1981	1985
Professional	7.8	11.7	12.5	19.7	26.3	26.5	12.4	16.7	15.9
Administrative	4.8	3.0	3.4	12.4	9.1	8.7	8.5	5.8	6.7
Clerical	18.0	20.0	20.7	22.0	23.0	22.1	17.2	20.7	21.8
Sales	7.0	7.8	8.4	10.4	10.6	9.1	10.1	10.6	9.2
Transport	6.9	6.3	5.8	4.0	3.2	3.1	5.7	4.6	3.9
Crafts/ laborers	48.5	42.3	39.6	25.2	19.6	21.0	39.1	34.3	33.8
Services	7.0	8.8	9.3	6.2	7.7	9.5	6.9	8.4	8.7

Sources: 1971, 1981 figures from Australian Bureau of Statistics, Population Census. 1985 figures are for May 1985, are subject to sampling error, and are taken from Australian Bureau of Statistics, *The labour force: Victorian regions,* quarterly.

concentrated in western Melbourne (Table 5.1 shows them to form a higher percentage of the labor force of the western region than of the other regions shown). Therefore the decline in manufacturing jobs in the traditional industries and the reappearance of new manufacturing jobs elsewhere (beyond the labor market accessible to western suburbs workers) will be influencing the traditional, male–dominated working class of this region disproportionately. (Table 5.2 measures the sex segregation of different occupational classifications and shows male dominance of the manufacturing sector.) The proportion of jobs held by women in the

Table 5.2 Percentage of each occupational group that is women.

Occupation	Region								
	West			Inner east			Outer east		
	1971	1981	1985	1971	1981	1985	1971	1981	1985
Professional	37	44	44	37	45	40	38	42	36
Administrative	13	8	14	11	9	16	12	7	16
Clerical	67	69	66	55	69	70	61	64	76
Sales	53	51	50	41	44	54	46	45	56
Transport	12	12	15	17	20	29	13	16	6
Crafts/ laborers	17	15	19	17	16	11	14	14	11
Services	62	60	62	60	58	59	58	58	56

Sources: 1971, 1981 figures from Australian Bureau of Statistics, Population Census. 1985 figures are for May 1985, are subject to sampling error, and are taken from Australian Bureau of Statistics, *The labour force: Victorian regions,* quarterly.

Table 5.3 Percentage of paid employees that are full-time (paid work for more than 35 hours per week).

Full-time workers			Region			
	West		Inner east		Outer east	
	1981	1985	1981	1985	1981	1985
Males	90.0	95.7	88.0	91.3	90.0	96.1
Females	69.4	69.0	63.0	60.4	60.8	60.8
All	82.2	85.6	77.4	78.6	79.2	83.1

Sources: 1981 figures from Australian Bureau of Statistics, Population Census. 1985 figures are for May 1985, are subject to sampling error, and are taken from Australian Bureau of Statistics, *The labour force: Victorian regions,* quarterly.

region, which are part-time, is growing over the brief period for which figures are available (Table 5.3). This, the growth of women's outwork and the continuing high unemployment of male (especially manufacturing) workers (Table 5.4, Wilson 1985) indicate a changing structure of the region's working class. Perhaps marginalization from class practices is increasing with the removal of men from the collective labor processes of the factory floor and the concentration of women in isolating, long homework. Certainly, within households and communities in which males have traditionally held paid positions and women have not, there will be confrontations of class places and gender roles if men are now the unpaid workers.

Second, in the outer eastern region of Melbourne there is a substantial but declining proportion of the labor force in manufacturing jobs, and growth in the proportion of the workforce employed in services, clerical, and professional categories (Table 5.1). The services sector is dominated by women workers, and the clerical sector even more so (Table 5.2). Through the 1980s, the same proportion of the female workforce has remained part-time, though the number of part-time jobs in the region is increasing. Women's participation rates and unemployment rates in this

Table 5.4 Unemployment rate and labor force participation rate, May 1985.

Workers			Region			
	West		Inner east		Outer east	
	Unempl. rate %	Partic. rate %	Unempl. rate %	Partic. rate %	Unempl. rate %	Partic. rate %
Male	9.8	76.8	5.3	78.3	N/A	79.2
Female	11.6	47.4	6.9	47.5	8.9	50.1
All	10.4	62.1	6.0	62.4	5.7	64.8

Sources: Australian Bureau of Statistics, *The labour force: Victorian regions,* quarterly (and subject to sampling error).

region are the highest of those shown (Table 5.4). In the inner east there is a lower and comparatively stable proportion of the population in manufacturing jobs, higher and relatively stable proportions of the workforce in sales, clerical, and services categories, and much larger workforce segments in professional and administrative jobs (Table 5.1). Women dominate services and clerical occupations, as elsewhere. In the inner east there has been a larger increase in the proportion of the female part-time workforce (as a proportion of the overall female workforce) than has occurred in the other regions shown (Table 5.3); this is probably mostly accounted for by the growing proportion of clerical workers in offices.

There are higher labor force participation rates and lower unemployment rates for both men and women in the east than in the west. A higher percentage of women in the workforce works part-time there. And participation rates for women are increasing – probably because of the documented, growing need for households to have two incomes to afford house mortgages. But it is likely that few new part-time service sector or clerical positions are in unionized occupational sectors, and that many jobs are casual or contract-based rather than salaried, permanent, and superannuated (Wajcman & Rosewarne 1986, Brennan & O'Donnell 1986, 115–16). Class relations may be changing with the appearance of a slightly different paid workforce, but class formation is probably not resulting from it, at least at this early stage. Whilst women are joining men in the paid labor force in these locations, also (though their unemployment rates are everywhere higher than men's), their job experiences may not be of equal "quality" to those of men. Even in examining the social relations of homework when men and women are conducting family businesses from their homes, Walker (1987) has found the jobs of women often lower in status than those of men.

In sum, we find national paid labor force trends exhibited in eastern and western Melbourne in different fashion. Some variation in the class and gender relations associated with these labor force trends have been hypothesized. These, of course, are general expectations. The precise experiences of class and gender in the different locations and workplaces of Melbourne are not indicated adequately by listing trends in occupational and industrial sector data. A range of other factors is relevant, their particular combination to be determined in field work. But one institutional feature that, along with the changing local labor market, differentiates class and gender relations across the city is the local state, and to it we now turn.

Class and gender relations in the local state

A major contributor to the modes of social and economic reproduction in local areas is the local state. Though it does not have complete control over the range of public services available or affordable in an area or investment in jobs there, the apparatus of the local state can influence aspects of economic and social opportunities. For example, local governments may take up or decline federal funds available to set up child day-care centers; they may start local employment creation initiatives or ignore this possibility. As one recent description of local government revenue and expenditure patterns explained: "some councils may decide it is not their role to house needy groups," and "not all municipalities take full advantage of their local entitlements" (City of Camberwell 1987, 2, 5). The local state is also an arena of local social and political practices that may be channels for changing local class and gender relations – community support networks that help sustain people in paid work are often formed using the resources of local government and local state. Even participation in local community organizing against the local state might weld participants into new groups with new class insights.

There is considerable variation across Melbourne in local government spending on matters that could form better community networks and better prospects for employment. But spatial differentiation in spending is more easily noted between outer and inner municipalities than between east and west (Figure 5.1 shows the usually accepted outer, fringe, middle, and inner "rings" of the metropolitan area). The size and adequacy of amounts spent on "education, health and social welfare" depends in part on factors like the age structure of the local population (numbers of elderly, children, needy), income and employment levels. However, "quality standards and priorities adopted by Councils will also influence expenditure," as noted already (City of Camberwell 1987, 3). Important points to note about the revenue and expenditure of Melbourne's local government councils are:

(1) There is considerable variation in rates (property taxes) as a proportion of total revenue: between 35 per cent and 65 per cent. Most councils receive 14 per cent to 24 per cent of their revenue in government grants. (Other sources are user charges, interest earnings, licences, fees, and fines.) Municipalities containing significant numbers of industrial and commercial properties collect higher rates per rateable property than those with primarily residential land. Rates are the most significant source of revenue, though outer and fringe councils rely on them less than do inner and middle ring councils, and rely more on government grants.

(2) Inner and outer municipalities make a higher proportion of their expenditures on roads than do middle ring councils. Other high expenditures are recreation and administration. Expenditure on health, education, and welfare is relatively low: it includes "support for pre-schools, adult education, health inspections, immunization programs, and the provision of social welfare services to the aged, disabled and others" (City of Camberwell 1987, 2). It does not include schools or the funding of general health care, which are state and federal responsibilities. Middle ring councils spend more on housing and community amenities (town planning, garbage disposal etc.) than do inner and outer ring councils, as a rule.

It is clearly impossible to tell from the recorded categories of local government expenditure whether sustained expenditure on "health, education and welfare" means sustained expenditure on community support for people in paid work. A look at the policies of senior levels of the state apparatus towards releasing funds to local governments for use in such services as child day-care and domiciliary care for elderly (or other) dependents in working households, is more helpful in providing a general overview of the options open to local governments.

Since 1983 the federal Labour Government has accepted major responsibility for increasing the supply of child day-care places, in contrast to administrations of the previous decade which emphasized that child care was a family responsibility and which directed most children's services expenditure to pre-schools (Jones 1985). The federal Office of Child Care estimated "that access to Commonwealth funded day care services . . . increased by 65% from 1982 to 1985" (Jones 1985, 451). A range of strategies enabled this increase in the supply of child day-care places. The major federal focus was on expansion of places in local government controlled child-care centers. Family day-care services (where individuals look after others' children in their own homes) were continued, but not emphasized. Government funding of private sector, for-profit centers was avoided altogether, although funding of non-profit, non-government centers was introduced; the possibility of federal support of workplace day-care has been broached only. Jones (1985, 453–4) adds to this:

> All forms of Commonwealth provision of child care involve elements of partnership with other institutions and groups. The most important of these have been local government and non-profit organizations, and they continue to play a central role. . . . Its (the federal Labour Government's) major strategy for expansion, however, has involved the creation of a new partnership with the State and Territory Governments. Joint agreements have been negotiated with all but one . . . whereby the two levels of government share the capital and establishment costs of building new child care centres, with recurrent expenditure remaining primarily a Commonwealth responsibility.

These partnerships of state and federal governments, through which planning can target centers to the most needy areas, replaces the previous approach to the allocation of federal funds to local centers which had been based on assessment of proposals from individual local centers. Note that there is no assumption of the right to universal service provision here, though that is long what activists have called for (Brennan & O'Donnell 1986).

The Labour Government has recently retreated from its promising start to increasing the supply of day-care places, and its commitment to the growth of non-profit day-care centers. Since mid-1985, apparently in response to the unanticipated size of the recurrent costs of centers (especially staff wages), a new agenda has been emerging. The concern for the provision of affordable, quality care seems to have been replaced with a concern for tight control of funding: previously, federal subsidies were based on the award wages of qualified staff, thus setting staff standards, and center fees were limited by a maximum recommended fee (White 1986). Now, the subsidy of staff has been replaced by subsidy of child places, reducing "the operational subsidies by approximately 50%, and . . . [removing] the Federal legislative basis for the employment of qualified staff" (White 1986, 38). A second measure has been the replacement of ceilings on center fees by a ceiling of a certain weekly rate above which no federal subsidy will be paid. Service quality, usually accepted as related to staff experience and training, must now be maintained at the discretion of the centers and the state governments. If centers are to cover their costs, they must either use cheaper (less qualified) staff, or charge higher income parents higher fees than are necessary to cover the cost of the service those parents receive. More recently still, in the mid-1987 budget announcement, the federal government indicated that family day-care services (cheaper, but long criticized for their exploitation of care givers and the lack of quality control possible) would be relied upon for further increase in the supply of child day-care places.

The implication of recent changes in federal financial commitment is that the quality of centers will vary according to users' ability to pay for them. Centers will perhaps become the day-care services of higher income users; family day providers and informally organized care will perhaps cater to lower income users. Responsibility for enforcing standards will rest more with local and state governments, and will no doubt have to be prompted by continual lobbying: day-care services, their maintenance and extension, have been thrown back into the arena of political competition. It is also, of course, most important to combine this insight with analysis of the workplaces where day-care of children is being done, and their class and gender relations. As family day-care is favoured by government funding strategy, with its exploitative, outwork characteristics for women carers, a scenario can be depicted of women working outside the home exploiting women working as child carers within it.

In 1984 new guidelines for access to government-funded centers specified for the first time that children whose parents were working or seeking work could be given priority for the scarce places available, along with needy children (those at risk of abuse or neglect, and the children of single parents, isolated, Aboriginal, ethnic or low-income families) (Jones 1985, 461). This indicates some recognition of the importance of child day-care services in a time of increasing female labor force participation. However, figures do not show the large increase in the numbers of children having access to day-care that would be needed to offset the demands of the increasing number of parents with jobs in the paid labor force:

The preliminary figures (from a survey of child care arrangements conducted by the Australian Bureau of Statistics in 1984) indicate that *approximately* 119,200 children aged 0–5 years used child day care services and family day care schemes in 1984. This compares with 90,500 in 1980. While this is a substantial increase it still means that more than nine out of ten children below school age have no access to formal child care whether government funded or commercially operated (Brennan & O'Donnell 1986, 17).

There remain many more pre-schools having half-day sessions for young children than there are child-care centers open for long hours. (In Australia in 1983 there were 4306 sessional pre-schools compared with 512 day-care centers (Brennan 1983, 12).) Of course, if locally provided children's services are primarily pre-schools, where children aged 3 to 4 years attend for several half days per week, this means a parent working more hours than that must make additional child-care arrangements. Clearly, the parent responsible for child-care, who cannot find full-time affordable care and must construct a web of arrangements to make do, will be less likely to hold down secure employment.

One must conclude that the recent withdrawals of the federal government from its previously high level of commitment to child day-care is not likely to prompt renewed investment in that service on the part of local government. Local governments are likely still to regard major day-care center funding as a federal responsibility: it would be politically expedient for a local government to take on the funding of something which other levels of government have been accustomed to paying for. However, should "community management" models of child day-care provision emerge as a way of coping with federal withdrawal, then Mowbray's (1984) comments must be kept in mind as well. He has argued that community-based and self-help strategies, that place the burden of social service provision on the "community," usually in fact set that burden squarely on the shoulders of women and "the family": they are often very conservative strategies. They are just what is not needed to complement increased female participation in the labor force. Government provisions for adults who care for elderly dependents make this clear.

For the last two decades federal government policy in Australia has been to encourage community care of the dependent elderly. Concerns to deinstitutionalize the elderly, however, have been underlain by economic as well as humanitarian motives: some analysts have interpreted the federal government's 1956 Home Nursing Subsidy Act and 1973 Domiciliary Nursing Care Benefit as less an indication of the government's willingness to pay for home-based care than of its desire to encourage more care within the community and the family (Kinnear & Graycar 1982). The family, they say, is increasingly performing as a "hidden welfare service", especially as community care becomes more favourably viewed in the public eye than institutional care (Kinnear & Graycar 1983, 79–80). Its costs to care-givers are hidden, as they are in family day-care provision. The 1984–5 federal budget, announced a new Home and Community Care Program to coordinate the range of domiciliary care services to groups including the elderly, and to assist families providing care. Yet

The amounts specified remain rather small. Further, they aim to consolidate rather than supplement some of the previous expenditure on home care and related services. While additional coordination in community support funding and provision is welcome, this move also means that the increase in overall public expenditure is not as great as might first have appeared (Rossiter 1985, 7).

Accompanying federal government efforts to encourage family or community-based care, it is argued, have been shifts in the burden of providing care and a withdrawal of government support of provision (Kinnear & Graycar 1983, 80). This is partly because, rhetoric notwithstanding, the bulk of federal dollars for care of the elderly continues to go to institutions: between 1975 and 1982 ten federal dollars have gone into institutional care for every one going into home support provisions (Kinnear & Graycar 1983, 80). Total federal funding of public sector nursing home care still acts as a disincentive to state governments to expand their support of community care services (Howe 1986, 12).

The general implication of a continuing government focus on community care as the most appropriate form of care for elderly dependents, and yet the lack of expanding financial support for this care, is no doubt clear. "The community" will assume the costs and burden of caring for its elderly residents; in practice, studies have revealed, this means "the family" will have this responsibility, and in turn this means that women in the family will be care providers. A recent Australian survey of families caring in their homes for elderly relatives echoed the findings of overseas research that family care is usually synonymous with care by women, with little support from spouses, children, and the extended family (Kinnear & Graycar 1982, 55). A range of changes in carers' lifestyles were observed, with most problems for carers associated with declining leisure activities, and deteriorated work performance, relationships with spouse and siblings, and sleeping patterns. Astoundingly, the survey showed that over 50 per cent of the carers interviewed, all women, had given up jobs in the paid labor force in order to provide this care (Kinnear & Graycar 1982). Clearly "current concepts of community care build on traditional sex roles, and the practice continues a sexual division of labour which makes it a viable and cheap care alternative for the state" (Graycar & Harrison, 1984, 6).

Now, there are considerable local spatial variations in access to care for elderly people, and also in access to services for family members caring for dependent relatives. These local variations occur despite the considerable financial involvement of the federal government in service provision which affects the whole of Australia. Howe (1986) comprehensively reviews the myriad institutions providing institutional and home-based care for elderly people in the state of Victoria, showing how the presence of different sorts of providers (public, private, or voluntary sector) in localities dictates the availability of care there, as do the decisions of provider agencies to participate in particular programs. Non-institutional services, that is, community care, are overwhelmingly delivered by local government, and so are potentially universal in their provision. They are funded under the States Grants (Home Care) Act and include meals on wheels, home help and senior citizens centers (Howe 1986, 8). But, as noted already, local councils are not required to provide these services or to avail themselves of the appropriate federal or state funds. And matters are even more complex:

> The decision made by local government as to whether to establish any particular service and the scale of service to be conducted affects the distribution of allocations made by commonwealth (federal) or state governments, although the resource inputs from local government itself are very variable. In many cases, the impact of the local government contribution on the distribution of expenditure by other levels of government is out of all proportion to the amounts involved, yet the want of those few local resources may preclude any other assistance; this relativity may best be described as the local tail wagging the commonwealth/state dog (Howe 1986, 8).

It seems evident, and Howe observes, that the range and quality of services in a municipality may depend on its (local government's) financial ability (and willingness) to match funds available from senior levels of government. This is especially true of services giving care to residents outside institutions, which are delivered by local governments. As with the trend in provision of child day-care services, it appears that the trend in community care for the dependent elderly is one where higher socio-economic groups will obtain better service levels and where federal government policy will exacerbate this situation. The class and gender implications of this situation, for people working in the paid labor force who have dependent children or elderly relatives, are not entirely clear. For higher incomes and secure jobs do not guarantee that accessible services for dependents will be available to their holders in a locality; this is up to decision-makers in service-providing agencies and organizations. Some local councils accord high priority to provision of these services for their residents, even if those residents are not articulate lobbyists; other councils need to be prodded continually and remain convinced that care of dependents needs council support only to give women householders some time off. However, it does seem very clear that, across all regions of the city and income groups, women bear the major responsibility for dependents. Brennan & O'Donnell (1986) note in the case of child day-care, and Kinnear & Graycar (1982) in the case of care of elderly dependents, that women are the members of the household who most often take on less secure, less well paid and therefore more flexible work in order to meet the need for care of dependents. Federal expenditure on services for adults in the paid labor force, and policy to encourage the "community" basis of care, seems to be reinforcing this reality.

There are groups within the labor force that have particularly low access even to those few services available, however. Non-English speaking immigrants, for example, have been found to make less use of government-sponsored children's services than have families in Australia as a whole (Pankhurst 1984, 7). This is attributed primarily to the limited information available about centers, to their hours of opening and locations being inappropriate – especially for shiftworkers. Surveys have dispelled the myth that recently immigrated families prefer informal care, and have exposed the anxiety felt by workers whose children have been left alone at home or who have had to make complex child-care arrangements that are constantly breaking down (Pankhurst 1984). It seems to be the case, even in areas where formal day-care places are relatively numerous, that they are primarily patronized by middle-income, English-speaking Australians (Jamrosnik 1987). Immigrant women have been less able to use one of the alternatives open to English-speaking women – community-controlled child-care cooperatives. These sorts of centers are only sustained by successful written submissions and frequent meetings, work that requires English language and writing skills (Loh 1987). The class and gender relations of child day-care provision in the traditionally industrial western suburbs of Melbourne, whose workforce is ethnically very diverse, would reflect this special situation.

Local government provision of services like child day-care and domiciliary care for the dependent elderly, which support adults participating in the paid labor force, are not even in their spatial distribution. The question of how to explain this unevenness is interesting: why is it that some municipalities have better local government provision than others, some relying more on private or informal provision? Rather than just assuming service levels are set to meet "needs" or depend on the whim of individual bureaucrats, one promising alternative has been suggested by Mark-Lawson *et al.* (1985) in Britain. They have examined the history of women's involvement in the political arena of the local state, drawing links between this and the provision of community support services. Though local differences may be revealed and explained in Melbourne, federal government changes in social policy threaten every-where to reduce the accessibility of care for young and old dependents of workers. As these collective consumption goods take on "user pays" characteristics, they will be less affordable. The implications for women, the traditional "carers," may be that they are less able to hold down secure employment: their home, community, and workplace circumstances will combine to alter their experiences of class and gender.

Conclusion

This chapter has suggested ways in which the emerging characteristics of Melbourne's local labor markets and the reactions of local states to the changing priorities of federal social policy, may be affecting the class and gender relations experienced in different parts of the metropolitan area. (The degree to which local residents, workers, and decision-makers are creating distinct local variations in these changing circumstances has yet to be documented.) In concentrating on the local labor market and the local state as two institutional features structuring locally experienced class and gender relations, it is easy to think of the labor market as the realm of profit-making firms and the large bureaucracies of the public and private sectors. Equally, it is tempting to regard the local state, especially that part of its apparatus that is local government, as the provider of collective consumption goods for social reproduction. There are two ways in which matters are more complex than this, and I want to raise them as important foci for further research.

First, the local state, far from being only the site of the social reproduction of labor power, is also having some direct effect on the class and gender relations of employment in certain locations. British discussions, particularly, illustrate the activity of the state in the formation of urban enterprise zones and freeports, encouraging certain forms of enterprise to locate in certain places (*International Journal of Urban and Regional Research*, 1982). (Of course, the state's inactivity is relevant too: in so far as it does not act to prevent exploitative outwork practices in the employment of women locally, the state turns a blind eye to the growth of such practices.) The involvement of the local state in local employment initiatives, like those of London's Greater London Council, can improve the experience of employment and its availability for people in some localities (albeit, so far, on a very small scale: see Lowe (1987)). Australian local employment initiatives are summarized by Gow (1986). Again, they are on such a small scale that they will have little impact on the overall nature of employment-based social relations. But local employment initiative schemes, especially those based around notions of socially useful production, have very interesting implications for the way in which different workplace experiences can facilitate different class and gender relations. There are hints in the Australian ones that special efforts may be made by branches of the state apparatus involved to improve the work opportunities and work experiences of women workers there (Gow 1986, Ch. 8), though affirmative action is not the same as socially useful employment. In any case, the involvement of the state in employment in places, and the degree to which its workplaces can be made the sites of better class and gender relations, is something which should not be overlooked in studies of the local state.

The second point is related to the first. Provision of community services to adults who have dependents and who need those services in order to participate in the paid labor force, is a matter of workplaces as well as of service use. Services are produced as well as consumed. So the other side of the coin that is marked with declining accessibility of child care, is the side that is marked with the exploitation of women outworkers who are picking up the task of caring for workers' children on a less formal basis. That there seems to be a gender bias in both aspects of the situation, that women are often the ones needing the care for their dependents and that women workers are the ones doing the caring under increasingly difficult conditions, has often been remarked and needs to be explained. Class and gender relations are everywhere intertwined, within institutions like the local state as well as in their outcomes.

Note

This project is being supported (in part) by a Research Development Grant from the Faculty of Arts, University of Melbourne.

References

Blakers, C. 1986. Local employment initiatives: summary of thinking and developments, *Scan* 1, 136–43.

Bleitrach, D. & A. Chenu 1979. *L'usine et la vie*. Paris: Maspero.

Boddy, M. & C. Fudge 1984. *Local socialism*. London: Macmillan.

Bradbury, J. 1984. The impact of industrial cycles in the mining sector: the case of the Quebec-Labrador region in Canada, *International Journal of Urban and Regional Research* 8, 311–31.

Brennan, D. 1983. *Towards a national child care policy*. Melbourne: Institute of Family Studies Background Paper.

Brennan, D. & C. O'Donnell 1986. *Caring for Australia's children*. Sydney: Allen & Unwin.

Bureau of Industry Economics 1979. *Employment of demographic groups in Australian industry*. Research Report 3. Canberra: AGPS.

Bureau of Labor Market Research 1985. *Who's Who in the labour force?* Research Report 7. Canberra: AGPS.

Cass, B. & P. Garde 1984. Unemployment in the western region of Sydney: job seeking in a local labour market, SWRC Reports and Proceedings No. 47, University of New South Wales, Sydney: Social Welfare Research Centre.

City of Camberwell 1987. *Melbourne inter-council comparison*. 3rd edn. Melbourne.

Collins, J. 1984. Immigration and class: the Australian experience. In *Ethnicity, class and gender in Australia*, G. Bottomley & M. de Lepervanche (eds.). Sydney: Allen & Unwin.

Cooke, P. 1986a. The changing urban and regional system in the United Kingdom. *Regional Studies* 20, 243–51.

Cooke, P. (ed.) 1986b. *Global restructuring, local response*. London: Economic and Social Research Council.

Dear, M. & J. Wolch 1987. *Landscapes of despair*. Princeton N.J.: Princeton University Press.

Edgington, D. 1987. Melbourne, the next ten years – Victoria's new metropolitan strategy. Paper presented at the Annual Meeting of the Institute of Australian Geographers, Canberra, August.

Fincher, R. 1987. Space, class and political processes: the social relations of the local state. *Progress in Human Geography* 11, 496–515.

Foord, J. & N. Gregson 1986. Patriarchy: towards a reconceptualisation. *Antipode* 18, 186–211.

Fothergill, N. 1984. *Metropolitan employment analysis, 1961–81*. Melbourne: Melbourne and Metropolitan Board of Works.

Fothergill, N. 1987. *Metropolitan office development*. Working Paper 3, Ministry for Planning and Environment, Melbourne.

Giddens, A. 1984. *The constitution of society*. Cambridge, UK: Polity Press.

Government of Victoria 1987. *Shaping Melbourne's future*. Melbourne: Victorian Government Printer.

Gow, H. 1986. A national overview of local employment initiatives in Australia.

Canberra: National Conference on Local Employment Initiatives, Background Paper 2.

Graycar, A. & J. Harrison 1984. Ageing populations and social care: policy issues. *Ausralian Journal on Ageing* **3**, 3–9.

Harvey, D. 1975. Class structure in a capitalist society and the theory of residential differentiation. In *Process in physical and human geography: Bristol essays*, R. Peel, M. Chisholm & P. Haggett (eds.). London: Heinemann.

Harvey, D. 1987. Flexible accumulation through urbanization: reflections on "post-modernism" in the American city. Paper presented to a symposium at the Yale School of Architecture, February.

Howe, A. 1986. Aged care services: an analysis of provider roles and provision outcomes, *Urban Policy and Research* **4**, 2–20.

Howe, A. & K. O'Connor 1982. Travel to work and labor force participation in an Australian metropolitan area. *Professional Geographer* **34**.

Huws, U. 1985. Terminal isolation: the atomisation of work and leisure in the wired society. *Radical Science* **16**.

International Journal of Urban and Regional Research 1982. Urban enterprise zones: a debate. **6**, 416–46.

Jamroznik, A. 1987. Household economy and social class. Paper presented to the workshop/conference "The future of the household economy and the role of women", University of Melbourne, Centre for Applied Research on the Future.

Jones, A. 1985. The child care policies of the Hawke Labor government, 1983–1985. *Proceedings of the 27th Annual Conference of the Australasian Political Studies Association* **1**, 446–79.

Katznelson, I. 1981. *City trenches*. Chicago: University of Chicago Press.

Kinnear, D. & A. Graycar 1982. *Family care of elderly people: Australian perspectives.* University of New South Wales, Kensington: Social Welfare Research Centre, Research Paper 23.

Kinnear, D. & A. Graycar 1983. Non-institutional care of elderly people. In *Retreat from the welfare state: Australian social policy in the 1980s*, A. Graycar (ed.). Sydney: Allen & Unwin.

Livingstone, D., M. Luxton & W. Seccombe 1982. *Steelworker families: workplace, household and community in Hamilton, Ontario.* Research Proposal to the Social Sciences and Humanities Research Council of Canada.

Loh, M. 1987. On-the-job child care. *Australian Society*, April, 38.

Lowe, M. 1987. Book review of "Very nice work if you can get it: the socially useful production debate" (ed. by Collective Design/Projects (1985) Spokesman Books, Nottingham), *Environment and Planning D* **5**, 352–4.

McDowell, L. 1986. Beyond patriarchy: a class-based explanation of women's subordination. *Antipode* **18**, 311–21.

Mark-Lawson, J., M. Savage & A. Warde 1985. Gender and local politics: struggles over welfare policies, 1918–1939. In *Localities, class and gender*, L. Murgatroyd et al.

Massey, D. 1978. Regionalism: some current issues. *Capital and Class* **6**, 106–25.

Massey, D. 1984. *Spatial divisions of labor.* New York: Methuen.

Massey, D. & R. Meegan 1982. *The anatomy of job loss.* London: Methuen.

Mishrah, R. 1984. *The welfare state in crisis.* Brighton, Sussex: Harvester Press.

Mowbray, M. 1984. Localism and austerity: the community can do it. *Journal of Australian Political Economy* **16**, 3–14.

Murgatroyd, L., M. Savage, D. Shapiro, J. Urry, S. Walby & A. Warde with J. Mark-Lawson 1985. *Localities, class and gender.* London: Pion.

Nelson, K. 1986. Labor demand, labor supply and the suburbanization of low-wage office work. In *Production, work, territory*. A. J. Scott & M. Storper (eds.). Winchester, Mass.: Allen & Unwin.

O'Connor, K. 1981. The development of suburban labour markets. Paper presented to the Regional Science Association, North American meeting, Montreal.

Pankhurst, F. 1984. *Workplace child care and migrant parents*. Canberra: National Women's Advisory Council Research Report. AGPS.

Rossiter, C. 1985. Policies for carers. *Australian Journal on Ageing* **4**, 3–8.

Sandercock, L. 1986. Economy versus community. *Australian Society*, July, 12–15.

Savage, M. 1987. Understanding political alignments in contemporary Britain: do localities matter? *Political Geography Quarterly* **6**, 53–76.

Storper, M. 1985. The spatial and temporal constitution of social action: a critical reading of Giddens. *Environment and Planning D: Society and Space* **3**, 407–24.

Urry, J. 1981. Localities, regions and social class. *International Journal of Urban and Regional Research* **5**, 455–74.

Urry, J. 1985. Social relations, space and time. In *Social relations and spatial structures*, D. Gregory & J. Urry. London: Macmillan.

Wajcman, J. & S. Rosewarne 1986. The feminisation of work. *Australian Society*, September, 15–17.

Walker, J. 1987. Home-based working in Australia: issues and evidence, Urban Research Unit Working Paper no. 1. Canberra: Australian National University.

Walker, R. 1985. Class, division of labor and employment in space. In *Social Relations and Spatial Structures*, D. Gregory & J. Urry (eds.). London: Macmillan.

White, M. 1986. Child care funding: a change of direction. *Australian Journal of Early Childhood* **11**, 38–41.

Wilson, K. 1985. *Manufacturing industry in the western metropolitan region of Melbourne*. Vol. 2 *An introductory analysis of firm data*. Melbourne: Western Region Commission.

Wright, E. O. 1985. *Classes*. London: Verso.

6

A feminist perspective of employment restructuring and gentrification: the case of Montréal[1]

DAMARIS ROSE

For a number of years, the urban-geographical literature on the gentrification[2] of inner-city neighborhoods suffered from what became a somewhat sterile debate between "consumption-side" and "production-side" explanations of the phenomenon. The first wave of "back to the city" studies, most of them American, was firmly within the neo-classical land economics tradition. They focused on how changes in "lifestyles" were reshaping patterns of consumer demand, changing the trade-off of residential preference between inner city and suburb, and thus leading to reinvestment in central city neighborhoods (see, e.g., Alonso 1980, Laska & Spain 1980).

As a corrective to this "consumer-sovereignty" orientation, neo-Marxist writers, inspired by David Harvey's work on urban land markets in Baltimore in the early 1970s, insisted on the primacy of the economic forces shaping the historical production of, and change in, urban land values. It was argued that the resulting cyclical patterns of divestment and reinvestment in the built environment determined the supply conditions of gentrifiable housing, which in turn structured the demand (Smith 1982).

Dissatisfaction with such dichotomous orientations has led authors of diverse persuasions to seek out more comprehensive explanations focusing on the interaction of production and consumption. In this respect the notion of the "post-industrial" or "information" city has become popular. These essentially descriptive concepts situated urban change with reference to broader economic trends, and in particular the increased specialization of major urban centers in "advanced tertiary" activities (Cossette 1982, 71–89, Simmie 1983). This evolution also entailed the production of a new terrain devoted to consumption activities – the gentrified urban landscape.

This approach in its turn spurred the development of a Marxist variant, based on the analysis of the interactions of broader processes of economic restructuring (new spatial divisions of labor), patterns of capital investment in the central city, and the production of a gentrifier fraction within the "new middle class" (see, e.g., Williams & Smith 1986). This approach explicitly attempted to overcome the production-consumption dichotomy, by focusing on the production of social groups who themselves become

118

agents of urban change, albeit within a set of broader economic forces not of their creation. There would also seem to be a link between this approach and a "humanist" variant of the post-industrial city thesis: Ley (1981), for instance, emphasizes the role of the new middle class in generating the urban reform movement, and thus shaping the gentrification process.

Without ignoring this debate between neo-classical and neo-Marxist approaches to gentrification, a few writers have insisted on the necessity of taking on gentrification from a feminist perspective. It was pointed out, for instance, that the displacement effects of gentrification were not gender-neutral, and in particular that female-headed families (who form a large proportion of low-income inner-city populations) were often the hardest hit by gentrification; at the same time, these very families were especially dependent on the combination of inexpensive housing, access to jobs and to a network of supportive community services that inner cities could offer (Holcomb 1984). It was further argued (notably by Markusen 1981) that gentrification had to be seen in the context of an urban structure which has historically reflected and reinforced the separation of the "spheres" of waged work and reproduction (q.v. Larsen & Topsoe-Jensen 1984; Mackenzie & Rose 1983). Thus, middle-class women with both "career" jobs and children might well opt to become gentrifiers, in order to reduce the time-space constraints resulting from their dual roles; for this type of woman, too, the inner city would be a more supportive and convenient environment than the low-density suburb geared toward the traditional nuclear family (Rose 1984, Rose & Villeneuve 1987, Wekerle 1984).

It is worth noting that this latter line of argument has been developed primarily by Canadian authors; this is probably because Canadian cities are, by and large, seen as more attractive places to raise children than most of their American counterparts (cf. Goldberg & Mercer 1986, 154–6). One contributory factor to this reality has been the long and ongoing tradition of involvement of women in urban reform movements in Canadian cities, their efforts centering on struggles for community services and facilities (Andrew & Moore-Milroy 1987).

This emergent feminist orientation rejects the notions of consumer sovereignty inherent in neo-classical perspectives. Yet it upholds and provides conceptual structure to the view articulated by some neo-classicists (see Alonso 1980), that analyses of gentrification must be concerned with demographic trends and with the increase in women's labor force participation and its effects on household structure and the roles of family members. The analytical separation of production and consumption, which has plagued urban geography in particular, is seen by feminists to be artificial; it is argued that a dynamic urban theory must focus on their interrelationship. The analysis of reproduction[3] of people and social life thus has to become central to urban analysis (Andrew & Moore-Milroy 1987; Mackenzie 1987; Women and Geography Study Group of the IBG 1984, 43–66). Changes in women's employment situations thus have to be related to recent rounds of economic restructuring (cf. Massey 1984, *passim*) as these are mediated through the particular employment structure of a city. All the same, one should not subsume these changes in women's situation within the general logic of economic restructuring. In contrast, neo-Marxist writers still accord the gender component in labor force restructuring an inferior conceptual status (see, e.g., Williams & Smith 1986, 208). The dynamic interrelation of women's employment situations with their family situations, and the consumption practices of the households they live in, is thus obfuscated (Walby 1985).

It will be argued in the present chapter that gender relations can be actively *constitutive* of the new fractions of the labor force, and of the formation of some of these fractions into "gentrifiers," depending on the particular characteristics of the advanced tertiary employment structure of the city in question. First, we focus on professionals living in inner-city neighborhoods in Montréal: we shall examine how this labor force was restructured during the 1970s, showing that some key elements of this restructuring can be understood only by analyzing the labor force simultaneously by gender and by economic sector. We shall also see that some of these key elements illustrate the specificity of the Canadian economic context, as opposed to that of the United States. Second, in a more theoretical vein, we sketch out the nature of the dynamic interrelation between changes in the sexual division of labor, employment restructuring, and the types of households in which professionals live. It would appear that what is being produced is not a single group of gentrifiers with homogeneous cultural practices and ways of using space, but several different, perhaps even divergent fractions.

Urban hierarchies and the gender division of professional employment: the position of Montréal

In recent years, the Marxist-influenced urban literature has increasingly come to study how, at national and international levels, recent transformations of the economy, of power relations and decision-making structures, are "mapped" into a restructured functional urban hierarchy (Cohen 1981, Lipietz 1986, cf. Pred 1978, and Walker 1985, 245–7, for a sympathetic critique). In this spirit, some students of gentrification have tried to define more precisely the new fractions of managerial and professional labor that are becoming so prevalent in "global" and "regional cities" (Fainstein & Fainstein 1982, Williams 1986). These broad shifts are related to urban revitalization and gentrification by simultaneously looking at how the existing built environments of central cities are reshaped.

The research developing this perspective on restructuring and gentrification has been influenced by the experience of global cities such as New York and London, whose employment structures are very starkly polarized (see, e.g., Greater London Council 1986, 45–63; Sassen-Koob 1984). Not surprisingly, then, this work has emphasized the role of the central city as corporate control center, and, concomitantly, as a place dedicated to a "conspicuous consumption" that helps reproduce the strata of highly privileged managerial and professional personnel associated with the decision-making structures of major corporations. Gentrifiers are seen as being, by and large, recruited from these fractions, which are seen as occupying a class position that very strongly polarizes them away from working-class groups. Furthermore, it has been cogently pointed out that an increasingly important fraction of the working class is employed as the "new servant class" that "prepares the gourmet take-out food for the wealthy, stitches their designer clothes and helps manufacture their customized furniture" (Sassen-Koob, quoted in Ehrenreich 1986, 62). Many of these very same workers (of whom a large proportion are women on very low incomes) are at risk from displacement by gentrification.

This approach represents a significant theoretical advance. However, as far as Canadian cities are concerned, we should be somewhat wary of such images of the new middle class and of gentrifiers, since they stem, to a considerable extent, from empirical generalizations (q.v. Sayer 1984, 217–19), based on notions of the global, corporate city. With the partial exception of Toronto, this type of city does not exist in Canada. Moreover, as we shall see with reference to Montréal, even the concept of *regional* city may need some refinement in as much as this notion tends to evoke images of private sector control centers and fails to take account of the importance of the public sector in the urban economy.

If we are interested in coming to grips with the specificities of labor force restructuring and its relationship to gentrification in Canadian cities, a useful point of departure may be the work of sociologists Black & Myles (1986). These authors explore from a new angle the influence of Canada's history of "dependent industrialization" on the class structure. They compare the occupational division of labor and particularly the authority and control structure within extractive and transformative sectors (where the "pace-setters" are US multinationals), within privately-provided services (dominated by a mid-sized indigenous capital) and the state sector. They show that in general, occupational structures are most polarized in the sectors dominated by US multinationals, where, furthermore, upper-level white-collar workers are most likely to be oriented toward or incorporated into management power structures within the labor process. In the other sectors, and most especially in the state sector, the upper fractions of the middle class are more autonomous and hold more "neutral" positions in the labor process.

It is the comments of these authors on the positioning of *professionals* (whose power is related most particularly to knowledge), rather than that of managers (who have a strong decision-making power, but are much smaller in number),[4] that is of particular interest for our present purposes, professionals being the fraction most clearly associated with gentrification in the literature. For instance, from the point of view of urban change, their findings raise the interesting question of whether these differences between new middle-class fractions according to economic sector could influence, at the level of the neighborhoods where these groups live, their cultural and political practices.

Taking the argument of Black & Myles one step further, we could put forward a second proposition. It is possible that professionals in the "corporate production" sector in Canada might be on a somewhat higher career trajectory and have better financial prospects than those in the public and parapublic sectors. For, since the late 1970s, the recession and economic climate of the new Right (q.v. Rocher 1984) have effectively blocked the prospects of large numbers of younger public sector professionals (Morgan 1985), as well as marginalizing many of those in other sectors heavily dependent on state funding (community organizations, arts and culture etc.).

In the same way, the important but fragile gains which Canadian and Quebec women have made, with respect to careers and to the recognition of their needs by unions (Maroney 1983), risk being particularly hard hit by roll-backs in the state sector. For women professionals not only tend to be younger, and thence to hold less seniority, than their male counterparts; they are also overwhelmingly concentrated in economic sectors undergoing retrenchment after having experienced spectacular growth in the 1970s – notably education, health, and welfare (traditionally "female ghettoes"). Furthermore, the proportion of women among professionals in public administration also increased significantly over this period (Armstrong 1984; Rose & Villeneuve 1985). We might well conclude, therefore, that in the 1980s women professionals are relatively "marginal" with respect to employment incomes, job prospects, and job security (even if they manage to develop a career as well as having a family).

Taken together, these arguments – though in need of further research and refinement – suggest some important lines of demarcation within the "new middle class," depending in part on the economic sectors in which fractions of this "new class" are embedded, as well as upon the age at which they entered the labor market (in relation to the onset of the present period of crisis). Gender relations are related to both of these elements, but not reducible to either of them.

Furthermore, we could use these arguments to make some tentative propositions about differences in the "new middle class" between two Canadian cities that have experienced different kinds of economic and labor force restructuring – Montréal and Toronto. The 1970s saw a move from Montréal to Toronto of many Canadian head offices, especially in finance, in natural resources, and in the manufacturing sector (Semple & Green 1983). Over the same period, Montréal strengthened its position *vis-à-vis* the province of Quebec as a *regional* center for business services; but, as regards employment, its tertiary sector has a marked specialization in education, health, and social services (Lamonde & Polèse 1984). Since there is a much lower proportion of women among professionals in private sector tertiary jobs than in the public and parapublic sectors (Armstrong 1984), we may suppose that the different sectoral specialization of these two cities is mirrored in the gender composition, and filters through to the consumption practices of the professionals who live in them.

In addition, for reasons not unrelated to the divergent recent economic

Figure 6.1a Professionals, division by economic sector and gender, 1971 and 1981. Montréal Census Metropolitan Area.

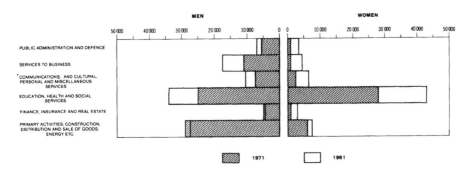

Figure 6.1b Three inner-city neighborhoods in transition.

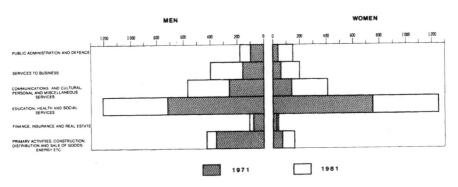

trajectories of the two cities, gentrification in Montréal has taken place in a somewhat less "hot" real estate market than that experienced by Toronto (q.v. Howell 1987). Moreover, the gentrifiable housing stock of the two cities is very different: in Toronto we find a majority of single-family dwellings with a fair number of mid-sized apartment buildings; while Montréal's housing stock is overwhelmingly constituted of rental units, mainly superimposed flats in buildings (called "plexes") with from two to six units. In comparison to Toronto, it is probable that gentrifiers of inner-city and formerly working-class neighborhoods in Montréal are a more diverse group with respect to occupation, income, and household type (fewer two-earner couples, more female-headed households, for instance) and housing tenure (more tenants and owners of legally undivided or "informal" condominium units).

We do not have access to data that would allow exploration of these ideas through a comparative analysis of the gender and sectoral division of the "new middle class" of these two cities. All the same, by breaking down the labor force restructuring that occurred in Montréal between 1971 and 1981 according to sector and gender, we can illustrate the linkages between the specificity of a city's economic sectoral structure, the importance of women professionals, and the human face of its gentrification.

In the Montréal metropolitan area (CMA) the number of jobs held by professionals increased by 40.8% from 1971 to 1981.[5] Yet the increase was much less marked in sectors related to the production and distribution of goods (9.9%), and in finance, insurance and real estate ((16.2%)[6] (see Fig. 6.1a). There was an actual *decrease* in jobs occupied by male professionals in these sectors, which is probably accounted for by the aforementioned migration of certain head offices. Thus, in these sectors, the increase in the total is due entirely to the increase in the female workforce; this coul well mean that the positions in question were lower-level and not linked to decision-making structures. In contrast, we see large absolute and relative increases in professional jobs in cultural and communications services, services to business management, and public administration. At the same time, the education, health and welfare sector remains relatively very important due to a huge absolute increase, which in turn is responsible for the major part of the increase in numbers of professional jobs held by women.

Sectoral and gender divisions of labor among professionals living in three inner-city Montréal neighborhoods undergoing transformation

We now explore a particular case of differentiation of an occupational group by economic sector, gender, and employment income. We focus on the change between the census years of 1971 and 1981 for the Montréal CMA and for three neighborhoods (designated zones A, B and C) situated within the Greater Plateau Mont-Royal district (see Fig. 6.2). These neighborhoods, traditionally home to working- and lower-middle class groups (Lussier 1984, Mathews 1986), have undergone a slow process of gentrification which has – so far – left them with considerable socio-economic diversity. Zone A (the western part of the St. Louis and Mile End neighborhoods, and the traditional immigrant "corridor" of Montréal) also retains very considerable ethnic diversity.[7]

In general, as one might expect, the Montréal CMA labor force became more polarized during the 1970s between managers and professionals at one end, and low-skilled white-collar, service, sales, and industrial workers at the other end (for more details, see Villeneuve & Rose 1985). In our three selected neighborhoods, the proportion of residents with a professional occupation increased more rapidly than in the metropolitan area as a whole. Already, in 1971, zone C (the closest to downtown) had proportionately twice as many profesionals as the CMA; by 1981, zone B had also achieved this status, while in zone A the proportion of professionals had become similar to that of the CMA as a whole. Also, in

124

Figure 6.2 Percentage division of professionals by economic sector and gender, 1971 and 1981. Location of study area.

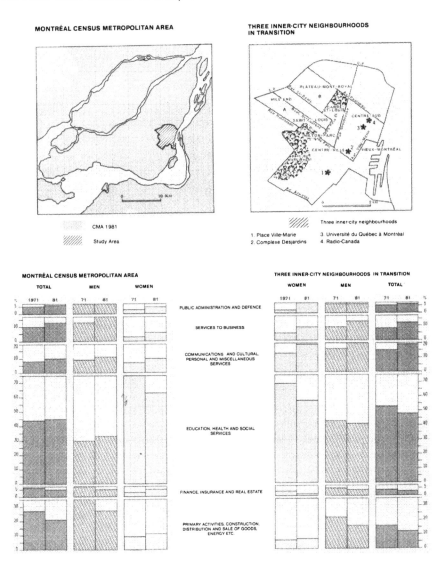

1981 these three neighborhoods had a higher proportion of women among their resident professionals than was to be found in the metropolitan area.

In 1971, the three neighborhoods we have selected, the proportion of professionals employed in the education, health and welfare sector was higher than in the CMA (see Fig. 6.2). This difference was even more pronounced for the culture and communications sector[8] and became still greater over the period studied, for both sexes: by 1981, this sector employed one out of every five professionals living in these areas, as compared with one in ten at the scale of the metropolitan area.

In contrast, in 1971 and even more so in 1981, the proportion of professionals in finance, insurance, and real estate was smaller in the three inner-city zones than in the CMA as a whole. Business services became relatively more important, though still less so than in the CMA as a whole. Finally, the public administration sector figured more strongly among professionals in the CMA in 1971 than in the three neighborhoods; by 1981 the situation was reversed.

We thus see living in these neighborhoods in 1981, a population of professionals whose existence and living conditions are strongly linked to the development of the public sector, and that of the communications and cultural sectors (often dependent on public funding). The findings of other researchers help flesh out this picture: people in artistic and literary occupations, often economically marginal, are especially over-represented on the Plateau Mont-Royal (Dansereau and Beaudry 1985). Furthermore, this precariousness is likely to be reinforced by the fact that women are over-represented among professionals in the neighborhoods under study, compared to the CMA.

Our data also allow for the inclusion of another dimension. The relationship between the location of places of employment and those of residence leads us to believe that there is a symbiotic linkage between the growth of public and parapublic sector employment, especially of women, and the gentrification of the neighborhoods studied. In general, public administration jobs are highly concentrated in the eastern (and very strongly francophone) part of Montréal's downtown, while private sector jobs remain in the original (traditionally anglophone but now linguistically-mixed) CBD further to the west. If we look at the employment/home separation of professionals (these data cannot be further broken down by economic sector), we find that in 1981 12.8% of women (but only 6% of men) who work in the eastern part of downtown live in the Greater Plateau district. No other neighborhood in Montréal is anywhere near as important a source of female professional labor for jobs located in either section of downtown. We should note that downtown-east, as well as being the center of public administration, also contains the Université du Québec à Montréal, the Québec headquarters of the Canadian Broadcasting Corporation and of Radio-Québec, and several major hospitals.

The concentration of residents of the Plateau Mont-Royal among women professionals working in the eastern part of the downtown area is in fact one expression of a broader tendency: women employed downtown are more likely to live in the inner city than their male counterparts (Rose 1986). The Plateau area may be particularly attractive to women professionals in the public, non-profit, and cultural sectors due to its proximity to downtown, its well-developed social and information networks which, for example, might help contractual workers to obtain employment, its modest-priced housing stock, and extensive provision of a diverse range of community services.

Our census data do not permit an analysis of the relationship of mean employment income and economic sector. Nevertheless, it seems clear (see Table 6.1) that a large proportion of the professionals living in the three neighborhoods examined here could be called "marginal" with respect to incomes. In relative terms, this is as true of men as of women, if we compare their incomes to those of their counterparts in the metropolitan area; in absolute terms, the marginality of women stands out, given that they comprise a greater percentage of the resident professional workforce than in the CMA as a whole.

We can thus see that if we are interested in the implications of economic restructuring for the reshaping of neighborhoods, we need to examine the interrelation of the gender composition of a city's professionals and their division by economic sector. For gender divisions influence employment incomes and job security of "new middle-class" fractions from which gentrifiers are produced. Yet in order to explore how this interrelation might influence gentrification, we also need to break down the professional population of these neighborhoods in terms of the types of household they live in. Household type is likely to influence the effect of occupational position on consumption practices, and on where these take place (through the number of persons employed and the number of dependents, for example) and, more broadly, on ways of living and forms of reproduction. This brings us to the second main theme of the present chapter.

Gentrification, household structure, and reproduction

If we now look at the professionals living in our three neighborhoods in terms of household type (here again, a disaggregation by economic sector is not available), we see (from Table 6.2) that in 1981 three types of households were more strongly represented there than in the CMA: persons living alone, those living with unrelated adults, and single parents. According to our calculations (not presented in Table 6.2), professionals of both sexes are more likely to live alone than are other workers; although this is also for the CMA as a whole, it is particularly pronounced for males in the three neighborhoods. The fairly large proportion of gay men in

Table 6.1 Professionals: mean employment incomes (1980) in three Montréal neighborhoods experiencing gentrification.

Zone		Total	Men	Women
A	St-Louis/Mile End (western part)	$14,467	$16,957	$11,664
B	Plateau Mont-Royal (central part)	$15,609	$17,089	$13,954
C	Milton-Parc (east)/Carré St-Louis	$20,296	$23,260	$15,001
CMA	Montréal metropolitan area	$22,297	$26,961	$16,960

Source: Statistics Canada, Census of 1981, special tabulations.

St-Louis and Mile End could in part explain this finding (Séguin 1984) –
although this influence should not be over-stated since this is not the main
"gay area" of Montréal. As to women professionals, we find them over-
represented among persons living with unrelated adults, compared to the
whole metropolitan area.[9]

Married couples or those living together (for most purposes, the
Canadian census does not distinguish between these two categories) are
clearly under-represented among the resident professionals in these
neighborhoods. In most cases, both spouses are in the labor force; it is all
the same noteworthy that the proportion of married women who are the
sole earners is higher (14%) than in the CMA as a whole. It has been noted
elsewhere that among families where a professional woman is the
"reference person" (census term that replaced "household head" in 1981),
only a small proportion have children living at home (Rose & Villeneuve
1987). However, since the 1981 census, there are indications of a baby
boom among "career women" in their thirties, at least in Mile End, where
flats tend to be larger and a more stable neighborhood ambience exists
(Rose & Villeneuve 1987).

This brief glimpse at household structures, combined with our
discussion of the forces which render women's new-found autonomy
fragile and precarious, highlights the importance of gender relations as an
element structuring ways of life in these gradually gentrifying neighborhoods. With this perspective, and taking into account the specificities of
Montréal's housing stock and tenure structure, we may suggest some
hypotheses about the types of housing and housing tenure to which these
gentrifiers might be drawn.

On the basis of exploratory studies, it seems that the presently
"unofficial" tenure form of *copropriété indivise* (undivided co-ownership –
not unlike the "equity co-ops" found in the United States) is particularly
appealing to women professionals whose financial future seems stable or
on the upturn (Choko & Dansereau 1986, Mondor 1984). Women whose
situation is more marginal or precarious are more likely to opt for a
housing cooperative (non-profit, no-equity) in a centrally located neighborhood (Rose 1984, Rose & Le Bourdais 1986). In both cases, access to
social networks and to community facilities and services seems to be just
as important in neighborhood choice as the characteristics of the housing
unit as such. Among male professionals, those who are more financially
secure and "established" are likely to be drawn toward rental apartments at
the "luxury" end of the market (a wave of speculation has considerably
increased the number of such units available) or to condominium
developments. Two-earner families might in certain cases convert
duplexes into a single-family dwelling; more often than not, however,
those with the financial means might be drawn to other less central,
though still conveniently located neighborhoods where single-family
houses are more readily available.

The relationship between restructuring and consumption practices is,
however, neither direct nor unilinear. The internal composition and the
increased diversity of the new middle class are influenced by changes in

Table 6.2 Professionals by sex and by type of household lived in, three Montréal neighborhoods experiencing gentrification, 1981.

	Men				Family households			Others[c]
	Total	Non-family households[a]		Lone parent[a]	Total	Couple (M-F)		
							Spouse	
Zone		1 Person	>1 Person			in L.F.	not in L.F.[b]	
A St-Louis/Mile End (western part)	785	295	80	30	310	71.0%	29.0%	70
B Plateau Mont-Royal (central part)	970	325	80	15	430	74.4%	25.6%	120
C Milton-Parc (east)/Carré St-Louis	1180	670	95	20	355	70.4%	29.6%	40
Sub-total of the three zones	2935	1290	255	65	1095	72.1%	27.9%	230
	100.0%	44.0%	8.7%	2.2%	37.3%	26.9%	10.4%	7.8%
CMA Montréal metropolitan area	102070	13320	2755	1465	73105	60.9%	39.1%	11425
	100.0%	13.0%	2.7%	1.4%	71.6%	43.6%	28.0%	11.2%

continued overleaf

Table **6.2** *continued*

Women Zone	Total	Non-family households[a]		Lone parent[a]	Family households Total	Couple (M-F)		Others[c]
		1 Person	>1 Person			in L.F.	Spouse not in L.F.[b]	
A St-Louid/Mile End (western part)	690	170	80	40	280	83.9%	16.1%	120
B Plateau Mont-Royal (central part)	835	275	90	65	300	93.3%	6.7%	105
C Milton-Parc (east)/Carré St-Louis	670	315	70	55	220	77.3%	22.7%	10
Sub-total of the three zones	2195 100.0%	760 34.6%	240 10.9%	160 7.3%	800 36.4%	85.6% 31.2%	14.4% 5.2%	235 10.7%
CMA Montréal metropolitan area	68100 100.0%	12645 18.6%	2530 3.7%	4730 6.9%	38625 56.7%	94.5% 53.6%	5.5% 3.1%	9570 14.1%

Source: Calculated from *Statistics Canada*, Census of 1981, special tabulations.
[a] Only household reference persons are included in the data corresponding to this heading.
[b] This category was calculated by subtracting those whose spouse was not in the labour force from the total number of married persons.
[c] This residual category includes those who are neither married nor "reference persons" (according to census terminology). A good number of women living with unrelated adults, as well as a certain number of single parents, are likely to be "concealed" within this category.

men's and women's ways of living, which are not reducible to the "economic." In order to grasp these relations in all their complexity, and in particular their effects on inner-city neighborhoods, we need to turn once again to the notion of production or reproduction of the labor force, this being viewed as a process structuring residential and community space. This notion has, of course, been central to much Marxist-inspired urban analysis, but feminists argue that it should be refined and enriched. Rather than continually conflating, as does much urban political-economic analysis, the activities of consumption and reproduction, a feminist perspective would explicitly discuss "reproductive work," much of which is carried out in the domestic sphere and thus remains invisible, but without which goods purchased could not be consumed; child-care must also be included within this concept (see, e.g., Andrew & Moore-Milroy 1987, McDowell 1986). In fact, it may prove useful to enlarge the notion still further, to take into account the waged work of people employed in public and private service provision, in so far as their work allows certain households to replace part of their domestic and household labor by the purchase or use of services and "ready to consume" goods. Adding to the concept of reproductive work in this way could help us develop appropriate frameworks for understanding the ways of life of so-called non-traditional households: lone-parent families, households formed of unrelated adults, men and women living alone, two-earner couples, and so on.

If further research can explore in more depth the links between reproduction and consumption touched on here, we may perhaps uncover some of the missing elements in existing conceptualizations of the relationships of economic restructuring, demographic trends, household formation, forces structuring housing demand in relation to the supply of gentrifiable housing, and the diversity of ways of living among gentrifiers. In particular, the residential preferences of some types of non-traditional households for inner-city neighborhoods may in part be explained by the following factors: these *mileux* enable a diversification of ways of carrying out reproductive work; they offer a concentration of supportive services; and they often have a "tolerant" ambience (see, e.g., Mills 1986, Rose & Le Bourdais 1986, Watson 1986).

Obviously, this type of reasoning should be qualified by taking account, for instance, of the fact that inner-city areas will not necessarily be able to continue over time to support such diversity and to facilitate such "alternative" consumption practices. This will depend in part on the economic future of the central cities in question, which will condition both the rhythm of production of potential gentrifier fractions of the labor force and the development of local housing markets. Much will also depend on the extent of intervention in the private housing market by the state and non-profit housing sectors, as well as turnover rates among a neighborhood's "incumbent" owner-occupiers (this last question being particularly important in strongly ethnic areas of Canadian inner cities).

In a schematic manner, we may put forward a conceptual framework which draws together the complex sets of interrelationships sketched out

Figure 6.3 Schematic representation of production-reproduction gentrification relations.

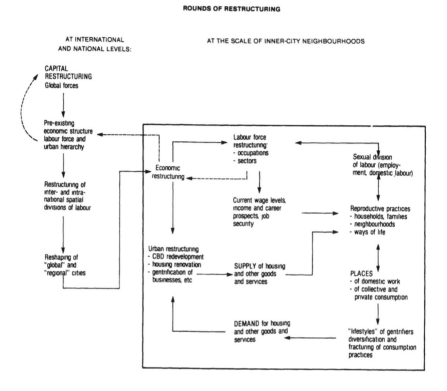

ROUNDS OF RESTRUCTURING

AT INTERNATIONAL
AND NATIONAL LEVELS:

AT THE SCALE OF INNER-CITY NEIGHBOURHOODS

CAPITAL
RESTRUCTURING
Global forces

Pre-existing
economic structure
labour force and
urban hierarchy

Restructuring of
inter- and intra-
national spatial
divisions of labour

Reshaping of
"global" and
"regional" cities

Economic
restructuring

Labour force
restructuring:
- occupations
- sectors

Current wage levels,
income and career
prospects, job
security

Urban restructuring
- CBD redevelopment
- housing renovation
- gentrification of
businesses, etc

SUPPLY of housing
and other goods
and services

DEMAND for housing
and other goods and
services

Sexual division
of labour (employ-
ment, domestic labour)

Reproductive practices
- households, families
- neighbourhoods
- ways of life

PLACES
- of domestic work
- of collective and
private consumption

"lifestyles" of gentrifiers
diversification and
fracturing of consumption
practices

so far; this is summarized in Figure 6.3. This preliminary effort is offered as a contribution to the debate, and in the hope of fostering empirical work aimed at a better understanding of, for instance, how the diversification of occupational structures affects neighborhood consumption practices. Our framework suggests that the evolution of modes of reproduction mediates such practices (e.g., use of non-profit day-care centers and participation in their management); it also influences the gender division of labor and may even help reshape gender relations within the wage workplace. Thus, for example, women's struggles for equal pay for work of equal value may take on increased personal cogency for the growing number of women who are heading households:

Conclusion

This chapter has argued that changes in women's employment situation, as mediated by other aspects of gender relations, may often be *constitutive* of gentrification processes, rather than being derivative or secondary elements. In this respect the analysis stands also as a sympathetic critique of some of the more interesting of recent trends in research on urban restructuring and gentrification. Through our empirical focus on a particular city, Montréal, we were also forced to evaluate critically the pertinence of the concept of the "corporate" city for a different national context and political economy from that in which the notion was developed. Furthermore, it seemed important to "unpack" the very general concept of "new middle class," which in fact includes a spectrum of social groups with different residential location patterns and consumption practices. Further exploration of the links between class, occupational position in contexts of economic restructuring, gender, and consumption practices, seems warranted if we are interested in a deeper understanding of gentrification and its impacts in different kinds of cities. It may well prove worth embarking on the far-from-easy task of integrating approaches which could be complementary. For instance, certain French sociologists (Bidou *et al.* 1983) have developed far more subtle and inclusive analyses of the "new middle class" than those that pervade the North American literature, but without considering the urban context; while others have focused on the role of culture, symbolism, and image in structuring the personal experiences and needs for particular types of habitat among those involved in gentrification, yet have ignored gender (Collective 1979).

Our analysis has suggested that changes in gender relations may influence how potential gentrifiers are constituted into households, and the ways that they appropriate and alter particular segments of the housing stock and neighborhood infrastructure. Gender-blind analyses – both of gentrifiers and of those threatened by gentrification – are likely to leave us with, at best, a partial understanding of these processes. In cities such as Montréal, strong polarizations of occupational structures certainly do exist within the labor force resident in the inner city. Yet when these structures are analysed by economic sector, by gender and by household type, the effects of occupational polarization may be muted to some degree while other fault lines, based more on gender than class, come to the surface. It is not inconceivable that some types of "new urban lifestyles" may prove more compatible with the needs of a neighborhood's less-privileged populations than others. We remain broadly in agreement with the political-economic view that labor force restructuring resulting from wider economic change is the underlying motor of gentrification. But such restructuring does not take place on a genderless planet, and this should make a difference not only to our modes of analysis, but also to the range of possible political strategies one might adopt in the face of the growing presence of a fractured, divergent, and still incompletely understood force in urban change.

Notes

1 This article was originally published in French in *Cahiers de géographie du Québec* **31**, September 1987; this translation appears by permission. An earlier version was presented in English at the Annual Meeting of the Canadian Association of Geographers (Ontario Division), Ottawa, October 1986. My thanks to all those who made comments on previous drafts. The research presented here was funded by the Social Sciences and Humanities Research Council of Canada. (I have been a tenant in the St-Louis and Mile End neighborhoods since moving to Montréal in 1982.)

2 Expression first used in Britain in the 1960s, making reference to the traditional elite, the "gentry." Although a plethora of other terms have been developed, this one has the widest currency.

3 There is an active debate over the possibilities of rehabilitating this term from its "functionalist" past in urban analysis of the 1970s (see, e.g., Pinçon-Charlot *et al.* 1986, 131–43; Watson & Austerberry 1986, 83–7).

4 Using the four-digit classification of the *Canadian Dictionary of Occupations*, we have attempted to regroup census categories into a seven-fold occupational classification, based on power relations rather than on notions of prestige and social status (cf. Hunter & Manley 1986). Thus, for instance, nurses are classed as technicians rather than as professionals. For more details, see Villeneuve & Rose 1986.

5 Unless otherwise indicated, all statistical data are drawn from special tabulations from the 1971 and 1981 censuses, compiled by Statistics Canada for a larger research project jointly undertaken by Paul Villeneuve (Université Laval) and the author.

6 We developed a classification into eight economic sectors, based on a Marxist-inspired division of the formal economy into production and distribution of goods, exchange, consumption, and regulation. For present purposes, the number of categories is reduced to six. For more details, see Villeneuve & Rose 1986.

7 The neighborhoods selected are neither the first nor the only ones to undergo gentrification. But during the period studied, they experienced a more marked transformation than others: for instance, the proportion of people with a university degree (a classic index of "social upgrading") increased very considerably there (author's calculations from published census data; cf. Ley 1986, 73). Moreover, their housing stock (mainly plexes) is more representative of that typically found in Montréal than that of the first areas to be gentrified.

8 This group also includes personal services, but very few professionals are to be found in this latter category.

9 We believe that official statistics under-estimate the scale of cohabitation among unrelated people. Furthermore, data on household composition are only available according to the occupation of the "reference person." (In its "household" file, Statistics Canada does not recognize for coding purposes that a household can have more than one "reference person.") See also the notes to Table 6.2.

References

Alonso, William 1980. The population factor in urban structure. In *Internal structure of the city*, 2nd edn, Larry S. Bourne (ed.), 540–51. New York: Oxford University Press.

Andrew, Caroline & Beth Moore-Milroy 1988. Introduction. In *Gender relations and Canadian urban structure*, Caroline Andrew & Beth Moore-Milroy (eds.). Vancouver: University of British Columbia Press.

Armstrong, Pat 1984. *Labour pains: women's work in crisis*. Toronto: The Women's Press.

Beauregard, Robert 1986. The chaos and complexity of gentrification. In *Gentrification of the city*, Neil Smith & Peter Williams (eds.), 35–55. Boston: Allen & Unwin.

Bidou, Catherine, Monique Dagnaud, Bruno Duriez, Jacques Ion, Dominique Meal, Monique Pinçon-Charlot & Jean-Paul Tricart 1983. *Les couches moyennes salariées: mosaïque sociologique*. Paris: Ministère de l'Urbanisme et du Logement.

Black, Don & John Myles 1986. Dependent industrialization and the Canadian class structure: a comparative analysis of Canada, the United States and Sweden. *Canadian Journal of Sociology and Anthropology-Revue canadienne de Sociologie and d'Anthropologie* **23**, 157–81.

Choko, Marc & Francine Dansereau 1986. *Restauration résidentielle et copropriété au centre-ville de Montréal*. Research report submitted to Canada Mortgage and Housing Corporation. Montréal: INRS-Urbanisation, appendices.

Cohen, R. B. 1981. The new international division of labor, multi-national corporations and urban hierarchy. In *Urbanization and urban planning in capitalist society*, Michael Dear & Allen J. Scott (eds.), 287–315. London and New York: Methuen.

Collective 1979. Revalorisation des espaces anciens. *Espaces and sociétés*, July–Dec. 30–1.

Cossette, Alfred 1982. *La tertiarisation de l'economie québécoise*. Chicoutimi: Gaetan Morin.

Dansereau, Francine & Michel Beaudry 1986. Les mutations de l'espace habité montréalais: 1971–1981. *La morphologie sociale en mutation au Québec, Cahiers de l'ACFAS* **41**, 283–308.

Ehrenreich, B. 1986. Is the middle class doomed? *New York Times*, sec. 6, magazine. 7 September, 44, 50, 54, 62, 64.

Fainstein, Norman I. & Susan S. Fainstein 1982. Restructuring the American city: a comparative perspective. In *Urban policy under capitalism*, Norman I. Fainstein & Susan S. Fainstein (eds.), Urban Affairs Annual Reviews, vol. 22, 161–89. Beverly Hills: Sage Publications.

Goldberg, Michael & John Mercer 1986. *The myth of the north American city: continentalism challenged*. Vancouver: University of British Columbia Press.

Greater London Council, Industry and Employment Branch 1986. *The London Labour plan*. London: GLC (distributed by the London Strategic Policy Unit).

Holcomb, Briavel 1984. Women in the rebuilt urban environment: the United States experience. *Built Environment* **10**, 18–24.

Howell, Leigh 1987. The affordable housing crisis in Toronto. *City Magazine* **9**, 25–9.

Hunter, Alfred & Michael C. Manley 1986. On the task content of work. *Canadian*

Journal of Sociology and Anthropology-Revue canadienne de Sociologie et d'Anthropologie **23**, 47–71.

Lamonde, Pierre & Mario Polèse 1984. L'évolution de la structure économique de Montréal, 1971–1981: désindustrialisation du reconversion? *L'Actualité Economique. Revue d'analyse économique* **60**, 471–94.

Larsen, Vivi & Hanne Topsoe-Jensen 1984. *Urban planning and the everyday life of women.* Horsholm: Danish Building Research Unit, Urban and Regional Planning Division.

Laska, Shirley Bradway & Daphne Spain 1980. *Back to the city: issues in neighborhood renovation.* New York: Pergamon.

Ley, David 1981. Inner-city revitalization in Canada: a Vancouver case study. *The Canadian Geographer-Le Géographe canadien* **25**, 124–148.

Ley, David 1986. *Gentrification in Canadian cities: patterns, analysis, impacts and policy.* Ottawa: Canadian Mortgage and Housing Corporation.

Lipeitz, Alain 1986. New tendencies in the international division of labor: regimes of accumulation and modes of regulation. In *Production, work, territory: the geographical anatomy of industrial capitalism*, Allen J. Scott & Michael Storper (eds.), 16–40. Boston: Allen & Unwin.

Lussier, Robert 1984. *Le Plateau Mont-Royal au 19e siècle.* Montréal: Comité Logement Saint-Louis.

Mackenzie, Suzanne 1987. Women's responses to economic restructuring: changing gender, changing space. In *The politics of diversity: feminism, Marxism and Canadian society*, Roberta Hamilton & Michèle Barrett (eds.), 81–100. London: Verso.

McDowell, L. 1986. Beyond patriarchy: a class-based explanation of women's subordination. *Antipode* **18**, 311–21.

Mackenzie, Suzanne & Damaris Rose 1983. Industrial change, the domestic economy and home life. In *Redundant spaces? Studies in industrial decline and social change*, James Anderson, Simon Duncan & Ray Hudson (eds.). Institute of British Geographers, Special Publication no. 15, 155–99. London: Academic Press.

Markusen, Ann 1981. City spatial structure, women's household work and national urban policy. In *Women and the American city*, Catherine R. Stimpson, Elsa Dixler, Martha J. Nelson & Kathleen B. Yatrakis (eds.), 20–41. Chicago: University of Chicago Press.

Maroney, Heather 1983. Feminism at work. *New Left Review* **141**, 51–71.

Massey, Doreen 1984. *Spatial divisions of labour.* London: Macmillan.

Mathews, Georges 1986. L'évolution de l'occupation du parc plus ancien de Montréal de 1951 à 1979. *Etudes et Documents* no. 46. Montréal: INRS-Urbanisation.

Mills, Caroline 1986. Changes in family and household type: their impact on inner city landscapes. Paper presented at the annual meeting of the Canadian Association of Geographers, Calgary, June. University of British Columbia, Department of Geography.

Mondor, Françoise 1984. La mobilité des femmes monoparentales en rapport au phénomène de "gentrification" ou de "retour en ville." Montréal: Université de Montréal, Institut d'Urbanisme.

Morgan, Nicole 1985. Nowhere to go. *Policy Options politiques* June, 14–16.

Pinçon-Charlot, Monique, Edmond Preteceille & Paul Rendu 1986. *Ségrégation urbaine: classes sociales et équipements collectifs en région parisienne.* Paris: Editions Anthropos.

Pred, Allen 1978. *City-systems in advanced economies.* London: Hutchinson.

Rocher, François 1984. Le crise de l'état-providence: éléments d'un débat théorique. *Notes de recherche* no. 14, Université de Montréal, Département de science politique.

Rose, Damaris 1984. Rethinking gentrification: beyond the uneven development of Marxist urban theory. *Environment and Planning D: Society and Space* **2**, 47–74.

Rose, Damaris 1986. Transformations de la structure de l'emploi féminin dans la region metropolitaine de Montréal. Draft papers, INRS-Urbanisation, Montréal.

Rose, Damaris & Céline Le Bourdais 1986. The changing conditions of female single parenthood in Montréal's inner city and suburban neighborhoods. *Urban Resources* **3**, 45–52.

Rose, Damaris & Paul Villeneuve 1985. Women and the changing spatial division of labour in Montréal. Paper presented at the Annual Meeting of the Association of American Geographers, Detroit (April 22–25); Montréal, INRS-Urbanisation.

Rose, Damaris & Paul Villeneuve 1987. Women workers and the inner city: some implications of labour force restructuring in Montréal, 1971–1981. In *Gender relations and Canadian urban structure*, Caroline Andrew & Beth Moore-Milroy (eds.). Vancouver: University of British Columbia Press.

Sassen-Koob, Saskia 1984. The new labor demand in global cities. In *Cities in transformation: class, capital and the state*, Michael P. Smith (ed.), Urban Affairs Annual Reviews **24**, 139–71. Beverly Hills: Sage Publications.

Sayer, Andrew 1984. *Method in social science: a realist approach*. London: Hutchinson.

Seguin, Jean-Marc 1984. *St-Louis du Parc en 1981: étude des caractéristiques du quartier à partir des données de Statistique Canada*. Montréal: CLSC St-Louis du Parc.

Semple, Keith & Milford Green 1983. Interurban corporate headquarters relocation in Canada. *Cahiers de géographie du Québec* **27**, 389–406.

Simmie, James S. 1983. Beyond the industrial city? *Journal of the American Planning Association* **49**, 59–76.

Smith, Neil 1982. Gentrification and uneven development. *Economic Geography* **58**, 139–55.

Villeneuve, Paul & Damar Rose 1985. Technological change and the spatial division of labour by gender in the Montréal metropolitan area. Paper presented at a meeting of the International Geographical Union, Commission on Industrial Change, Nijmegen, Netherlands, Aug. 19–24. Université Laval: Département de géographie.

Villeneuve, Paul & Damaris Rose 1986. De la place des femmes dans la division spatiale du travail: le cas de Québec entre 1971 and 1981. In Les genres de vie urbains: essais exploratoires, Rodolphe de Koninck & Linda Landry (eds.), *Notes et documents de recherche* no. 26, 71–92. Université Laval: Département de Géographie.

Walby, Sylvia 1985. Spatial and historical variations in women's employment and unemployment. In *Localities, class and gender*, Linda Murgatroyd et al., 161–76. London: Pion.

Walker, Richard A. 1985. Technological determination and determinism: industrial growth and location. In *High technology, space and society*, Manuel Castells (ed.), Sage Urban Affairs Annual Reviews, vol. 25, 161–89. Beverly Hills: Sage Publications.

Watson, Sophie with Helen Austerberry 1986. *Housing and homelessness: a feminist perspective*. London: Routledge & Kegan Paul.

Wekerle, Gerda 1984. A woman's place is in the city. *Antipode* **6**, 11–20.

Williams, Peter & Neil Smith 1986. From "renaissance" to restructuring: the dynamics of contemporary urban development. In *Gentrification of the city*, Neil Smith & Peter Williams (eds.), 204–24. Boston: Allen & Unwin.

Women and Geography Study Group of the Institute of British Geographers 1984. *Geography and gender*. London: Hutchinson.

7

The mobility of capital and the immobility of female labor: responses to economic restructuring

GERDA R. WEKERLE & BRENT RUTHERFORD

Introduction

Distributional issues were the concern of the 1970s; economic develop-ment has been the issue of the 1980s (Healey 1986). As we near the end of the 1980s, distributional issues are again coming to the fore. In the analysis of the impacts of economic restructuring and the creation of new forms of inequality, the spatial dimensions of production and reproduction are the focus of a growing literature. A body of work which has received relatively little attention to date links several previously unrelated issues: labor market restructuring, women's employment, and spatial inequalities. The literature comes from two quite distinct sources: behavioral journey-to-work studies which focus on the relationship between access to transportation and women's employment opportunities; and structural analyses of spatial divisions of labor which document the development of local economies based on cheap female labor. These studies converge around two related concerns: an interest in why women work closer to home than men do, and the implications of this for women's employment opportunities; a focus on the development of gender-specific local labor markets. While framed differently, each approach examines the roots and implications of the development of local pools of predominantly female labor. The two approaches differ substantially in the unit of analysis chosen, the theories of explanation favored, and the proposals for change and intervention that follow from them. Table 7.1 provides a highly schematic presentation of the two approaches, their similarities and differences.

This chapter begins with a brief discussion of the restructuring of the Canadian labor force focusing specifically on the gender segregation of the labor market. This is followed by a review of the literature emerging from the structural and behavioral approaches to the study of women's employment and the development of gender specific urban spatial labor markets. Both of these literatures cover a very broad and complex body of work, each of which is frequently discussed in isolation from the other

Table 7.1 Approaches and responses to gender, space and urban labor markets.

	Behavioral	Structural
Level of analysis:	Primarily individual	Primarily regional and/or firm and/or aggregate occupational categories
Focus:	Accessibility by individual to employment, services	Regional location of activities
Key findings:	Women work closer to home than men Women's travel opportunities are limited Workers' incomes increase with increased distance	Industry seeks to minimize labor and land costs by strategic locational choices Labor market is gender-segregated
Key assumption:	Individuals make choices regarding employment, residence and transportation	Firms seek to maximize profit
Analysis/explanation:	*Choice model*: women choose to work closer to home because of dual role obligation *Constraint model*: women are forced to work closer to home because of limited transportation options and the spatial concentration of female-dominated jobs Patriarchy reinforces dual role obligation Men control transportation capital	*Radical model*: women in gender-segregated local labor pools because of limited alternatives *Feminist model*: women's opportunities are constrained by industrial practices which reinforce job segregation and lower pay Patriarchy underlies industrial decisions which profit from women's low paid labor Women seen as secondary wage earner
Systemic discrimination:	Women's dual worker role, lower earnings, greater part-time employment, less access to private transport Transit system designed to service full-time workers, choice riders in CBD and employment centers	Maintenance of gender-segregated labor market; firm location to exploit pools of cheap female labor

Table 7.1 *continued*

	Behavioral	Structural
Policy prescription:	Improved transportation increases employment opportunity	Improved transportation increases worker competition and may reduce wage rate. Improved transportation increases competition among employers
Policy goals:	Land use policy to locate employment close to residences Improved transportation service to increase accessibility for transit dependent workers	State intervention to limit firm flight; economic development programs pay equity legislation

(with the exception of recent works by Hanson & Johnston 1985, Hanson & Pratt 1988, Holcomb 1986, Rutherford & Wekerle 1988), without any indication that they may be complementary in their approach. The focus of the chapter is on the commonalities and contrasts between the two approaches and on some contradictions we see within and between them.

Restructuring of the labor force

A primary interest of scholars in the 1980s has been the analysis of economic restructuring and its impact on the labor force. A substantial body of research has documented the relationship between economic restructuring and changes in women's labor force participation. One focus of this work is the obstacles to equality of opportunity within the labor market. A second focus is to examine how the continuing, and in some cases increasing, gender segregatation of the labor force functions in the accumulation of capital.

Over the last few decades, a fundamental restructuring of the labor force has occurred in Canada, as in other industrial countries. Two key aspects of this are the increasing feminization of the workforce and the continuing segmentation of the labor market. In the past decade, sectors employing large numbers of women – retail trade, accommodation and food services, recreational services, public administration and finance – have expanded. Between 1966 and 1982, the female labor force grew by 119.4%, with the largest growth in employment of women over the age of 25; the male labor force grew by 35.6% (Abella 1984, 56). At the same time, deindustrialization has resulted in a growth in two sectors, professional and low skilled blue- and white-collar jobs, and a decline in the number of

141

skilled white-collar jobs in the service sector (Abella 1984, 62–70, Gunderson 1985, Villeneuve & Rose 1985). These processes have increased the segmentation of the labor market.

In labor market analyses, the professional and managerial sector are referred to as the core labor market or primary sector. This is where we find "High wages, good working conditions, employment stability, chances of advancement, equity, and due process in the administration of work rules" (Doeringer & Piore 1971, 165). Unskilled blue- and white-collar jobs, classified as being in the secondary market, "[tend] to have low wages and fringe benefits, poor working conditions, high labour turnover, little chance of advancement, and often arbitrary and capricious supervision" (Doeringer & Piore 1971, 165). According to Reich *et al.* (1980), "These groups seem to operate in different *labour markets* with different working conditions, different promotional opportunities, different wages, and different market institutions." Gordon *et al.* (1982) argue that it is in the interests of capital to treat workers differently to enhance competition among them.

Gender is a primary basis for labor market segmentation (Armstrong 1984, 31). Women are concentrated in particular industries; they are slotted into a limited number of jobs within these industries (Armstrong & Armstrong 1984). Despite women's increasing labor force participation, they remain much more concentrated than men in a few occupations. While the male labor force is characterized by a heterogeneous occupational structure, with no major occupational categories predominant, women are still largely concentrated in clerical, sales, and service occupations (Abella 1984, 67).

An analysis of 23 occupation groups in Canada, by a recent federal commission, the Abella Commission on Employment Equity, determined that women were dominant (70% +) in only two – medicine/health and clerical – while men were dominant in 15 groupings. Only three occupational categories – clerical, sales, and service occupations – covered 62% of all women workers, the same categories in which women predominated in 1901 (Abella 1984, 245). These are all in the secondary labor market where women continue to cluster.

Over the past 50 years women's gender segregation in the labor force has remained relatively stable. An examination of the ten occupations in which women dominated in 1971 shows that, in seven of these, women's domination had increased even further by 1981. Fox & Fox's (1987) research finds that in both 1931 and 1981 nearly 75 per cent of women were in occupations in which women were in a majority; in 1981, 50 per cent of women worked in occupations that were 75 per cent female. In 1981, over 60 per cent of men or women in the labor force would have had to change occupational categories for the two genders to have the same occupational distribution, i.e., for occupations to be integrated rather than segregated.

Armstrong (1984, 29) criticizes dual systems theory for being more descriptive than analytical, for it does not explain how labor markets came to be structured initially to segregate women from men. These theories also ignore the relationship between the labor market and the household economy, since they focus on structural factors rather than power relations within the household.

Discussions of gender–segregated labor markets also neglect the pervasive theme in recent geographical analysis of spatial inequalities in economic growth and development (Browett 1984). They neglect the evidence that points to the function of spatial differentiation in contributing to capital accumulation; the research on the spatial differentiation of labor markets in urban areas (Rose & Villeneuve 1985); or the empirical findings that position in the division of labor is a major determinant of the distribution of spatial as well as social inequalities (Coates *et al.* 1977, 253). Particularly in the suburbs and periphery of North American cities, where there is a growth in white-collar secondary sector employment, much of it female-dominated, one must examine to what degree occupational segregation by gender is spatial, given problems of low density, accessibility to transport, and zoning constraints on employment location.

The structural approach: the mobility of capital and the development of gender-segregated local labor markets

Spatial analyses have contributed a missing ingredient to discussions of labor force restructuring by insisting that spatial restructuring goes hand in hand with economic restructuring (Storper 1981). In the clearest statement of this perspective, Doreen Massey (1983, 68) argues that space is neither a passive surface nor a negative constraint; it is used as an active element of accumulation as capital makes positive use of distance and differentiation in the drive to rationalize production. She argues that shifts within a country of the location of major sectors of production to peripheral regions involve "the development and reorganization of . . . spatial structures of production" (Massey 1983, 7). These new spatial divisions of labor, as she calls them, "represent whole new sets of relations between activities in different places, new spatial patterns of social organization, new dimensions of inequality and new relations of dominance and dependence. Each new spatial division of labour represents a real, and thorough, spatial restructuring" (Massey 1983, 8).

Much of the recent literature on industrial location and the spatial division of labor seeks to describe and explain the distribution of employment over space through a structural analysis of labor market changes. The unit of analysis may be large population aggregates – the paid labor force, sectors of the economy such as manufacturing, or changes in the location of individual firms or subunits of large corporations. One focus is the changes occurring at the regional and metropolitan level as sectors of the economy and firms seek out local pools of cheap labor. According to Browett (1984), capital seeks out reserve armies of labor in underdeveloped regions to stabilize wages when labor is scarce and capital expands, and to provide a cheap supply of labor during periods of recession. In a period of economic restructuring, Hudson (1982) notes that companies attempt to boost profit by exploiting fresh labor reserves as a way to avoid or overcome their profitability crisis: "because of changes in labour processes and the emergence of a new spatial division of labour, companies often seek locations with large masses of relatively unskilled labour-power."

Historically, single industry and single employer towns, such as are found in mining regions, have been seen to have a captive labor force. Competition among employers is low or absent, and the accessibility of workers to other sources of employment is limited. (In some Canadian resource towns, part of the contract with the employer is employer-paid periodic trips out of the community to "civilization.") More recently, scholars have observed an urban version of this theme where corporations relocate from the aging metropolis to peripheral areas offering a large pool of unemployed or underemployed labor. Industrial siting decisions may be motivated by a search for workers who have not been organized, or are likely to prove "loyal" and reliable workers. Spatial inequalities are created as firms locate to exploit these cheap labor pools.

What is lacking from many of the current structural/spatial analyses is any reference to gender, even though the pervasive gender-segregation of labor markets is a dominant characteristic of capitalist economies. The recent volume, *Geography and gender* (Women and Geography Study Group of the IBG 1984, 79) notes that Marxist and political economic analyses try to specify mechanisms underlying change in location and labor requirements of industry, but fail to explain why there is a demand for *female* labor or why women's labor is cheaper. This approach tends to ignore divisions within the labor force, particularly pervasive gender divisions, relating to changes in the reorganization of employment.

A few researchers are beginning to link industrial location analyses with a feminist critique of how gender-segregated labor markets are created and maintained. Industrial restructuring is viewed as a response to the relative immobility of female workers and the availability of cheap local labor pools (reserve armies of labor) which are created by gender-segregated labor markets. This work (Massey 1983, Lewis 1984, Lewis & Foord 1984, Hudson 1982) demonstrates that industries often move from center city to periphery, or from one region to another within a country, to exploit local labor markets of often unskilled and semi-skilled workers which are frequently predominantly female and captive in the sense that there are few alternative employment options or available transport to seek employment elsewhere. By locating on cheap land on the periphery, employers also shift transportation costs to workers, who are often required to purchase an automobile rather than capital absorbing transportation costs by locating in more accessible locations.

Various British studies have documented that since the 1960s there have been changes both in the inter- and intra-regional distribution of jobs for women (Women and Geography Study Group of the IBG 1985, 67). In Britain, between 1951 and 1971, the decentralization of female employment coincided with the decline of traditional manufacturing industries and a corresponding growth of assembly and service sector industries in the suburbs and peripheral locations (Women and Geography Study Group of the IBG 1984, 73, Massey 1983, 225, Lewis, 1984). With the growth of semi- and unskilled mass production and assembly work, Jane Lewis (1984) argues that female labour became more important in the locational decisions of firms which separate managerial and technical functions from processes of direct production, often in different locations of a city, country, or even globally. Lewis & Foord (1984) report that female employment was decentralized to peripheral regions in the north and west of Britain, areas with traditionally low levels of female participation in the labor force which had been dominated by heavily male-dominated sectors of the economy such as mining, shipbuilding, steel, and heavy engineering.

The new jobs created for women were frequently in the secondary sector, providing jobs at low pay and with poor working conditions, but posing only minimum conflicts with home demands, and frequently located either within walking distance or within easy access of home (Lewis & Foord 1984). In some cases, industrial economic development projects have introduced low paying female employment to depressed regions, thereby decreasing or eliminating skilled jobs for men in traditional manufacturing industries (Lewis 1984).

Several studies find that firms employing women have tended to concentrate in new towns and on post-war industrial estates. Hudson (1982) reports on the relationship between economic development policies to create new employment in Washington New Town in the north-east region of England, and forces which push married women into relatively low paid employment in these new industries.

> Wage levels in those industries that have located in the town are often relatively low, both for men and for women. The existence of potentially cheap labour-power was a strong incentive for many capitalists to locate there. At the same time, the conjunction of low wages and the costs of housing served to push married women into wage-labour to meet these and associated living costs, as higher rent or mortgage levels were not translated into higher male wages.

Hudson further reports that public transportation is poor and expensive in this new town which was planned on the assumption of almost 100 per cent car ownership. Since most families have only one car, usually used by the husband to get to work, women's employment opportunities are further restricted to working within the new town or on certain adjacent industrial estates.

Lewis & Foord's (1984) case studies of Peterlee New Town and East Kilbride in Britain add to this picture. They argue that new town planning policy in Britain has assumed and reinforced traditional gender divisions of labor by creating concentrations of unemployed married women with limited job opportunities which attract multi-plant manufacturing functions and branches of government departments. As in Hudson's report of factors which push married women into the labor force in Washington

New Town, Lewis & Foord (1984) also find that women enter the paid labor force to help pay the high rents for new state-built housing in the new town. At the same time, married women's employment serves to keep demands for wage increases down in the mines (which employ predominantly men). Paradoxically, they report that the new role of women in the labor market is not reflected in the planned environment of this new town, which is oriented to the concept of the mother at home and around the assumption of high levels of car ownership. A lack of public transportation to women's jobs and a lack of available child-care for employed mothers has forced some women to accept part-time shift work at night when children can be cared for by their fathers.

In interpreting these findings, the Women and Geography Study Group of the IBG (1984) argues that capital is not gender-blind in industrial relocation decisions. Patriarchy is pervasive in industrial decisions which are meant to profit from women's lower pay and segregation in female employment ghettos. Women workers' greater attachment to local labor pools occurs as a result of patriarchal structures, since women's competing family commitments and concomitantly less available time to travel to work make them less available for employment than male workers. These geographers argue that women manage their dual role by taking jobs close to home; by taking non-demanding jobs in the secondary sector; by working in part-time and shift work; and by interrupting their work at intervals to care for children and other family members such as the elderly (Women and Geography Study Group of the IBG 1984, 71). By drawing upon the concept of patriarchy as this affects employers' decisions at the macro-scale and women workers' options at the micro-scale of the household economy, this approach goes beyond a strictly structural explanation of spatial divisions of labor. It suggests a potentially more sophisticated theory which attributes the development of local labor markets and spatial inequalities not only to the mobility of capital but also to the immobility of labor due to inequalities within the household.

The work by Damaris Rose and Paul Villeneuve on industrial restructuring and changes in gender segregated labor markets in Montréal between 1971 and 1981 is a variant on these themes (Rose & Villeneuve 1985, Villeneuve & Rose 1985, 1987). They are interested in the spatial differentiation of labor force characteristics within metropolitan areas and argue that the feminization and polarization of the labor market are reflected in a new spatial sorting (Rose & Villeneuve 1985). This work uses census data to examine the development of spatially defined labor pools with specific characteristics by comparing changes over a decade in the distance between home and workplace locations of male and female workers both in specific occupational groups and in specific industrial sectors. The basic mode of analysis compares the distance from place of residence to the CBD (Central Business District) for, say, women employed in service occupations between 1971 and 1981. Any change in the decade could be attributed to a redistribution of such jobs in terms of their urban centrality versus suburbanization. A similar mode of analysis is used for industrial sectors.

Between 1971 and 1981, they find that all occupational groups increased the distance of their journey to work as jobs became more centralized and required longer work trips. In documenting two forces at work within the metropolitan area – centralization of professional and managerial jobs, and the decentralization of unskilled blue-collar and white-collar employment – this research provides an insight into the spatial operation of the dual labor market. This supports Scott's (1981) findings that in US SMSAs (Standard Metropolitan Statistical Areas) there have been recent strong increases in white-collar jobs in central cities and suburbs, with core areas of major metropolitan regions becoming more and more specialized centers of labor-intensive management and control functions, while production functions are dispersed to the hinterland.

Villeneuve & Rose (1987) also found that the distances traveled to work decreased between the genders in the top five occupational categories, but distances increased for unskilled white-collar and blue-collar workers. They interpret this as an indicator of the increasing ghettoization of female blue-collar workers. To evaluate the influence of women's household responsibilities on work trip length, the percentage of married women in each occupational grouping is correlated with changes in average work trip length over the decade. They find that the correlation is not as strong in 1981 as it was in 1971, and conclude that household responsibility is losing ground to forms of labor market participation in its effects on work trip length. They suggest that local labor markets which employ women in specific occupations may be more important than women's family responsibilities in explaining married women's shorter journey-to-work trips.

There are some difficulties with this interpretation. Marital status is a very imperfect measure of household responsibility, since there are women with household responsibilities who are not married, i.e., they may be living in common-law relationships or may be single parents. Further, degree of household responsibility is very much a function of division of labor within the family, which cannot be adequately measured by a single variable such as percentage of married women.

Kirsten Nelson's (1986) study of the location of back offices in the San Francisco-Oakland region changes the focus of analysis within the structural model from labor markets or occupational groupings to the decision-making of individual firms which relocate low level clerical functions to suburban locations. While still focusing on the question of the link between women's gender-segregated labor and spatial locations, this study uses interviews of office managers rather than aggregate analyses of labor market changes and treats the firm as a key actor in this process. Nelson finds that corporations choose sites within easy highway access of single-family neighborhoods of young families. The office managers she interviewed look for, and find, a high quality but docile female labor supply to fill highly skilled but lowly paid clerical jobs in suburban residential communities. The managers view these workers as relatively immobile and report that these workers will tolerate only short journeys-to-work since their family responsibilities necessitate a tradeoff between home and work demands and result in limited career aspirations. Nelson (1986) argues that lowering transportation costs by working closer to home is one way these women justify taking an outside job.

Nelson (1986) emphasizes that the major reason for back office relocation is not cheaper land costs, but a search for a preferred labor supply. "A shift to a demand for these married, secondary-earner women ('homemakers') further increases the importance of proximity to the desired labour supply, since it has been demonstrated that married women, particularly with small children, have a much shorter journey to work on average than single women" (Nelson, 1986).

Firms seeking out a well-educated, docile (i.e. non-unionized) and low paid workforce, exploit the relative immobility of this reserve army of labor of married women by locating nearby and creating new clerical employment in suburban sub-centers. For corporations, a female labor supply in new single-family housing areas translates into lower labor costs due to less turnover, lower training time per worker, increased productivity, a longer working day for employees, and less chance of a unionized workforce. Nelson (1986) concludes that:

> As female-dominated industrial and occupational sectors continue to grow in the advanced capitalist economies, the conjunction of the household and the wage economies in the local labour market becomes increasingly important to industrial location analysis.

The behavioral approach: spatial constraints and the immobility of female labor

Studies of journey-to-work are alternative sources of information on the linkage between female employment and spatial constraints, but, for the most part, they have been lacking in any structural analysis of labor market restructuring. Empirical journey-to-work travel surveys are behavioral in approach, relying on either sample surveys or census data for information on the trip behavior of individuals or households, usually including data on the location of employment, the time or distance to work (or both), the mode traveled, and sometimes information on alternative options.

The studies comparing men's and women's transportation patterns build upon an earlier tradition of transportation research popular in the 1960s in the United States (Altshuler 1979) which addressed the problems arising from a previous phase of decentralization of employment from center city to suburban industrial parks. They found that poor and visible minorities living in US center city areas who did not have access to a private automobile were disadvantaged in obtaining and maintaining employment in these areas of new employment growth. Originally framed as a civil rights issue and part of the agenda of the US anti-poverty program of the 1960s and early 1970s, interest in the linkage between transportation and access to employment opportunities has largely remained dormant in the past decade.

Today there is a renewed interest in this linkage, but the emphasis has shifted to a focus on women's access to transportation and the role this may play in access to employment, and, at a macro-level, the creation and maintenance of gender-segregated labor markets. Empirical research based on journey-to-work surveys, time budget studies, and census analyses of location of residence and workplace compare male and female transportation patterns and reveal substantial inequalities in modal choice, travel times, and stress on employed women who are transportation disadvantaged because they are disproportionately reliant on public transportation in a primarily car-oriented urban system (Hanson & Johnston 1985, Rosenbloom 1978, Michelson 1983).

A key finding has been that women work closer to home than men and that this has implications for the size of women's job search area, and, ultimately, the degree to which female labor can move farther afield for higher wages. Unlike earlier studies of transportation and employment which established simplistic linkages between improving workers' physical mobility and enhanced job prospects (which were later rejected by empirical findings; Altshuler 1979), feminist analyses have been both more sophisticated and more complex.

The literature on women's travel patterns has shown that, compared with men in the labor force, employed women work closer to home, spend less time getting to work and use public transportation more (Fox 1983, Michelson 1983, Howe & O'Connor 1982, Rosenbloom 1978, Ericksen 1977). If women work in the suburbs, they tend to live closer to work than men and are less likely to outcommute (Madden & White 1978). An early study (Kaniss & Robins 1974) speculated that short journey-to-work trips limited the size of women's job search area and their ability to compete in the job market to find the highest paying jobs or employment commensurate with skill levels. These empirical findings suggest that women workers are more sensitive to trip length than male workers, and a substantial body of work has developed to examine the implications of this relative immobility and its root causes.

A consistent finding of journey-to-work research is that shorter work trips are associated with lower incomes (Madden 1981, Hanson & Johnston 1985, Pucher *et al.* 1981, Millar *et al.* 1986). If female workers are the group which works closest to home, this disadvantages them. Yet the causality is unclear. Is it because women workers live closer to home that they have lower incomes? Or do women who earn low incomes work closer to home because they do not gain additional income by traveling farther?

An explanation forwarded by some researchers is that women *choose* to work closer to home because they are secondary wage earners with less attachment than men to the labor force (Taaffe *et al.* 1963, Madden 1981). An alternative explanation emphasizes the interpretation of household economy and market economy in understanding women's inequality. Researchers such as Ericksen (1977), Madden & White (1978),

149

Rosenbloom (1986) and Michelson (1983) de-emphasize the choice implied in the first explanation, and focus instead on a theory which portrays women as constrained to work closer to home because of their additional household and child-care responsibilities. They link women's domestic roles within the home and family, including decisions regarding the length of the journey to work, to behavior in the labor market. The concept of patriarchy explains women's more limited options: patriarchal relations within the home force women to adapt and limit their employment options, often by shortening the time and distance traveled to work because they are still primarily responsible for the home and children; gender inequalities within the family mean that men generally have access to the family automobile (when there is only one) and women are forced to rely on slower and less flexible public transportation. This focus on women's dual worker role links transportation studies more closely to feminist theories which examine the interrelationships between patriarchy and capitalism in contributing to women's inequalities within the home and workplace.

Several recent studies emphasize the constraints on women's employment (Hanson & Johnston 1985, Hanson & Pratt 1988, Rutherford & Wekerle 1988, Howe & O'Connor 1982) by focusing on their limited transportation options and the concentration of female-dominated jobs in specific locations. This research adds a structural analysis to gender and journey-to-work studies insofar as it is concerned both with labor market questions, such as the location of female-dominated occupations, and with the behavioral issues of how actual women respond to specific environmental opportunities or constraints. It is within this perspective that we locate our own work.

These studies question the predominant emphasis on gender dynamics within the family to explain women's differential travel-to-work patterns. Based on their analysis of a Baltimore travel survey, Hanson & Johnston (1985) conclude that dual roles cannot explain women's shorter journey-to-work trips, since, in their sample, the presence or absence of either children or husbands does not affect women's work trip lengths. They cite their greater reliance on public transportation, their travel on slower modes, and their employment in gender-segregated occupations which, in some cities, are closer to concentrations of female employment. Women may stay in female-dominated occupations, according to this study, because of accessibility problems and because some sectors which employ large numbers of women may locate near the neighborhoods where there are high concentrations of skilled female workers.

An earlier study by Howe & O'Connor (1982) in Melbourne, Australia, focused directly on the link between women's journey to work and the locations of women's jobs, and provided the behavioral parallel to Nelson's research on firms' locational behavior. They concluded that a strong force in the expansion of suburban labor markets is the availability of a female work force either unwilling or unable to travel far to work. Melbourne suburbs with a high concentration of light industry employed an above average number of local women. In these suburbs, there was a high correlation between low average wages and a high female industrial workforce. High wage suburbs, on the other hand – with jobs in chemicals, metals, and machinery-sectors dominated by males – tended to offer few opportunities for women. They note that "the local availability of employment may be as important as the actual type of employment in inducing women to enter the workforce." Accessibility to jobs is an additional, and overlooked, factor limiting women's employment:

> Sex bias in employment may arise in part from difficulties in job accessibility of suburban women. That occurs because suburban residential growth has outpaced the dispersal of employment and, in single-car households, job access for a wife is constrained to a limited local area job market. This situation has worsened in recent years as suburban unemployment among young women has risen rapidly. This trend reflects in part the fact that many families made residential location decisions in terms of accessibility of the father's job; these locations create problems for young women in both finding and holding employment (Howe & O'Connor 1982).

They conclude that the overconcentration of women in a limited range of occupations may be compounded by the spatial concentration of female-dominated jobs.

The behavioral journey-to-work studies began with simple analyses of the differences between men's and women's trip behavior. The very dramatic and substantial differences recorded called for an explanation which initially was derived from studies of household divisions of labor with particular reference to the tradeoffs employed women are constrained to make due to their dual worker role. As this explanation is proving inadequate to explain empirical findings, scholars are seeking to combine a behavioral focus with a structural explanation emphasizing the development of local labor pools in a gender-segregated labor market.

Captive riders and captive labor: an empirical example

The studies by Howe & O'Connor (1982) in Melbourne, Australia, and another by Black (1977) in Sydney, Australia, both focus on accessibility, or the ease of reaching employment opportunities, as a key concept in studying the relationship between women's journey to work and resultant inequalities of income and access to employment. This focus relates back to earlier research conducted in the 1960s and early 1970s which linked access to transportation to the employment chances of visible minorities living in core areas of US cities (Altshuler 1979). Accessibility is a concept with a variety of somewhat different operational definitions, but any of them is developed from the perspective that the shorter the distance or time or both from origin to destination, the greater the accessibility. Although the accessibility of an area may be expressed as a weighted function of the time/distance to jobs in all areas (Wachs & Schofer 1972), a more easily understood expression of accessibility is the number of jobs within a given number of minutes from a residential area for any given classification of workers (Wachs & Kumagai 1973). When time is utilized as the expression of accessibility, presuming that car travel is more rapid in many instances, the relative disadvantage of workers totally dependent on public transport to get to work may be expected.

Gender differences in work trip length[1] in a suburban municipality of Metropolitan Toronto are presented in Figure 7.1 which portrays the percentage of workers traveling a given distance to work that are men and women. Women clearly form the majority of those traveling short distances to work and are, increasingly, the minority gender for those

Figure 7.1 Percent of men and women by distance.

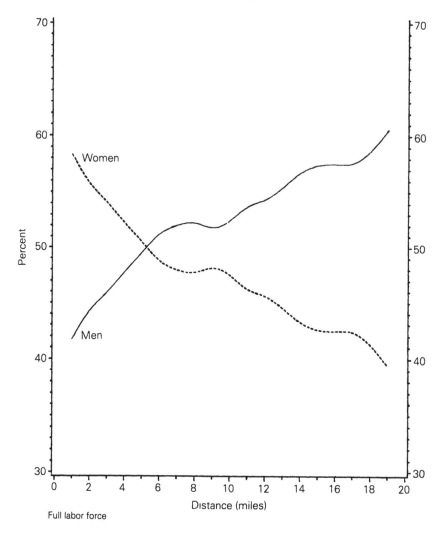

Full labor force

Table 7.2 Travel mode and choice, by gender.

	Female	Male
Automobile (Percent of N)	62.0	87.6
Public transit (Percent of N)	38.0	12.4
Transit captive (Percent of transit users)	56.6	30.4
Transit captive (Percent of N)	19.0	4.2
Total N	2858	4511

traveling longer distances. This basic relationship shows the strong gender difference not only in work trip length, but in the concomitant size of the area of presumed job search.

Accessibility to transportation is a necessary precondition to accessibility to the workplace. Although the usual distinction in modal choice is between car and transit, an important sub-group of transit users are those who use transit not by choice but because they have no other choice. We define "captive riders" as persons who either have no driver's license, do not own a car, or report that no car is available to them. "Choice riders" are persons who either own an automobile or have one available to them but choose to use transit for reasons of speed, convenience, or economy.

Substantial gender differences are shown in Table 7.2 between mode of transportation used for work trips when the worker had a choice of mode or no choice. While the majority of both men and women workers use the car, women are more than three times as likely to use public transit than men. Of those using public transit, nearly 60 per cent of women have no choice but to use public transit. This is almost twice as many as men. When transit captives are considered in proportion to the total group, nearly one in five women workers are transit captives[2] as compared with approximately one in 25 for men.

To clarify the role of choice/captivity in modal choice, Table 7.3 presents average incomes, journey-to-work distance, and travel times to work by gender and choice of mode. Marked income differences exist between the genders even when controlled for mode. The largest income differences are between car and choice riders (approximately $8,000), while men and women who are captive riders are more similar, with an income difference of $3,500.

Women who are car users or choice transit users work closer to home than men (an average of 2 and 1.5 miles, respectively). Because over one-half of the choice users work in the CBD and use subway services, their travel distances are higher than for car or captive riders. Little gender difference exists between male and female captive riders.

Table 7.3 Income, distance, and time, by mode and gender (averages).

	Female	Male
Income*		
Car	$20,050	$28,700
Transit Choice	$16,600	$25,900
Transit Captive	14,500	17,000
Miles to Work		
Car	6.8	8.8
Transit Choice	10.9	12.4
Transit Captive	9.1	9.0
Travel Time to Work		
Car	20.9	24.2
Transit Choice	52.3	56.2
Transit Captive	50.6	52.2
Total N	2858	4511

* In Canadian dollars.

Car users spend the least time in their work trips due to the faster speed and the shorter distances traveled. Transit users take twice as much time as car users. While car users take on the average 44 minutes per day traveling to and from work, nearly half of all women must spend over 100 minutes per day using public transit to jobs which pay much less than men earn.

Because of the persistent finding of gender differences in incomes and journey-to-work distances (Millar *et al.* 1984, Pucher *et al.* 1981, Hecht 1974), one conclusion which seems to follow is that if women worked farther from home their incomes would increase. Better to resolve the relationship between distance and income, we previously conducted (Rutherford & Wekerle 1988) an analysis which examined the strength of this relationship. Although distance and incomes are correlated, we found that the strength of the association is modest with a correlation of 0.132 for women and 0.191 for men. Even when these relationships were separately studied for mode and choice, the magnitude of the relationship was little changed. However, when gain in income (as a regression beta) is predicted by an additional mile traveled, the value was $407 for men and $182 for women.[3] Although beta values varied somewhat by mode, in all cases they were less for women, and considering the lower speeds by which women using transit get to work, the yearly time required to travel the additional distance per year resulted in hourly gains which were less than their current hourly wage rate.

This analysis provides the basis for suggesting that a plausible reason for women to work close to home is simply that little income gain is possible, even when a woman is willing to travel longer distances: the marginal gain is not worth the marginal costs. This "rational" reason is distinct from the "dual-role" explanation. The structural basis for this rational model is that most women work in occupations which not only have low wage rates, but that those wage rates vary little from area to area. For example, clerical and service worker salaries can be expected to be relatively constant anywhere within an urban area.

Gender distribution of male and female workers by distance and income

A further approach to understanding the distance and income linkage is to examine the distribution of men and women workers over a surface

Figure 7.2 Female worker density by distance and income.

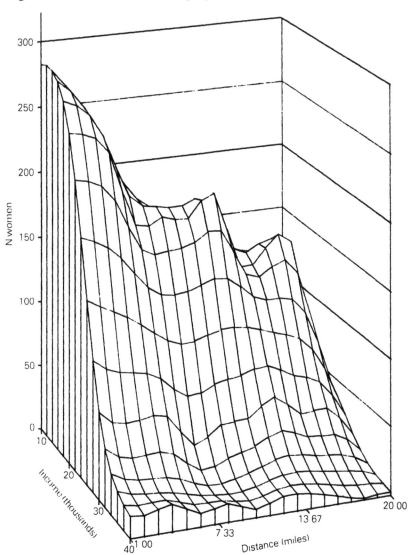

Figure 7.3 Female worker density by distance and income (contour).

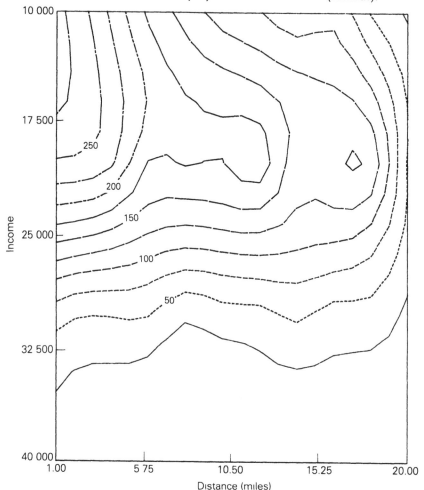

representing the length of the distance to work and income. Figures 7.2 and 7.3 portray the number of women at combinations[4] of distance and income, while Figures 7.4 and 7.5 show the same relationship for men. The highest concentration of women is where distances are short and incomes are low. The relative number of women declines with longer trip lengths, but even more strong is the decline in women workers at higher income levels, regardless of distance traveled. Thus, it is not so much that women work closer to home, which they clearly do, but that the proportion of women is very low in the higher income area. Many women do travel long distances, but receive little or no income gain for their efforts.

While a considerable number of men work close to home and at moderately low incomes, the distinguishing feature of their distribution is

Figure 7.4 Male worker density by distance and income.

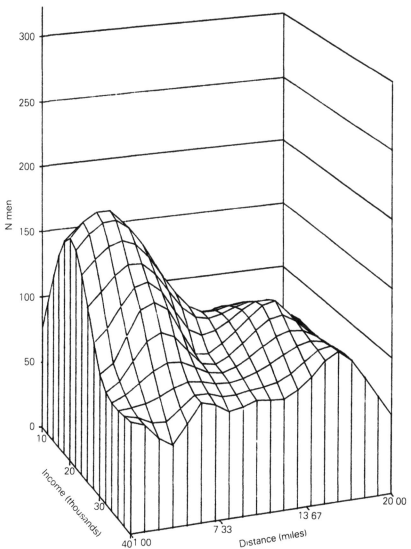

that men with higher incomes are found at all distances, with a tendency for higher income males at longer distances. Overall, the distribution for males is much more uniform across the income and distance surface.

These distribution figures contribute to an understanding of the reasons why the correlational and regression analysis discerned weak distance-to-income linkages. For a correlation to be high, men or women must be distributed along a "ridge line" from low incomes and low distance to high incomes and high distance. Clearly, this is not the case for women and only appears as a weak form for men. Many men have high incomes

Figure 7.5 Male worker density by distance and income (contour).

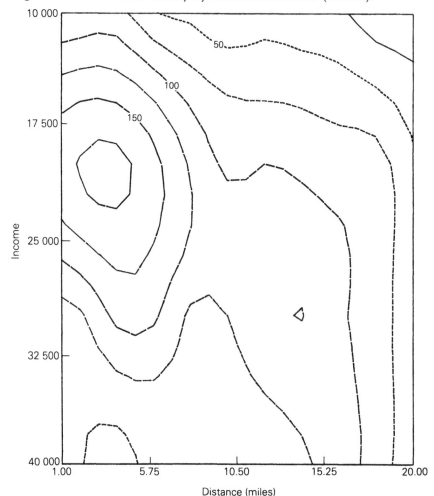

and work close to home.[5] Thus, while the regression and correlational analysis has accurately reported the low relationship between distance and income, the values could only have been high if it were the case that few workers have high incomes and live close to home (which would appear to be a perfectly reasonable goal for many workers), and that few would travel long distances with low incomes – a reality for many workers.

Captive labor force

A taxonomy of levels of access to employment opportunities based on journey-to-work distance and income was constructed to explore these relationships further. In this analysis our focus shifts from the individual to the area in which he or she works. The first defining characteristic of a

Figure 7.6 Taxonomy of employment accessibility by income and distance.

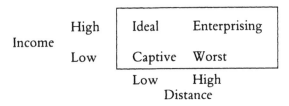

model of a captive labor market would appear to include the concept of average trip length at the work zone level. If the argued consequence of captive labor markets holds, then average income at the zone level is either a result or a defining characteristic of captive labor markets. A typology was constructed for the ultimate purpose of assigning each workzone into one of four categories: low income earners who travel low distances to work, low income earners who travel high distances to work, high income earners traveling low distances, and high income earners traveling high distances.

The "captive" type characterizes a local labor market where individuals receive low incomes but work close to home. Although zones of this type will be termed "captive," nevertheless, some proportion of workers might simply value time over income, or simply not have the willingness to seek employment and possible high wages by a longer journey to work.

While "captive" zones have received considerable attention, the "worst" typed zones are those where wages are low, but workers must travel long distances to obtain low wages. A structural interpretation of this condition might be one of high competition of labor for employment: workers must pay high travel costs if they are to be employed anywhere.

The "ideal" situation is to have high wages and low journey-to-work distances. These might have high proportions of self-employed high earners such as those described by Feldman (1977) who have the lowest journey-to-work. This situation is the image held by some feminists who argue that land use changes should decrease the separation between home and work, and that mixed land uses should encourage local employment (Hayden 1984, Leavitt 1985, Wekerle 1985).

Finally, the fourth type is termed "enterprising" as it contains workers who receive high incomes and travel long distances to work.

An examination of these concepts related them to places of work within Scarborough. For each of 68 zones,[6] the average distance to the place of work was calculated and then dichotomized at the median. Likewise, the average income was calculated for each work zone, and dichotomized at the median.

Following the model above, each of the 68 zones was assigned to one of the four types.[7] Table 7.4 presents characteristics of the typology. The "high" distance zones differ from the "low" distance zones by only 3 miles in distance traveled to work – a modest difference; nevertheless, the farther distance zones show twice the average work trip distance than the shorter distance zones. Although the income dichotomy was based upon all

Table 7.4 Distance to work and income by captive labor typology (full-time workers).

	"Captive"	"Ideal"	"Worst"	"Enterprising"
Distance type	Low	Low	High	High
Income type	Low	High	High	Low
Distance (average miles)	3.0	3.1	5.6	6.1
Income* (average)				
Female	$16,300	$19,300	$15,100	$21,000
Male	$25,600	$29,300	$21,700	$28,600
Number of zones	17	17	17	17
Number of workers	940	645	364	547

* In Canadian dollars.

worker incomes, both male and female incomes vary as expected by type, however, the "worst" type shows substantially lower incomes for both males and females than the "captive" type.

Table 7.5 presents characteristics of workers and their journey to work by the captive labor typology. The largest percentages of women are found in the "captive" and "worst" typed zones. Although this is partly a result of the method of constructing the typology, as areas with a large proportion of women workers would be expected to have lower zonal average incomes, it also shows that the "worst" type zones contain large proportions of women who travel longer distances. This is in sharp contrast to the presumption that low-paid female-dominated zones are characterized by short travel distances.

Moderately higher proportions of part-time workers are found in zones which are characterized by low average trip distances, but an equal number of part-time workers are found in both the "captive" and "ideal" type.

Although it has been argued from the dual-worker role perspective that women with young children tend to work closer to home, in this zonal analysis, the zones with the largest numbers of women with young children are actually those with longer work-trip distance, the "worst" and "enterprising" zones.

When controlled for gender, in each zone type women spend longer times on their work trips, reflecting their greater transit dependence, and for both genders, the difference between the nearer and farther work zones is approximately eight minutes. The zone with the highest number of transit users, as well as transit captives, is the "worst" type rather than the "captive" type.

In considering these results, both conceptual and empirical, it appears that in addition to the "captive" zonal type, the "worst" type is even a more troublesome work zone, because not only are relatively low wages

Table 7.5 Worker and mode characteristics by captive labor typology.

	"Captive"	"Ideal"	"Worst"	"Enterprising"
Female (percent)	51.2	43.2	56.4	46.8
Part-time (percent)*	21.2	20.6	14.4	11.9
Any child(ren) in house-				
hold less than 12 years				
(percent)				
Female	24.6	23.1	36.8	38.8
Male	31.7	32.1	35.8	29.9
Travel time (average				
minutes)				
Female	16.4	16.1	24.6	25.5
Male	15.3	12.4	20.8	19.3
Transit use (percent)				
Female	21.9	17.9	32.7	15.9
Male	9.6	4.7	16.5	5.4
Transit captives (percent				
of transit users)				
Female	75.8	78.8	81.8	63.7
Male	64.3	45.5	71.4	34.0
Transit captive (percent				
of N)				
Female	16.6	14.1	26.7	10.1
Male	6.2	2.1	11.8	1.8

* Part-time workers not included in other values.

paid, but workers in these areas suffer the travel costs which workers in the "captive" zones do not. As well, on measures of both male and female income, percent female workers, transit use, and captive riders, these are the least fortunate work zones, and also have long travel times and many women with young children. This is in no way to dismiss concern and attention to "captive" labor zones, but serious concern must also be given to those zones which we here have termed the "worst," and indeed they appear to be. These zones may reflect the increasing decentralization of unskilled blue- and white-collar jobs, and parallel a trend in Montreal reported by Villeneuve & Rose (1985). Furthermore, it is probably the case that such zones are those typified by new industrial parks relatively far from residential areas and poorly served by public transit.

Re-examining behavioral and structural approaches

The behavioral approach

A focus on accessibility makes it possible to link questions arising from the behavioral and structural approaches to the study of women's employment and transportation. It permits attention to the behavior of individuals and groups within the labor force, at the same time as it allows an examination of the environmental context to which they are responding. Traditionally, the literature on journey to work, including recent studies which focus on women's travel patterns, has viewed behavior in isolation – as if land uses, characteristics of the transportation system, or the location of firms had little effect on workers' decisions or day-to-day behavior. Behavioral studies of journey to work have presumed that any observed differences are attributable to individual differences rather than to structural characteristics in the distribution of appropriate employment opportunities. A focus strictly on the individual's journey to work, without such linkages, lends itself to a liberal choice model whereby workers are viewed as choosing their mode of travel and place of work to accord with lifestyle or lifecycle needs.

The addition of a feminist analysis linking the household economy and the political economy added greater depth to theories explaining journey-to-work patterns, but were often limited by a frame of reference which overemphasized dynamics within the individual household. Reliance on existing data sources, such as standard travel surveys or the census, often did not provide sufficient information to permit researchers adequately to test the dual worker role hypothesis against competing explanations. For example, Hanson & Johnston's (1985) and Villeneuve & Rose's (1985) use of marital status as an indicator of women's dual role is quite inadequate as a measure of family responsibility. Behavioral studies which provide a very rich source of data on household dynamics (often based on detailed time budgets) often have rather small samples (Rosenbloom 1986, Michelson 1983). When these are also sample surveys drawn from a large metropolitan area, as Michelson's study (1983), we lose the ability to examine the effects on women's employment of specific locational constraints or opportunities.

The structural approach

What the industrial location model contributes to our understanding of women's journey to work and employment linkages is that the local availability of employment may make it possible for some women, who might not otherwise be able to take paid employment, to work at all. When gender-segregated jobs are located close to or far from residential areas where women live, this affects their ability to take advantage of employment opportunities. In suburban areas, there is frequently a mismatch in so far as women may be highly educated and skilled but few managerial and professional jobs locate there. The location of back office clerical employment or electronic plants near residential areas where educated and skilled women workers live, provides low paying unstable employment and perhaps considerable underemployment if women's education or skill levels exceed the requirements of the jobs.

The spatial division of labor model emphasizes the mobility of capital to find cheap labor. This structural approach presumes that behavior is determined by the needs and characteristics of capital. By ignoring the journey-to-work behavioral literature, this model ignores the parallel adjustments by workers who travel to job opportunities offering better pay or working conditions or who change residence to decrease their journey to work. Our emphasis on captive labor markets focuses attention on the conditions under which some workers have greater mobility than others within the context of the regional economy. We also provide data on the impact of limited mobility on the employment opportunities of certain classes of workers. As for women, we cannot assume that all women living in a local area will work in local jobs, as, somehow, the discussions of local labor pools imply (Lewis & Foord 1984, Lewis 1984, Massey 1983). Nor can we assume that a wide range of employment opportunities in a region will be effectively available to women if there is limited investment in public transportation.

At the same time, while increasing workers' access to transportation may increase job opportunities and result in higher wages, at least one author (Feldman 1977) argues that this can have the opposite effect – by keeping wages low as more workers compete for the few available jobs. Improving transit may merely serve employment growth centers by bringing in cheap labor. Similarly the feminist proposals for restructuring urban space through mixed land uses, less separation of home and work, and the creation of local employment (Hayden 1984, Levitt 1985, Wekerle 1985), do not take into account what kind of employment this might be, and the possible effects of captive labor markets on women. From this review and critique of the literature, we conclude that journey-to-work studies and structural analyses of gender-segregated labor markets, while based on very different premises, are complementary in providing both behavioral and structural explanations for women's labor market subordination.

Structural analyses of spatial divisions of labor enrich our understanding of the relationships between economic restructuring and the spatial differentiation of the labor force, especially the creation of gender-segregated local labor pools. According to Doreen Massey (1983, 4), this approach represents an advancement in geographical analysis as it is a shift away from a purely spatial analysis of the geographical distribution of jobs to a focus on the spatial organization of production. This approach links geographical changes, which are often quite site-specific, to shifts in the national political economy or wider international system.

Notwithstanding the utility and recent popularity of these approaches, critiques of structuralist urban theory emphasize some of its shortcomings. Both John Browett (1984) and Michael Peter Smith (1983) point out that much of current structural analysis makes space, regions, or capital active agents in the current economic restructuring. In speaking of regional shifts in sectors of the British economy, Massey (1983, 67), for example, writes:

> if such geographical patterns are the outcome of socio-economic processes (operating over space) then in order to understand a pattern we must go behind it and interpret it in terms of the structures and processes on which it is based.

Browett (1984) objects that this stance tends to personify spaces as exploiting, as, for instance, when authors speak of local labor pools as exploiting female low paid workers. Smith (1983) adds that individuals in these schema are acted upon by systematic constraints whether they be class, ideology, or access to opportunities, and in the process human agency is downplayed.

These criticisms apply not only to spatial/structural theories, but also to much of the literature on dual labor markets. In her analysis of gender-segregated labor markets, Pat Armstrong (1984, 29) states that dual labor market theory tends to shift the focus from the characteristics of workers to those of labor markets; women's behavior is seen in response to their labor market position. She writes:

> Jobs rather than people, structures rather than personal characteristics, were the focus of attention. From this perspective, it was not poor choices but lack of choice that explained the division of labour by sex (Armstrong 1984, 29).

Following this logic, the focus of recent research on spatial divisions of labor is on the increasing geographical mobility of capital, as sectors of the economy shift to escape a well-organized workforce or seek out cheap labor pools. There is not an equivalent focus on the mobility of labor seeking better employment opportunities at the intra-metropolitan or inter-regional levels.

Both Browett (1984) and Smith (1983) urge the development of structuralist analyses which view people as active agents shaping their social world. The Women and Geography Group of the IBG (1984) further questions the ability of radical analyses to deal with gender differences in the location of employment. They point out that the gender division of labor does not come about just through the invisible hand of market forces, but frequently is maintained by quite specific industry practices, including those of trade unions, which reinforce, and profit by, job segregation by gender. This suggests attention to union policies and practices as active agents in maintaining a gender-segregated workforce.

A related concern is that there is only limited empirical verification of structuralist theories (Smith 1983). The few empirical studies linking spatial differentiation and gender-segregated labor markets are pre-dominantly small case studies of specific towns, several of them British new towns (Lewis & Foord 1984, Lewis 1984, Hudson 1982), and only recently have studies used large data sets to examine changes over a whole metropolitan areas (Hanson & Pratt, 1988, Hanson & Johnston 1985, Villeneuve & Rose 1985 & 1987, Rutherford & Wekerle 1988). While theories, these empirical studies are still too few and too limited in scope to assess the implications for new theory building. We do not know, for instance, whether the experience of the development of local labor pools documented in three British new towns is an anomaly specific to Britain or to new towns in Britain, or whether it is indicative of a more widespread trend. For instance, do these processes explain employment decentralization to regions such as the US Sunbelt? Do they explain the expansion on the periphery of major North American cities of low skilled white-collar employment? Are there specific factors which would allow us to predict *which* peripheral areas would be attractive to capital seeking cheap pools of captive female labor?

The interpretation of empirical findings is also open to question in some of these studies. For instance, Lewis & Foord (1984) portray capital as *exploiting* cheap female labor in the two British New Towns they studied. Yet early community studies of British New Towns lamented the absence of employment opportunities for married women in these communities. Under what conditions is employment for women viewed as an opportunity? Under what circumstances is this a case of capital exploiting labor?

Similarly, in Nelson's (1986) study, the interpretation of the finding that back offices locate near concentrations of married female workers is based on a discussion by managers that these workers prefer to work closer to home. Workers' reasons are imputed from interviews with management only; there is no independent verification of women workers' reasons for working close to home. Is this because of competing responsibilities within the household economy? Are women's skill levels so low that regardless of how far they travel income will not go up? Or is the location of employers seeking female workers conditioned by women's pre-existing behavior to travel a specific distance and no farther? Specifically, there are rather limited discussions of the role of transportation, both personal and public, in expanding or limiting women's employment opportunities. A purely structural analysis cannot test these assumptions since part of the answer lies in the experience and perceptions of workers which can only be obtained through empiricist methodologies.

The concept of gender-segregated labor pools is being used to explain phenomena in very different geographical settings – isolated new towns as well as developments on the periphery of a major city such as San Francisco. It is problematic whether a concept of "captive labor force" can prevail in large urban settings. While the financial and time costs of transportation to work may be high for workers from their homes to certain locales of employment, the transportation infrastructure is present and, indeed, is one function of a metropolis. As well, any number of other employers are also present in large urban settings.

One way to dissolve captive labor markets, or at least minimize their effects, is to increase workers' mobility to other places of employment. However, in some of these cases, the location is relatively isolated.

Subsidized transportation has been used to supply local labor demand as in the British new towns reported on by Lewis & Foord (1984), or the paratransit operations which are subsidized by industrial parks in the Philadelphia area (Fox 1986). Presumably, a concern with the negative impacts of local labor markets on women implies some form of state intervention on questions such as pay equity or working conditions, to prevent employers from exploiting local pools of female labor. One conclusion we draw from our examination of structural theories is that explanations of gender-segregated labor markets, as these are manifested spatially, must go beyond purely structuralist theories to incorporate the perspective of women workers as active agents within the labor market.

Policy implications

From a policy perspective, empirical research which links women's journey to work and labor markets also has significant implications. Starting from the journey-to-work literature, we find that all users of the public transportation system are not equal. Some are considerably more disadvantaged than others and this inequality translates into lower incomes, more time spent in traveling to work, and a more limited choice of locations for employment. In addressing these inequities, subsidies could be heavily concentrated on routes with high transit dependencies, thereby targeting transit subsidies to the poor (Meyer & Gomez-Ibanez 1981, 251).

Although a major focus of public policy today is reducing discrimination against women, especially as this relates to inequalities of employment, proposed remedies are aspatial and do not include attention to physical accessibility to employment. According to economist Morley Gunderson (1985, 24), for the most part, employment equity programs emphasize policies to reduce barriers to women entering the core labor market. These focus on getting women workers into certain programs (affirmative action), or extending the benefits of the core sector through equal pay legislation.

At the same time, there is also a focus on systemic discrimination or behavior that has an arbitrarily negative impact on groups. A recent Canadian federal taskforce on employment equity states:

Systemic remedies are a response to patterns of discrimination that have two basic antecedents: (a) a disparately negative impact that flows from the structure of systems designed for a homogeneous constituency; (b) a disparately negative impact that flows from practices based on stereotypical characteristics ascribed to an individual because of the characteristics ascribed to the group of which he or she is a member. The former usually results in systems primarily designed for white able-bodied males; the latter usually results in practices based on white able-bodied males' perceptions of everyone else. In both cases, the institutionalized systems and practices result in

166

arbitrary and extensive exclusions for persons, who, by reason of their group affiliation, are systematically denied a full opportunity to demonstrate their individual abilities (Abella 1984, 9).

Based on our review of the literature on women's travel to work, we would argue that women have been negatively impacted by public transportation systems designed for full-time male wage earners, many of whom are assumed to have access to a private automobile. These are often the choice riders who choose public transport largely for reasons of speed and convenience. Captive riders, who include a high proportion of women, are negatively impacted by poorer service at off-peak hours or in the evenings and weekends, and by system design which provides transport largely to employment locations but not as well to shopping and recreational facilities.

When employed women are stereotyped as secondary wage earners or still largely in the home, this has a further "disparately negative impact." For instance, few transit systems take into account and plan for the double trip that working mothers often must make to take children first to child-care or school before they continue on to their own employment location. Free or inexpensive transfers with time allowed between transfers could mitigate some of the present financial burden. Locating child-care at major transit nodes could make connections easier, at least for some parents. This is not to say that improving the transportation system *per se* will respond to the basic inequities of lower incomes and more limited employment opportunities experienced by women working in gender-segregated occupations and sectors of the economy. However, accessibility to paid employment can be made easier or harder by the design of the transportation system.

The Abella Report on Employment Equity makes a distinction between "pre-employment conditions that affect access to employment" and "those conditions in the workplace that militate against equal participation in employment" (Abella 1984). Although not directly related to production, available child-care is viewed as facilitating women's employment and expanding women's choice of jobs (Abella 1984, 177), and thus is seen as a necessary part of an employment equity program. Yet accessibility to transportation and the spatial location of jobs, which are as much preconditions for employment as available child-care, are not considered as factors which might keep women out of the core labor market.

Hanson & Johnston (1985) argue that eliminating the occupational segregation of women goes beyond training for non-traditional jobs. Greater attention must be directed to the locations of job opportunities in male-dominated or integrated occupations and their accessibility to transportation. Female job ghettos, which draw upon local labor pools, can perpetuate segregation when public transportation is inadequate or unavailable. Greater attention must be devoted to planning the connections between residential areas, new employment centers and transportation infrastructure, particularly in peripheral areas. For instance, there is little recent research on transit captives to determine whether they are spatially concentrated in certain work or residential locations. Despite all the current work on the gender segregation of the labor market, there are very little data on the locations of female-dominated jobs and particularly whether new jobs created in peripheral locations are female-dominated, male-dominated, or integrated. There are questions concerning the influence the peripheral location of "high tech" industrial parks will have on the non-traditional job opportunities of women employed in declining sectors of the economy, who may be undergoing retraining from unskilled work, especially if these locations are only available by private automobile.

Sawers (1984, 226) is one of the few scholars in the Marxist geography tradition who link mobility to opportunities for employment in the core or secondary labor markets. He notes that workers with stable jobs in the core labor market can increasingly locate near jobs in suburbs, whereas the secondary labor market, which changes jobs frequently, lives in the core where adequate public transportation provides access to a wide range of job opportunities. This relates to current studies showing a higher concentration of employed women living in core areas of US cities like Baltimore (Hanson & Johnston 1985). Relocating residences to be closer to jobs is less functional for workers in the secondary labor market who change jobs more often and are lower paid (usually without relocation allowances). Especially if they are women, they are also less likely to own their own homes and most likely to rent – a form of tenure less available in the suburbs and therefore in higher demand and higher priced (McClain & Doyle 1984). Women in female job ghettos frequently require some form of low cost or subsidized housing, especially in cities with tight housing markets and high housing costs. As new employment moves to peripheral areas, the distance between such housing and sectors of employment growth will continue to increase.

Marion Fox's (1986) recent review of transportation problems associated with suburban industrial park development concludes that women stand to gain most from suburban paratransit and pooling programs which would make available higher paid employment and jobs which offer some career advancement. She shows that when there is a labor shortage, some employers are motivated to pay for van pools to bring in female labor which could not otherwise access these industrial parks. However, keeping in mind the arguments about the negative effects of captive labor markets, such employer paid benefits might only serve to keep women workers in female job ghettos in the absence of other options, much in the same way as some unions now argue that workplace day-care may keep workers tied to one employer.

In Canada, Section 15 of the Canadian Charter of Rights and Freedoms, which came into law through the Constitution Act 1982, includes not only equal protection under the law but also "equal benefit of the law without discrimination." This implies a potential emphasis on inequitable practices or policies especially by public agencies which have a disproportionately negative impact on women. Under this provision, women might argue that they have been subjected to systemic discrimination by transportation agencies which do not adequately serve captive riders, or that they have been discriminated against by industrial economic development programs which subsidize, with public funds, the formation of local labor pools of primarily low paid female employment. While so far this has not been the focus of litigation, the law, as it stands, does not rule it out in future.

Notes

The co-authors gratefully acknowledge the support of the Social Science and Humanities Research Council of Canada for supporting this research. The Toronto Transit Commission kindly provided the data on which much of the present paper is based. This research would not have been possible without the Research Development Fellowship Program awarded to Rutherford by York University and LaMarsh Research Program on Violence and Conflict Resolution.

1 To explore questions regarding the role of gender in transit behavior, we conducted a secondary analysis of an extensive survey conducted by the Toronto Transit Commission in autumn of 1983. The survey was designed to assist the planning of bus routes following from the completion of a new light rail extension of the existing subway line. At the time of the survey, the subway line extending West to the CBD ended at the west border of the study area. Because of the detailed information required to reconsider each route, the sample size was large, numbering 6,434 households and 16,249 individuals for which transit behavior and personal characteristics were gathered. The sample frame was taken to be nearly all of Scarborough, a suburban municipality approximately 10 miles east of the downtown center of Toronto. The sampled households constitute approximately 10 per cent of the population of the study area.

 The data were collected utilizing a self-administered questionnaire. The original mailing was followed by two subsequent reminders that resulted in a final response rate of 72%. The study design sought information from all members of the households who were 12 years of age or older, thus the basic design of the study focused upon the individual rather than the nature of the household, and included those who were family members but neither heads of households nor their spouses. Questions of the location of home and workplace mode and the amount of time travelled to work also permitted a comparison of work trip travel and distance travelled of male and female workers.

 Scarborough may be typified as a suburb largely planned and built in the 1950s and is still expanding in both housing and industrial development. Industries are generally light and modest in size. Housing is largely single-family, but with a considerable number of large apartment blocks along the main streets.

2 The number of captives may be even higher as the study did not include travel on foot and bicycle.

3 A lower correlation will produce a lower regression beta, presuming equal variances. The lower beta for women ($182CAN) is partly because of the lower correlation between distance and income. Whether the focus is on the correlation or the beta, the conclusion remains; women do not gain much in income by traveling additional distances.

4 To permit density comparison with subsequent figures for men, the number of

women was increased by a factor such that the total numbers of workers, and the areas under the surfaces are equal.

5   Indeed among women, a greater proportion of $40,000CAN income women work closer to home rather than farther.

6   The zones are approximately equivalent to census tracts, but more precisely have been previously defined by the Toronto Area Regional Modelling Study and are known as TARMS zones.

7   Zone assignments were necessarily based on the sample frame (residences of Scarborough) and could not include workers living outside of Scarborough, but traveling to work locations within Scarborough.

References

Abella, R. S. 1984. *Equality in employment: a Royal Commission report.* Ottawa: Ministry of Supply and Services.

Abella, R. S. (ed.) 1985. *Equality in employment: research studies.* Ottawa: Ministry of Supply and Services.

Altshuler, A. 1979. *The urban transportation system: politics and policy innovation.* Cambridge, Mass.: MIT Press.

Armstrong, P. 1984. *Labour pains: women's work in crisis.* Toronto: Women's Press.

Armstrong, P. & H. Armstrong 1984. *The double ghetto: Canadian women and their segregated work.* Toronto: McClelland & Stewart.

Black, J. 1977. *Public inconvenience: access and travel in seven Sydney suburbs.* Canberra: Australian National University.

Browett, J. 1984. On the necessity and inevitability of uneven spatial development under capitalism. *International Journal of Urban and Regional Research* **8**, 155–76.

Coates, B. E., R. J. Johnston & P. L. Knox 1977. *Geography and inequality.* London: Oxford University Press.

Doeringer, P. B. & M. J. Piore 1971. *Internal labour markets and manpower analysis.* Lexington, Mass.: Lexington Books.

Erickson, J. A. 1977. An analysis of the journey to work for women. *Social Problems* **24**, 428–35.

Feldman, M. 1977. A contribution to the critique of urban political economy: the journey to work. *Antipode* **9**, 30–50.

Fox, B. J. & J. Fox 1987. Occupational gender segregation of the Canadian labour force 1931–1981. Toronto: Institute for Social Research, York University.

Fox, M. 1983. Working women and travel: the access of women to work and community facilities. *American Planning Association Journal*, Spring, 156–70.

Fox, M. 1986. Faster journeys to suburban jobs. *Women and Environments* **8**, 15–18.

Gordon, D. M., R. Edwards & M. Reich 1982. *Segmented work, divided workers: the historical transformation of labour in the United States.* New York: Cambridge University Press.

Gunderson, M. 1985. Labour market aspects of inequality in employment and their application to crown corporations. In *Research studies of the Commission on Equality of Employment*, R. S. Abella (ed.). Ottawa: Minister of Supply and Services.

Hanson, S. & I. Johnston 1985. Gender differences in work-trip length: explanations and implications. *Urban Geography* **3**, 193–219.

Hanson, S. & G. Pratt 1988. Spatial dimensions of the gender division of labor in a local labor market. *Urban Geography* **9**, 173–193.

170

Hayden, D. 1984. *Redesigning the American dream: the future of housing, work, and family life.* New York: W.W. Norton.

Healey, P. 1986. Distributive consciousness in spatial planning. Paper presented at conference on Planning Theory in Practice. Torino, Italy.

Hecht, A. 1974. The journey-to-work distance in relation to the socio-economic character of workers. *Canadian Geographer* **18**, 367–78.

Holcomb, B. 1986. Geography and urban women. *Urban Geography* **7**, 448–56.

Howe, A. & K. O'Connor 1982. Travel to work and labor force participation of men and women in an Australian metropolitan area. *Professional Geographer* **34**, 50–64.

Hudson, R. 1982. Accumulation, spatial policies, and the production of regional labour reserves: a study of Washington New Town. *Environment and Planning* **14**, 665–80.

Kaniss, P. & B. Robins 1974. The transportation needs of women. In *Women, planning and change*, K. Hapgood & J. Getzels (eds.). Chicago: American Society of Planning Officials.

Leavitt, J. 1985. A new American house. *Women and Environments* **7**, 14–16.

Lewis, J. 1984. The role of female employment in the industrial restructuring and regional development of the United Kingdom. *Antipode* **16**, 47–60.

Lewis, J. & J. Foord 1984. New towns and new gender relations in old industrial regions: women's employment in Peterlee and East Kilbride. *Built Environment* **10**, 42–52.

McClain, Ian & W. Cassie Doyle 1984. *Women and housing.* Toronto: Lorimer.

Madden, J. F. 1981. Why women work closer to home. *Urban Studies* **18**, 181–94.

Madden, J. F. & M. J. White 1978. Women's work trips: an empirical and theoretical overview. In *Women's travel issues: research needs and priorities*, S. Rosenbloom (ed.), 201–42. Washington, DC: US Dept. of Transportation.

Massey, D. 1983. *Spatial divisions of labor: social structures and the geography of production.* New York: Methuen.

Meyer, J. R. & J. A. Gomez-Ibañez 1981. Autos, transit and cities. Cambridge, Mass.: Harvard University Press.

Michelson, W. 1983. *The impact of changing women's roles on transportation needs and usage.* Washington, DC: US Dept. of Transportation.

Millar, E., G. N. Steuart, D. R. Ross, R. Potvin & R. Ridout 1986. *An occupationally disaggregate analysis of census data for the Greater Toronto area, 1971–1981.* Toronto: University of Toronto/York University Joint Program in Transportation.

Miller, M., R. Morrison & A. Vyas 1986. *Travel characteristics and transportation energy consumption patterns of minority and poor households.* Argonne National Laboratory: Center for Transportation Research.

Nelson, K. 1986. Labor demand, labor supply and the suburbanization of low-wage office work. In *Production, work, territory*, A. J. Scott & M. Storper (eds.). Boston: Allen & Unwin.

Pucher, J. 1982. Discrimination in mass transit. *Journal of the American Planning Association* **48**, 315–26.

Pucher, J. R., C. Hendrickson & S. McNeil 1981. The socioeconomic characteristics of transit riders: some recent evidence. *Traffic Quarterly* **45**, 461–83.

Reich, M., D. M. Gordon & R. C. Edwards 1980. A theory of labour market segmentation. In *The economics of women and work*, A. H. Amsden (ed.). London: Penguin.

Rose, D. & P. Villeneuve 1985. Women workers and the inner city: some social implications of labour force restructuring in Montreal, 1971–1981. Paper presented at the Canadian Urban Studies Conference, Winnipeg.

Rosenbloom, S. (ed.) 1978. *Women's travel issues: research needs and priorities.* Washington, DC: US Dept. of Transportation, Research and Special Programs Administration.

Rosenbloom, S. 1986. The unique travel pattern of single heads of households with children in three Southwestern SMSA's. Washington, DC: Transportation Research Board.

Rutherford, B. & G. R. Wekerle 1988. Captive rider, captive labour: spatial constraints on women's employment, *Urban Geography* **9**, 173–93.

Sawers, L. 1984. The political economy of urban transportation: an interpretive essay. In *Marxism and the metropolis*, W. K. Tabb & L. Sawers (eds.). New York: Oxford University Press.

Scott, A. J. 1981. The spatial structure of metropolitan labor markets and the theory of intra-urban plant location. *Urban Geography* **2**, 1–30.

Smith, M. P. 1983. Structuralist urban theory and the dialectics of power. *Comparative Urban Research* **9**, 5–12.

Storper, M. 1981. Toward a structural theory of industrial location. In *Industrial location and regional systems*, J. Rees et al. (eds.), 17–40. London: Croom Helm.

Taaffe, E. J., B. J. Garner & M. H. Yeates 1963. *The peripheral journey-to-work.* Evanston, IL: Northwestern University Press.

Toronto Transit Commission, Scarborough Transit Services, 1984. *Improvement study. Report No. 2: Survey findings.* Toronto: Service Planning Dept.

Villeneuve, P. & D. Rose 1985. Technological change and the spatial division of labor by gender in the Montreal metropolitan area. Paper presented at a meeting of the Commission on Industrial Change, International Geographical Union, Nijmegen, The Netherlands.

Villeneuve, P. & D. Rose 1987. Gender, labour force restructuring and the journey-to-work in Montreal, 1971–1981. Paper presented at the American Association of Geographers, Portland, Oregon.

Wachs, M. & T. G. Kumagai 1973. Physical accessibility as a social indicator. *Socio-economic Planning Sciences* **7**, 436–56.

Wachs, M. & J. L. Schofer 1972. Public transit and job access in Chicago. *Transportation Engineering Journal* **98**, 351–66.

Wekerle, Gerda R. 1985. From refuge to service center: neighborhoods that support women. *Sociological Focus* **18**, April, 79–95.

Wekerle, G. R. & B. Rutherford 1986. Employed women in the suburbs: transportation disadvantage in a car-centered environment, *Alternatives: Perspectives on Society, Technology and Environment* ·**14**, 49–54.

White, M. J. 1977. A model of residential location choice and commuting by men and women workers. *Journal of Regional Science* **17**, 41–52.

Women and Geography Study Group of the IBG 1984. *Geography and gender.* London: Hutchinson.

Part IV

The state

8

Interpretive practices, the state and the locale

NICHOLAS K. BLOMLEY

"Woe unto ye also ye lawyers, for ye lade men with burdens grievous to be borne, and ye yourselves touch not the burdens with your fingers" Luke 11:46.

"The lawyer's truth is not truth, but consistency, or a consistent expediency" (Thoreau 1980, 239).

Introduction

This chapter will explore the role of one specific element of the state apparatus in the process of social reproduction. Specifically, those agencies charged with the enforcement and interpretation of law will be examined. While the state's fundamental implication in the process of socio-spatial reproduction is widely acknowledged, few geographers have turned their attention to the role of the law and legal interpretation *per se* (exceptions include Johnston 1983, 1984, Clark 1981, 1982, 1985, 1986, forthcoming; Blacksell *et al.* 1986, Blomley 1986a,b, 1987a,b, forthcoming). Predominantly, analysis of law from within the geographical literature has been sparse or theoretically uninformed, law being regarded as a "given" and analyzed in terms of impact assessment. This is lamentable given the very real importance of socio-spatial context in shaping law and its interpretation. Space is not simply a stage on which laws are applied, but serves to shape the very process of legal interpretation. There is, therefore, a need for analysis of law that is both theoretically and spatially informed.

This chapter will argue that law is an important agent of social reproduction, representing a conjoint expression of state ideology and instrumentality. Thus, on the one hand, law has a distinctive ideological dimension in that it summons up a mass of associations of naturalistic morality, democracy, and determinacy. Law is, however, not *just* ideology; it aims to *do* something.

The very instrumentality of law lies, in part, in its ideological nature (for example, its claims to "common morality") which encourages compliance given the social costs of deviancy. It is this assumption which underpins the concept of the "rule of law" (Dicey 1965). However, the

175

effectiveness of law also lies in its perceived determinacy as a text. The assumption is that law represents a body of determinate textual commands which, by their very nature, ensure that some specific goal is achieved.

Such determinacy is, however, as elusive as it is improbable. As Clark (1985) suggests, legal interpretation is far from unproblematic. On the contrary, it is inherently evaluative and potentially indeterminate. Crucially, he recognizes not only that law shapes space, but also that context may simultaneously shape law. This perspective is shared by Geertz (1983), who argues that "law is local knowledge not placeless principle and that it is constitutive of social life not reflective" (218). Thus, law is not an abstract, determinate text, but a social artefact, formed and interpreted by socially and spatially situated people (Thrift 1983, 1985).

Given the complexities of legal interpretation, law can generate contradictions both for the state and for the specific constituency in which it is applied. It will be argued that distinctive interpretive procedures are adopted by different state agencies, with specific standards of interpretive validity being given priority in each sphere. The law courts opt (formally at least) for a formalized and abstractive procedure of interpretation. This contrasts with the contextual and place-specific strategy of the local enforcement agency. These place-specific interpretations in turn give rise to social practices which differ among geographic locales. The "geography" of legal interpretation rests not only in these spatial differences, but more importantly in the structuring effect of a locale on interpretive practice in any one spatial region or level of the state.

In the section which follows, those state agencies charged with the interpretation and enforcement of the law are scrutinized. In particular, the intra-state differences in interpretive function and strategy are theorized. In the third section, the Shops Act (1950), which regulates the hours and conditions of retail employment and operation in England and Wales, is used to illustrate the different strategies adopted. Conflicts around the interpretation of the Act indicate that the various modes legal interpretation can have important consequences for intra-state tensions, for ideologies of the state and law, and ultimately for the efficacy of the state in shaping social reproduction in particular geographical locales.

Law and the state apparatus

Introduction: the ideal type

Law, then, is an important agent of reproduction for the state. How is law applied within the state? The orthodox ideal type is a formalized one (Dicey 1965). Law is *formulated* by the legislature (derived from common conceptions of morality), and then *applied* by a myriad of agencies, to whom responsibility is both delegated and monitored by the central state. Some autonomy is accorded these agencies, it being recognized that the practical details of often complex laws are best devolved to those expert in a specific field. The assumption is, however, that such application will be unproblematic, the text of a statute being related to any given situation. Any problems as to the specific meaning of law that may emerge are resolved in the courts, which are represented as being beyond or above the state. The courts' assumed function in this ideal type is to *interpret* law; to extend the original intent of those who drafted the statute to specific phenomena that were not extant at the time of codification and to do so in a manner that is "true" to the intrinsic meaning of the statute as a determinate text. Three finite legislative tasks for the state apparatus are thus identified; that of law-making, law-enforcing, and law-interpreting. Each of these tasks, in this idealized division of legal labor, is the dominant (or even sole) function of one agency; that of the legislature, enforcement agency, and court respectively.

Implicit in this ideal type are at least two conceptions. Firstly, this formalized division of judicial labor implies a *functional determinacy* such that specific agencies have clearly defined, separable functions with little (if any) intra-apparatus interplay. It is assumed that the law court does not "make" law, nor does the local agency "interpret" law. Secondly, there is an assumption of *statutory determinacy*, such that the law has meaning in and of itself. The state acts merely as a medium or conduit for the formulation, interpretation or application of law. Law is somehow derived "beyond" the state, in a formless field of naturalistic morality. It "enters" and "leaves" the state unsullied. The juridical practices of the state, in this conception, do not shape law. Law, if anything, shapes the state.

The concepts of functional and statutory determinacy are, however, deceptive. Firstly, the state as an apparatus cannot be divided so easily. Crucially, all agencies of the state are engaged in interpretive practices. Having said this, however, it is clear that interpretive "chains of command" exist within the state such that certain state agencies are obliged to acquiesce in the interpretations of higher state agencies. Secondly, the state is not simply a conduit through which law passes. It is in the very process of interpretation by those agencies within the state that statutory "meaning" is produced. These considerations can be observed when the interpretive procedures of both law court and enforcement agency are considered.

Interpretive practice and the law court

Formally, the function of the court is straightforward. Lord Devlin, one of the British Law Lords, argued that, in theory

> the judge is the neutral force between government and governed. The court interprets and applies the law without favour to either and its application in a particular case is embodied in an order which is passed to the executive to enforce. It is not the judge's personal order; it is substantially the product of the law and only marginally of the judicial mind (*The Sunday Times*, August 6, 1972, 14).

The judge, in this conception, must be "of the law," abstracted from society. As Griffith (1985) puts it, the judge must act like a "political, economic and social eunuch, and have no interest in the world outside his court when it comes to judgement" (191).

The interpretation of statutory meaning is unproblematic in this conception, given the determinacy of legal meaning. The perspective is a positivistic one; the statute can be read for meaning. Meaning is clear, and constant for all those who know how to interpret in a rigorous manner. Thus, Langan (1976) asserts that "a statute is an authentic expression of the legislative will, the function of a court is to interpret that document according to the intent of them that made it" (1). From this perspective, the interpretive process becomes a translative one, the statute being conceived of as pellucid.

How then, does the court uncover the intention of the legislature given that "it may only elicit that intention from the actual words of the statute" (Langan 1976, 1)? The "true" meaning of any statute, it is often implied can be uncovered via hard-nosed, formalized interpretive technologies (Twining & Miers 1982, Reynolds 1980). Frequent reference is made to such procedures as the syllogism, or a number of "rules" of interpretation. The "literal rule", for example, assumes that words and phrases are used in their ordinary meaning, and that phrases and sentences are to be construed according to the rules of grammar. The "mischief rule" holds that interpretation should proceed with reference to the mischief for which the law aimed to regulate. The so-called "Golden rule" extends the literal rule, such that "ordinary meaning" of text is to be adhered to "unless that is at variance with the intention of the legislature . . . or leads to any manifest absurdity or repugnance, in which case the language may be modified, so as to avoid any inconvenience, but no further" (*Becke* v. *Smith* [1836] 2 M & W 191, 195). Allied with these procedural technologies is the assumption that meaning is accumulative. This accounts for the frequent reference to leading cases. The principle of *stare decisis* ensures that the judicial decisions of higher courts have a binding force and enjoy a law quality *per se*. Such rulings can only be overturned by a higher court.

However, despite such "objective" procedures, the legal literature has long argued that determinacy is elusive. "Meaning" is not easily uncovered. Some blame the rules. Thus Cardozo (1924) argues that "in the complexity of modern life there is an increasing need for resort to some fact-finding agency which will substitute exact knowledge of factual conditions for conjecture and impression" (117). Others seem to recognize the fact that it is not the technologies that are to blame, but the interpreter. Thus, interpretation is seen at best as fluid, at worst indeterminate. Levinson (1982), for example, expresses anguish at the fact that "there are as many plausible readings of the United States Constitution as there are versions of *Hamlet*, even though each interpreter, like each director, might genuinely believe that he or she has stumbled into the one best answer to the conundrum of the texts" (391).

Others, while understanding the desire for interpretive determinacy, point to the fact that interpretation is not an abstract procedure, beyond the interpreter. Judges, despite their best intentions, are not *above* society.

Meaning is not uncovered by procedures alone (although, of course, one can make reference to such procedures). The interpretive process, on the contrary, is conducted by *social* beings, albeit elite social beings (Griffith 1985). As such, the interpretive process is a social process. Holmes (1881) puts the case clearly:

> the actual life of law has not been logic; it has been experience. The felt necessities of time, the prevalent moral and political theories, intuitions of public policy, avowed and unconscious, even the prejudices which judges share with their fellow-men, have had a good deal more to do than the syllogism in determining the rules by which law shall be governed (27).

This indicates a tension between the quest for interpretive determinacy and the potential indeterminacy inherent in relativism.

The local enforcement agency

A myriad of enforcement agencies serve to apply specific statutes in a given context (the police, housing authorities, pollution authorities, water boards, etc.). The assumption has again often been made that the operation of the agency is unproblematic, it being engaged simply in "application." The agency is, in other words, a medium and nothing more. Law that "enters" the agency will leave the agency unchanged. However, there is every indication that the picture is not so clear. Thus, for example, Selznick (1966) commented that any agency has to be related to its constituency of which it is, in some senses, a part. Others have recognized the specific institutional structure of the agency, and the effect this may have on interpretation (Richardson *et al.* 1982). Others have taken this further, and recognized the political context within which the agency is situated. Hawkins (1984) and Hawkins & Thomas (1984), for example, emphasize the constant internal concern for the legitimation of the agency and its actions. Hawkins goes on to emphasize the symbolic nature of much enforcement activity. With reference to the enforcement of pollution legislation, he argues that "what is really being sanctioned is not pollution, but deliberate or negligent law breaking and its symbolic assault on the legitimacy of the regulatory authority" (1984, 205; see also Manning 1977).

This has generated a number of informed studies which, if nothing else, demonstrate the complex interpretive procedures of the agency. Within certain constraints, the enforcement officer has an interpretive task to perform, such that the facts of the case have to be related not only to the substance of law, but to other considerations (such as those associated with the agency itself). Hawkins & Thomas (1984) argue for an interpretive perspective, recognizing that tasks and problems within the agency are, in large part, socially constructed. The ideologies associated with the law and those to whom the law is directed may shape the enforcement process itself (see also Kagan & Scholz 1984). There is a recognition of the variety of interpretive strategies that an agency can adopt (Reiss 1984), as well as the bargaining nature of much enforcement activity. Thus, Hawkins & Thomas (1984) argue that tactics of informal negotiation between agency and the regulated activity are mutually beneficial; "both organizations are concerned with conserving resources and minimising interference with established routines" (15).

The main thrust of the above is that enforcement of law by a state agency is not a simple imposition of law via an "abstract" agency. Quite the contrary; the process is interpretive, shaped by an organization with internally defined imperatives, with reference to a specific locality. This last point is addressed by Shover *et al.* (1984), who seek to understand the reasons that underlie marked regional differences in the enforcement of regulatory law within the coal mining industry. They argue for the importance of spatial context in the shaping of interpretive strategies, so that the place-specific form of personal ideologies, economic activity, and political environment influences the choice of a legalistic as opposed to conciliatory enforcement strategy. Thus, enforcement is not merely application. The agency is engaged in interpretive processes in a contextual and place-specific manner.

In sum, when interpretive procedures of both the law court and the enforcement agency are considered, liberal assumptions of functional and legal determinacy begin to break down, given an overlap in the interpretive function, combined with a specificity of interpretive practice. It is to differences in these practices that I know turn. Specifically, I suggest that differences in interpretive procedures are in large part political, representing an attempt at the legitimation of the role of the interpretive agency to a specific constituency.

Interpretation and interpretive communities

If the state apparatus is to intervene in the reproductive process via law, at least two conditions have to be satisfied; the one temporal, the other spatial. Firstly, law is formulated as a text. This text is drawn up at a specific time, in response to a specific socio-economic environment. That environment can change markedly. As a consequence, it is necessary that new conditions be evaluated with reference to the text; a time specific statute has to be related to changed and unforeseen conditions. This, formally at least, is the task of the judge. Secondly, the text has to be applied to a complex and shifting locality. The enforcement agency has again to make decisions as to the meaning and importance of the observed phenomena. This is a further interpretive task. Both court and local agency, in combination, provide for enforcement; both are engaged in interpretation.

To succeed both instrumentally and ideologically, however, the interpretive process at both levels of the state generates distinct procedures of interpretation, as well as internally defined standards of validity. This reflects the inherently political nature of interpretation.

Thus, by its very nature, the task of formal interpretation by the law court is a self-consciously abstracted one. Devlin (1979), one of the English Law Lords, refers to the essential quality of the judge as "impartiality and next after that the appearance of impartiality" (3). The stance of the court is therefore an ideological one (Tushnet 1983, Unger 1983, Clark 1985). It is clear that judicial definitions of meaning are potentially of great importance for a given constituency. Those that make the decisions are very often unelected, and thus find themselves in positions that can only be defended by example. As a result, the procedures of the court are imbued with a higher status to give them legitimacy, an interpretation being defined as a function of the statute, not the interpreter. For example, Devlin (1979) argues that although a judge should have some regard to the "intention" of the statute, "in the end the words must be taken to mean what they say and not what their interpreter would like them to say; the statute is the master and not the servant of judgement" (14). This encourages a tendency towards interpretations that appear to have been constructed on a higher plain. There is thus an inherent drive towards interpretive procedures that give an impression of abstraction from any context other than that simply of the text *per se*. Note that I am not saying that an "abstractive" strategy is *in reality* that which a court will adopt. The interpretive process is far too flexible and relative for that to occur (Willis 1938, Fish 1980, 1982). However, it is possible that the formalized adoption of such procedures, and the obeisance of the court towards all that lies beyond such procedures (Thompson 1975), will serve to direct and constrain interpretations such that they only have "validity" in so far as they conform (and can be reproduced by others using the same procedures) to a distinct abstractive and objective stance. Both procedures and the validity of an interpretation are, in this regard, internally defined. "Meaning" is not "objective" but internal to the interpretive agency itself.

When we turn to evaluate local agency interpretation, it is again apparent that distinct procedures and specific standards of validity are adopted. Such an interpretive strategy is, as with that of the law court, internally coherent and reflects the political nature of the interpretive task. Thus, for local application to proceed, it needs to be practical, pragmatic, and inherently locally responsive. I am assuming, as with the law court, that the local agency is obliged to legitimate its interpretation of law within a specific constituency. To achieve this, a formalized, abstractive procedure akin to that of the law court could be adopted. For example, every transgression of a specific statute could be prosecuted. However, to do so would not only place impossible strains on the local state, but lay the state open to charges of legalism and unresponsiveness; to deny, in effect, the place-specific form and needs of the locality. Given these two factors, the interpretive procedures adopted by the local state seem to reflect the specific character of the locality within which they are discharged. As indicated earlier, the procedure is a negotiatory and responsive one, a reaction to the dynamic and place-specific nature of a locality. The coherency of a local agency interpretation rests not so much on the assumed determinacy of law as a higher text (as with the law court), but in terms of the specific imperatives of the local state as an institution. Thus, a "valid" interpretation is not necessarily one which accords with the assumed intentionality of the legislature, but one which serves to advance or maintain an impression of the legitimacy and potency of the local state, and of its right to make such an interpretation.

How, then, can we contrast the interpretive strategies of the higher court and the local agency? In this context, the comments of the literary theorist, Fish (1980), are useful. He recognizes the relativity of textual interpretation, and the implausibility of what Dworkin (1986) terms "conversational interpretation," arguing that "linguistic facts . . . do have meaning, but the explanation of that meaning is not the capacity of syntax to express it but the ability of a reader to confer it" (8). However, he steers away from formless relativism and solipsism by arguing that what gives an interpretation coherence is that it is *shared* by an interpretive community who adopt similar perspectives and interpretive procedures in approaching any text. Such internally defined perspectives and procedures will be of the nature of "common sense" in so far as the interpretive community are concerned (for example, "God is love," "economic competition is good," "law applies to all"). However, *between* different communities, disagreement is possible or even inevitable. Thus, he comments:

> members of the same community will necessarily agree because they will see (and, by seeing, make) everything in relation to that community's assumed purposes and goals; and conversely, members of different communities will disagree because from each of their respective positions the other "simply" cannot see what is obviously and inescapably there (15).

To an extent, we can think of both the court and the enforcement agency as separate interpretive communities, especially if we adopt a rather broader perspective of the concept. Clearly, the court and the agency do share some interpretive common ground. There are, for example, presumably a number of shared ideologies in the broadest sense associated with the validity of law as a social institution. However, the different bodies view law, its interpretation, and the validity of those interpretations in a different manner given their distinct political positions. That is, the court may view law as something to be interrogated for "meaning" in an empiricist manner. The validity of a specific interpretation is couched in terms internal, in part, to the law court (how, for example, the ruling accords with precedence and internal logic). The local agency, on the other hand, is more concerned with the local significance of law. An interpretation of that statute, or an associated ruling, is valid in so far as it meshes with local conditions and locally expressed ideologies associated with law and the state. The definition of the cogency and legitimacy of interpretation, in both cases, is internal to the different agencies. Validity, in this conception, is not an abstract; it is internal to the interpreter. This tallies with Fish's (1980) arguments, where he asserts that an interpretation is "correct" in so far as it is "correct" for the interpretive community itself.

However, the interpretive community, at least in this study, is not a homogeneous mass. Although bound by some common assumptions, divisions are often evident. The disquiet within the legal literature as to the existence of different rulings on what is supposedly a determinate text, interpreted in a contextual manner, is a case in point. Merely because an interpretive community exists, does not mean that interpretation within that community is necessarily homogeneous, although it is more so than between communities. That community is bound together by a set of shared assumptions and ideologies which provide the reference point for interpretation. If internal differences outweigh consensus, the community divides.

This would suggest that interpretive strategies by all state agencies can best be viewed as lying on a continuum between the extremes of "abstraction" and "immersion" in any given geographic or temporal context (Fig. 8.1).

Two immediate points should be made concerning Figure 8.1. Firstly, both extremes of abstraction and immersion are impossible to attain. One

Figure 8.1 The interpretive continuum.

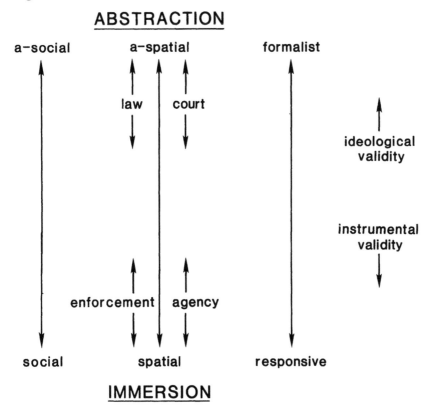

cannot somehow abstract oneself from any context in a Rawlsian sense (Clark 1986). By the same token, total immersion in any given context is not a practical possibility. An external frame of reference is always constructed. Secondly, we can think about the continuum as expressing an ideological and instrumental duality such that formally abstractive interpretations carry an ideological weight (as with the law court), yet a context specific interpretation may serve to be of much greater instrumental value in the discharge of state functions. This last point reflects a deeper contradiction within law, given state imperatives. If law is seen as policy that is to be implemented effectively, it needs to be applied via an immersive strategy, with due regard to the nature of its constituency. However, if law is to retain ideological legitimacy as a determinate "meaningful" text, it needs to be interpreted in an abstractive manner.

It is evident that different interpretive agencies adopt different positions on this continuum. However, the exact location is obscured by ideologies and imperatives associated with enforcement.

Thus, the law court might be envisaged to be located towards the top of the scale, given associated formalist and abstractive assumptions. This might be the position to which the court aspires. However, the inevitable social rootedness of an interpreter will as inevitably "pull" the court towards immersion. Similarly, the local agency may be seen to be located towards the base of the continuum, given its rootedness within a locality. However, the actions of the state have to acquire some formal legitimacy. As a consequence, enforcement may be legitimized by reference to formalistic positions. Both court and agency are thus in a state of continual tension, given a contradiction between place and legitimation. We would expect this tension to be worked and re-worked in the day-to-day operation of the agency.

This duality of abstraction and immersion would not be a concern were it not that court rulings acquire law-like status. If all judges do is interpret law, the logic runs, then any ruling must be an expression or extension of law. These rulings are of importance for local level application given the chain of command within the state – such that higher level rulings are binding on lower level interpreters. The danger is, however, that an abstractive ruling can have contradictory consequences when it comes to application by an agency further down the hierarchic scale of the state apparatus, given the very interpretive technologies that were purportedly used to reach that decision. We may expect to observe a tension between two interpretive communities, given the inappropriate exportation of a statutory interpretation reached by one set of internally defined procedures and standards of coherency, to a second agency, engaged in distinctive and locally-specific interpretive practices.

184

Interpretive communities; the Shops Act (1950)

We can see these interpretive communities at work, and the tensions that may set in both between and within different communities by considering the implementation of the Shops Act. Simply put, the Act is designed to regulate the hours and conditions of retail employment and operation in England and Wales.

The present Act, in so far as it consolidates at least four hundred years of similar statutes, has a long pedigree. Historically, the Act represents two broad concerns (*see* Whitaker 1933, 1940, 1973, Winstanley 1983, Wigley 1980, Blomley 1986b). Firstly, there is a (diminishing) concern for the continued observance and statutory protection of the Protestant sabbath. Interrelated with this ideological concern has been a more explicit productive concern, in that the state (through law) sought to intervene directly within the retail process in an attempt to ameliorate many of the problems generated by extended opening hours. Thus, the late night and Sunday opening policies of many 19th-century retailers, especially in the larger metropolises, were seen as inimicable to the interests of larger retail capital interests (and, ultimately, to retail capital as a whole), given competitive pressures and the increased overheads associated with extended hours. It was also implicitly recognized that the over-utilization of a predominantly full-time labor force was not conducive to productivity (Grindrod 1843).

In the last two decades, many of the ideological and instrumental concerns that found expression in the Act have been replaced, given marked changes in retailing and processes of socialization. Thus, some sectors of retail capital have changed their use of labor to the extent that it is possible for them profitably to meet the out-of-hours demand for shopping. This has induced many retailers, especially in the durable goods sector, to open illegally. Other sectors of retail capital, on the other hand, foresee increased overheads with few concomitant increases in turnover if they are – given competitive pressures – obliged to open late at night or on Sundays. As a consequence, many of the more "traditional" retail interests (for example, the department store sector) have bitterly opposed deregulation, and complained about the illegal actions of their competitors. This political pressure, coupled with parliamentary inertias, has prevented repeal or reform, despite some 20 attempts since 1950, the most recent leading to a Commons defeat for the Thatcher Government (Blomley 1985). Constrained by ideologies of the "rule of law," the state has been obliged to take action against offenders, despite an evident reluctance. Thus, legislation that was seen as being functionally valuable at the turn of the century has become contradictory and divisive by the 1980s. It is in this light that we need to consider interpretation of the Act by courts and local enforcement agencies.

Interpretation within the courts and the local agency

The higher courts have long been asked to interpret formally the substance of the Act. Since 1950, the *Scottish Law Book* has outlined 42 specific cases heard in higher courts (that is, above the level of the crown courts). Prior to 1950, there were several other landmark cases which are also relevant to the 1950 Act, given its consolidating nature. These disputes can be broken down into rulings concerned with the definition of specific clauses, phrases or even words within the Act (for example, *Betta Cars* v. *Ilford Corporation* [1960] where, on appeal, the plaintiffs argued that they were not "open for the serving of customers" within the meaning of the Act and were thus not liable for prosecution), and those that are more procedural (such as the landmark ruling in *Stoke-on-Trent City Council* v. *B&Q (Retail)* [1983] where, in part the plaintiffs argued that the justices had erred in the definition of the grounds sufficient to justify an injunction). Such rulings, as has been argued elsewhere (Blomley 1987b, forthcoming), have had important ramifications for the local state.

It is interesting to note the interpretive procedures that the courts adopt in their interrogation of the Act. There seems clear evidence of an attempt at conscious abstraction. Thus, Lawton LJ, in *Stoke-on-Trent* v. *B&Q (Retail)*, explicitly states that

> At the outset of this judgement I wish to make it clear what I regard as irrelevant considerations: first, that section 47 [the Sunday trading restrictions] of the Act is widely disregarded; secondly, that many people want it repeated; thirdly, that many people find it convenient to shop for non-exempt goods on Sunday; and fourthly, that with the resources of manpower and money which are available to local authorities many of them could not hope to stop unlawful Sunday trading except on a selective and spasmodic basis which would probably be regarded as unfair and oppressive. My judicial duty is to apply the law as laid down by Parliament, not to change it (84).

Given this formalized distancing of the court from the context of the case, a variety of judicial procedures have been adopted in an attempt to "uncover" the meaning of the Shops Act. This problem has been acute given the increasingly out-dated nature of the statute. Faced with such interpretive tasks, the courts have opted for a range of formalized procedures. An extensive body of leading cases has accumulated over the years, providing a touchstone for later interpreters. Similarly, syllogistic logic has been implicit in many rulings, as has the use of the "literal" or "Golden rule" (Blomley 1987b).

Despite such self-consciously formalistic procedures, we need to dispense with the notion of the judge as a higher interpretive being. On the contrary, the distinctive societal rootedness within time and space of the judge is evident. It is clear that specific justices, in interpreting the Act whether definitionally or procedurally, hold strong ideological convictions as to the law, the role of the state or the operation of the market. Thus, for example, repeated emphasis is placed on the "rule of law," especially given the blatant actions of many larger retailers. MANN J, in *Lewis and Another* v. *Rogers* [1984] quotes with approval the comments of LANE LJ, in the *Tisdall* case: "laws are there to be obeyed. The fact that an individual disapproves of a law does not mean that individual will escape prosecution. . . . It is a dangerous arrogance for anyone . . . to decide which laws they will obey and which they will not" (*The Times*, April 10, 1984). There is, therefore, clear evidence of both an abstractive and an immersive dimension to interpretation, with the court formally opting for rigorous, abstracted procedures, yet simultaneously evidencing socially derived ideologies of the law and the state.

The manner in which enforcement agencies, on the other hand, handle the Shops Act is also a fascinating one, in a number of distinctive ways. The relevant enforcement agency of the Shops Act is usually the environmental health department of a district council, coupled with a designated council committee. The Shops Act is usually only a minor part of this agency's total mandate, which centers on such concerns as pollution, occupational health, and working conditions.

Historically, there is interesting evidence of a tension between the two polar concerns of immersion and abstraction. This tension reflects the ambivalence towards the Act, and the role of the agency. Thus, when the Act was first proposed in the late 1940s, considerable debate ensued as to enforcement. Prior to this date, enforcement had been entrusted to a variety of agencies such as sanitary inspectors, weights and measurements inspectors, and the police. The Act obliges the local authority to appoint shops inspectors to apply the Act within their jurisdiction, although some degree of autonomy is accorded to the local authority as to the choice of action considered sufficient to secure compliance. This led to considerable debate within the officer community as to the virtues of local administrative autonomy as opposed to centralized (or regional) enforcement.

One school of thought argued for local level interpretation, given the advantages of place-specific knowledge in securing compliance. Thus, Wilson (1952) commented that local officers "have local knowledge and experience which the central departments cannot have. . . . Uniformity of practice, however desirable it may be, would be dearly bought if it meant a sapping of local initiative" (91). Others, on the other hand, argued for an abstractive interpretive strategy. In part, this reflected a concern for national geographic uniformity in enforcement. However, the political advantages of abstraction were argued for, given the limited moral mandate associated with the Act. Thus, Crookson (1951) argued for regional administration of the Act given that, "in the enforcement of a law which contains elements repugnant to public opinion the more impersonal and detached the Inspector is the better. The *anonymity* of the Inspector could not be obtained under a small authority" (37, my emphasis). The parallel with the law court is evident.

Thus, there appears clear evidence of the complex interplay of abstractive and immersive interpretive strategies with reference both to the higher law court and the local enforcement agency. This tension can also be observed when the interpretive strategies of one specific local agency are considered.

Examples

It is inevitable that, given the breadth of the comments outlined above, examples can only be illustrative of a few points. However, I wish to comment on two topics; firstly, the contextual nature of the interpretive practices of officers within a locality, and secondly, to outline some of the interpretive tensions between two distinct but interrelated interpretive communities: the local level and higher courts. The selected examples indicate that we can view the locally constituted state as an interpretive community such that interpretation and locality are mutually interdependent. However, interpretive tensions exist within the "community," in this case, between abstractive and contextual interpretive strategies.

Local level enforcement; the "law in action"

The process of local level interpretation of the Act by Bristol City Council, in the West of England, is a fascinating one. It clearly illustrates the tension between context and abstraction. On the one hand, interpretive strategies are specific to Bristol. Simultaneously, however, given a concern for the legitimization of the enforcement agency, the abstractive nature of law and legal rulings as "higher authorities" is used by Bristol's officers and councilors as a defensive resource in the face of criticism.

The procedure of enforcement in Bristol is a graduated task. Given an awareness of an offence (usually through a third party), officers will advise the offender of their transgression. This may be followed up by a formal notice. If the offense is repeated, officers may consider taking formal action. A summons will be drafted and then, after higher level approval is given, may be served. Repeat offenders may not receive the luxury of an advance notice, proceedings being instituted immediately.

This graduated process is, however, far from automatic. At every stage, complex interpretive decisions are made. Interpretations as to the importance of any offense, and the assumed intention of the offender are constantly being reached, on a case-by-case basis. Such decisions are as much a reflection of the coming together of officer ideologies and detailed local knowledge as they are of the Act as a formal text. Thus, for example, the various offenses detailed in the Act are informally graded by officers into "technical offences" and more serious transgressions. The former might include the failure to display requisite employee information (section 21), and are usually handled in a less formal manner through the serving of a notice on an employer. On the other hand, offenses relating to Sunday opening are evaluated as being more serious, reflecting, in part, external pressures from pro-regulation interests. Sunday trading offences may receive closer surveillance in the event of a subsequent non-compliance. It is noteworthy that all of the formal proceedings instituted by Bristol under the Shops Act since local government reorganization in 1974 were for Sunday trading offences.

Similarly, offenders are graded and evaluated in a deeply contextual manner. Officers are responsible for certain "patches" in the city, and have developed a detailed knowledge of the area and its habitual offenders. Individual cases are dealt with in a discretionary manner, with reference to the specific offender. Thus, one officer commented that

> one thing you find in this particular business; the results depend to a great extent on how you deal with the people. And you form an opinion when you meet someone as to the best way to deal with them. You don't treat everyone alike. You treat them the best way in order to get what you're after. (This and subsequent quotations relate to research interviews conducted in Bristol between 1983 and 1985, unless indicated otherwise.)

Some offenders, it was felt, would only respond to a formal approach. Others were more friendly or more respectful of the authority of the officer, and would respond to an advice giving strategy. Either way, responsiveness to context was seen as an important consideration; "we could walk into a premises; within five minutes, suss the thing out and decide that we'll have to deal with this particular man in this particular way."

To conclude, it is evident that the "ground-level" interpretation of the Act in Bristol is clearly shaped by the nature of that locale. It should also be noted that the interpretive procedures adopted in Bristol are clearly distinct from those of other local agencies. Nearby Bath City Council, for example, adopts a very lenient enforcement policy, not having proceeded against any offender since 1974 (Bristol has instituted proceedings against 41 stores between 1974 and 1985). Again, this can only be understood with reference to a set of place-specific political, social, and economic factors (Blomley 1986a,b). This tallies with Blomley (1987a), which reports on variations in enforcement in England and Wales for the period 1976–84, finding marked spatial variations in enforcement levels (many authorities taking no proceedings at all in the period), and different enforcement tactics (such as the use of injunctions against persistent offenders). It would seem that some local agencies adopt an active enforcement policy, while others seem reluctant to institute proceedings against offenders.

Video retailers and the Maidstone *case*

It is not the case, however, that the interpretive strategies of the local agency can simply be reduced to localized imperatives. The interplay of immersive and abstractive strategies, both within the state and between its various apparatuses, is evident with reference to the ramifications of the so-called *Maidstone* case in Bristol. In this instance, higher appellate judges ruled that the business of video tape rental from fixed retail premises came within the terms of reference of the Act. Thus, the rental of a tape on a Sunday constituted a formal offense under the Act. This ruling, reached via formalized interpretive procedures, had important implications for the local state, and its localized interpretation of the Act and its imperatives. In Bristol, a fascinating debate ensued as to the best interpretive strategy to adopt; a contextual or an abstractive one.

The appearance and continued growth of video rental outlets in Britain is a fairly recent phenomenon. Such outlets, often small and independently owned, or one component of a general convenience or petrol retail outlet, are generally open for longer hours to cater for the out-of-hours demand for their product. In April 1984, however, the status quo was upset by a High Court appeal ruling which held that the hire of a video tape from a shop constituted an offense (*Lewis and Another* v. *Rogers* [1984]). This followed an earlier case in a lower court. As argued elsewhere (Blomley 1987b), this ruling bore all the hallmarks of a formalized, abstractive interpretation. The case centered on whether the video hire business was a "retail trade or business," conducted in a "shop" within the meaning of the Act. The judges made explicit use of precedent, syllogistic logic and the "literal meaning" rule of statutory interpretation in reaching their interpretation.

Such a ruling effectively criminalized the operating policy of a large and growing sector of retail capital. The denial of out-of-hours trading is clearly of considerable financial import to the video sector, especially in the context of low margins and intense competition for customers. The chair of the Video Retailers Association claimed that the ruling would drive many of Britain's 10,000 video shops, who conducted up to 25 per cent of their business on a Sunday, out of business (*The Guardian*, April 3, 1984).

Bristol's response to the ruling was both complex and contradictory. Following the initial case, which prompted the *Maidstone* appeal, the decision was made to issue advisory letters to video traders, informing them of the ruling. This so alarmed the local trade that their organization, the Avon Video Retailers Association, petitioned the council to disregard the ruling and thus, in essence, to respond to local trade needs. However, the local state instigated action against those who continued to trade illegally, prosecuting 34 defendants (representing 31 outlets).

The manner in which Bristol accommodated the case law shift is informative. It is noteworthy that it "held fire" following the first ruling of the lower court until the High Court made the case a clear cut one on appeal. After that, the local state, conscious of its potentially vulnerable position, given the shift in the enforcement environment, proceeded in a somewhat abstractive fashion, arguing that the ruling was locally valid, and thus applied to all video stores regardless of local circumstance. The local state thus sought to ensure that its every step was "rule-bound," and thus formally defensible to external criticism. Hence, although reacting to local context, Bristol opted for a formalist, "letter-of-the-law" stance.

However, the video issue became more than a simple one of confrontation and response. The debates that were generated concerning enforcement policy and the video trade made reference to the interpretive strategies of the local state and to the interpretive chains of command within the state. Crucially, one local councilor took up the video case, and pressed for a re-evaluation of the manner in which the local state reacted to higher court interpretations and local demands. Ultimately, he was to prove unsuccessful, but his actions were to expose a number of important facets of enforcement.

The individual councilor, a Conservative, felt that "fundamentally this is wrong that the public have been able to purchase video tapes for seven years and suddenly they're told they can't do it." The concern of the councilor was to force a reinterpretation of law and case law that was, while within the broader constraints of law as a higher moral code, still responsive to local context. He argued against what he saw as the abstractive manner in which a "rule bound" local state served to alienate its constituency. He felt strongly, in conversation, that government should be much more responsive to change, and not so bound by its own internal logic. He went further, however, and criticized the "retreat to legalism" of officers and councilors; the tendency to adopt safe, "letter-of-the-law" positions if criticized. He questioned those who "climb and hide behind archaic laws or do not go as far as one might to obtain change. . . . I think the local authority should have a far wider viewpoint of things than just to hide within the law even if one knows the law is wrong." He criticized the "officer empires" in this context, with officers living "behind these laws and rules to maintain those empires." He also questioned the centrality of case law and its interpretation. In committee he criticized the very validity of the legal decision itself. He also questioned the manner in which legal interpretation soon became local creed, and the manner in which rulings were reached.

> Some lady complained about cars parking outside her shop on a Sunday. If that hadn't happened, the initial case would not have been attempted and therefore as there would have been no precedent, the video shops would still be open. And that's ridiculous. One should interpret it whether it's the *right* thing, or the *wrong* thing. It shouldn't be some random case now used as the basis for law.

In Committee, he repeatedly sought to find a loophole for the video trade. Clearly, such a flexible and pragmatic approach to the interpretation of rulings was antithetical to many orthodox procedures and ideologies within the state. If such a view held sway, the legal footing upon which enforcement was premised would be eroded, and the enforcement actions of the local state thrown into further doubt.

The issue, therefore, was more than a procedural one. The questions asked and challenges made, explicitly or implicitly, by the councilor, raised questions concerning the function of the local state, its interpretation of law, and the validity of "abstractive" rulings in a heterogeneous and changing locale.

The response of both officers and councilors to the claims and queries of the maverick in their midst was, unsurprisingly, a negative one. The committee and the department retreated to the case law, maintaining that it was valid, set a strong precedent, and as such, provided a basis for action. In effect, officers stonewalled the demands of the councilor and the video trade. To do otherwise would have led to disruptive redefinitions of policy, and affected the local state's image as an impartial agency. Councilors shied away as, in the words of a Conservative councilor, "they didn't want to act favourably to one section." To do so would have led to accusations of double standards. Far better, in those circumstances, to accept the legitimacy of the interpretation of another agency and write the situation off as anomalous or, perhaps, unfortunate. An officer commented that "we felt it was unfair to try and find a way out for the video traders. There's no reason why they should be any better off than any other Sunday trader. It's an unfair law, and that's that." Thus, faced with pressure towards a locally responsive strategy and a criticism of the interpretive chain of command within the state, the local state reacted by accepting the legitimacy of an abstractive ruling.

To summarize then, it has been argued that the interpretive strategies of the local state reflect a tension between an immersive and an abstractive strategy. When the actions of the local state are challenged, as in the *Maidstone* case, it appears that a formalized, abstractive strategy is adopted. In Bristol's case, this meant the acceptence of case law as valid, and thus locally relevant, with the consequent prosecution of offenders. It is interesting to note, however, that the impact of the ruling was somewhat shortlived. Following the first wave of proceedings, Bristol has not, to date, initiated further proceedings, despite evidence of continued offenses. This would suggest that the controversy of the ruling obliged action which, given its problematic consequences, was not repeated in later, less controversial times. The discretion accorded the local agency thus allows it (in some cases, at least) to distance itself from some higher court rulings. The top–down interpretive hierarchy is not necessarily so instrumental as may be implied. Local interpretive practices, in this regard, clearly have a role in shaping the process of social reproduction, and cannot simply be reduced to the technologies of the higher state.

The immersion/abstraction tension is an important one. The basis of the problem lies in the procedures by which the ruling is reached, and its assumed validity in a local context. In the *Maidstone* case, a formalized and abstractive ruling was applied to a locality that proved complex and specific. The coherence and validity of the higher court ruling was internally defined with reference to such things as previous court rulings, logical constituency, and textual "meaning." The coherence of local level interpretations, on the other hand, was only partially a function of law as an abstract code. Other, specific concerns were used to evaluate the validity of any interpretation. Locally defined concerns of legitimation and rationality were higher on the interpretive agenda. Given that court rulings were assumed to be valid for the local state, tensions arose when "meaning" was transferred between the two interpretive communities. Such tensions were evident within the local state, between the local state and its local constituency, and between local state and higher state agencies.

Conclusions

This chapter has sought to explore the question of legal interpretation, highlighting differences in interpretive practice both between and within state agencies. The divide between an abstractive and an immersive interpretive strategy has been considered. It would seem that the assumed determinate meaning of a statutory text can be altered or even lost when immersed within a locale.

A deep incoherence is, therefore, apparent within law. Law, by its very nature, is written as a text intended to cover all eventualities. In so doing, it inevitably abstracts from context. Clark (forthcoming) argues that the most powerful reason for judicial indeterminacy rests in the methodological separation between theory (principles) and practice (rules):

> Because principles are conceived as abstract analytical statements, empirical rules will always be distant from their original locations. Rules attempt to provide guidelines for action; principles eschew action for simple clarity (42).

While principles demand abstraction, given the legitimation imperative, that very abstraction makes principles difficult to relate to context. This dissonance is inevitable, given the nature of law.

This incoherency can be profoundly contradictory for state intervention, leading to an internal conflict between ideology and instrumentality. There is, therefore, a tension between the imperatives of law and those of the state. What, in other words, is law to be? Is it to be ideology or policy? Possibly the former requires an abstractive interpretation such that law and the legal system are imbued with a "higher level" status. The latter, on the other hand, demands application that is locally coherent. The state, seeking to intervene within the process of reproduction, may find itself in an impasse of its own construction.

Moreover, the immersive/abstractive divide can have implications for central/local state relations, given that the legal apparatus of the state operates at a number of levels. An interpretation, reached at a higher level of the state, may prove problematic when translated to the locale. This would suggest that the power of the higher level legal apparatus in structuring local practices is not unlimited. A higher court ruling has to be internally coherent for the local interpretive agency if it is to be readily adopted by the local state. Thus, although I would agree with Clark (1985) in recognizing the power of the courts in shaping the terms of reference of local level conflicts, it would seem that the potential contradictions that result from moving between the two interpretive levels can serve to constrain the actions of higher interpretive agencies. Given that any contradictions that may result from legal interpretation may ultimately affect the entire legal apparatus of the state, the central/local interpretive relation can therefore best be seen as one of biased reciprocity.

As well as tensions between interpretive communities, it has also been shown that tensions exist within agencies as to the relative merits of abstractive versus immersive strategies. Such tensions reflect the continual struggle for the legitimation of the agency coupled with the intrinsically place-responsive nature of interpretation. In the Bristol case, it was demonstrated that such internal interpretive conflicts could be of considerable importance. Clearly, the internal negotiation of interpretive strategies is a continual and potentially problematic process for the local state.

Another dimension, which has only been touched on in this chapter, is that of the interpretive differences that may exist between different localities. The spatial mosaic of enforcement strategies that can be observed between local government areas surely reflects the power of place in the shaping of interpretation. Such spatial differences may translate into tensions when, for example, a local agency is obliged to legitimize its specific interpretation of law in the light of comments concerning other, more lenient neighbouring agencies.

Finally, it should be clear that the implications of the argument raised in this chapter are of importance for those who seek to address the issue of legislative intervention in the reproduction process, as well as those concerned with wider issues of law, interpretation, and spatiality. It is clear that the inherently social nature of law and its interpretation is an urgent research task. An attempt has been made here to recognize the intrinsically contextual nature of that intervention, and its implications for the state. The concepts of functional and statutory determinacy have been shown to be elusive ones. Firstly, it has been shown that all those agencies of the state that come into direct contact with law do not act merely as a conduit for law, but may actively shape its immediate form. Moreover, this very multiplicity of interpretive functions may, given differences in interpretive technologies, lead to intra-state tensions. Secondly, the necessary socio-spatial rootedness of the interpreter ensures that interpretation is always contextual. Law is not a higher form of rationality, nor is it, by its very nature, intrinsically determinate. Law, in the last instance, is complex, subtle, potentially incoherent, and profoundly contextual.

Note

This paper benefited greatly from the informed comments of a number of people, including Gordon Clark, Terry Young, Ron Johnston and ex-California Supreme Court Justice Joseph Grodin, as well as Jennifer Wolch and Michael Dear. The field work in this study was made possible by a grant from the Economic and Social Research Council.

References

Blackwell, M., C. Watkins & K. Economides 1986. Human geography and law: a case of separate development in social science. *Progress in Human Geography* **10**, 371–96.

Blomley, N. K. 1985. The Shops Act (1950); the politics and the policing. *Area* **17**, 25–33.

Blomley, N. K. 1986a. Regulatory legislation and the legitimation crisis of the state: the enforcement of the Shops Act (1950). *Environment and Planning D, Society and Space* **4**, 183–200.

Blomley, N. K. 1986b. Retail law at the urban and national levels: geographical aspects of the operation and possible amendment of the Shops Act (1950), unpubl. Ph.D. dissertation, University of Bristol.

Blomley, N. K. 1987a. Retail regulation in England and Wales: the results of a survey. *Environment and planning a* **19**, 1399–1406.

Blomley, N. K. 1987b. Legal interpretation: the geography of law, *Tijdschrift voor economische en sociale geografie* **78**, 265–75.

Blomley, N. K. forthcoming. Law and the local state: enforcement in action. *Transactions of the Institute of British Geographers*.

Cardozo, B. N. 1924. *The growth of the law*. New Haven: Yale University Press.

Clark, G. L. 1981. Law, the state and the spatial integration of the United States. *Environment and planning a* **13**, 1197–232.

Clark, G. L. 1982. Rights, property and community. *Economic Geography* **58**, 120–37.

Clark, G. L. 1985. *Judges and the cities; interpreting local autonomy*. Chicago: University of Chicago Press.

Clark, G. L. 1986. Making moral landscapes: John Rawls' original position. *Political Geography Quarterly* **5**, 147–62.

Clark, G. L. forthcoming. Geography and law. In *The new models of geography*, R. Peet & N. Thrift (eds.). London and Boston: Unwin Hyman.

Crookson, G. M. 1951. Shops Act administration – a personal survey. *The Inspector* **6**, 32–7.

Devlin, P. 1979. *The judge*. Oxford: Oxford University Press.

Dicey, A. V. 1965. *Introduction to the study of the constitution*. London: Macmillan.

Dworkin, R. 1986. *Law's empire*. Cambridge, Mass.: Belknap Press.

Fish, S. 1980. *Is there a text in this class? The authority of interpretive communities*. Cambridge, Mass.: Harvard University Press.

Fish, S. 1982. Interpretation and the pluralist vision. *Texas Law Review* **60**, 495–566.

Geertz, C. 1983. *Local knowledge: further essays in interpretive communities*. New York: Basic Books.

Griffith, J. A. G. 1985. *The politics of the judiciary*. London: Fontana.

Grindrod, R. B. 1843. The wrongs of our youth; an essay on the evils of the late-hour system. Reprinted in *Demands for early closing*, K. E. Carpenter (ed.), 1972, British labour struggles: contemporary pamphlets. New York: Arno Press.

Hawkins, K. & J. M. Thomas 1984. The enforcement process in regulatory bureaucracies. In *Enforcing regulation*, K. Hawkins & J. M. Thomas (eds.), 3–22. Boston: Kluwer-Nijhoff.

Hawkins, K. 1984. *Environment and enforcement: regulation and the social definition of pollution*. Oxford: Clarendon Press.

Holmes, O. W. Jr. 1881. *The common law*. London: Macmillan.

Hopkins, A. & N. Parnell 1984. Why coal mine safety regulations are not enforced. *International Journal of the Sociology of Law* **12**, 144–79.

Johnston, R. J. 1983. Political geography of current events II; a reapportionment revolution that failed. *Political Geography Quarterly* **2**, 309–17.

Johnston, R. J. 1984. Residential segregation, the state and constitutional conflict in American urban areas. IBG Special Publication, no. 17. London: Academic Press.

Kagan, R. A. & J. F. Scholz 1984. The "criminology of the corporation" and regulatory enforcement strategies. In *Enforcing regulation*, K. Hawkins & J. M. Thomas (eds.). Boston: Kluwer-Nijhoff.

Langan, P. St. J. 1976. *Maxwell on the interpretation of statutes*, 12th edn. Bombay, Triparthi Private Ltd.

Levinson, S. 1982. Law as literature. *Texas Law Review* **60**, 373–403.

Manning, P. K. 1977. *Police work: the social organization of policing*. Cambridge, Mass.: MIT Press.

Reiss, A. J. Jr. 1984. Selecting strategies of control over organisational life. In *Enforcing regulation*, K. Hawkins & J. M. Thomas (eds.). Boston: Kluwer-Nijhoff.

Reynolds, W. L. 1980. *Judicial process in a nutshell*. St. Paul's, Minn.: West Publishing.

Richardson, G., A. Ogus & P. Burrows 1982. *Policing pollution: a study of regulation and enforcement*. Oxford: Clarendon Press.

Selznick, P. 1966. *TVA and the grass roots: a study in the sociology of formal organizations*. New York: Harper & Row.

Shover, N., J. Lynxwiler, S. Groce & D. Clelland 1984. Regional variations in regulatory law enforcement: the Surface Mining Control and Reclamation Act of 1977. In *Enforcing regulation*, K. Hawkins & J. M. Thomas (eds.), 121–46. Boston: Kluwer-Nijhoff.

Thompson, E. P. 1975. *Whigs and hunters; the origin of the Black Act*. London: Allen Lane.

Thoreau, H. D. 1980. *Walden* and *On the duty of civil disobedience*. New York: New American Library.

Thrift, N. 1983. On the development of social action in space and time. *Environment and Planning D, Society and Space* **1**, 23–58.

Thrift, N. 1985. Flies and germs: a geography of knowledge. In *Social relations and spatial structures*, D. Gregory & J. Urry (eds.), 366–403. Basingstoke and London: Macmillan.

Tushnet, M. N. 1983. Following the rules laid down: a critique of interpretivism and neutral principles. *Harvard Law Review* **96**, 781–827.

Twining, W. & D. Miers 1982. *How to do things with rules; a primer of interpretation*. London: Weidenfeld & Nicolson.

Unger, R. M. 1983. The critical legal studies movement. *Harvard Law Review* **96**, 563–675.

Whitaker, W. B. 1933. *Sunday in Tudor and Stuart times*, London: Houghton.

Whitaker, W. B. 1940. *The eighteenth century English Sunday: a study of Sunday observance from 1677 to 1833*. London: The Epworth Press.

Whitaker, W. B. 1973. *Victorian and Edwardian shopworkers: the struggle to obtain better conditions and a half holiday*. Newton Abbott: David & Charles.

Wigley, J. 1980. *The rise and fall of the Victorian Sunday*. Manchester: Manchester University Press.

Willis, J. 1938. Statute interpretation in a nutshell. *Canadian Bar Review* **16**, 1–27.

Wilson, J. P. 1952. Address to conference of the Institute of Shops, Health and Safety Acts administration. *The Inspector* **7**, 91.

Winstanley, M. J. 1983. *The shopkeepers world 1830–1914*. Manchester: Manchester University Press.

9

The shadow state: transformations in the voluntary sector

JENNIFER R. WOLCH

Introduction

The restructuring of metropolitan America has been characterized as a process of spatial and social change resulting from the reorganization of production and an evolving division of labor (Soja *et al.* 1983). But a second aspect of restructuring is the redefinition of state responsibilities for population welfare (Gilbert 1983, Mishra 1984, Dear & Wolch 1987). This redefinition, effectuated via public sector retrenchment and privatization of public services, has dramatically altered the organization of the post-1945 welfare state (Palmer & Sawhill 1984, Bawden 1984).

The extent of welfare state shrinkage should not be underestimated. For example, although total US federal spending has continued to grow primarily because of rising defense expenditures, domestic spending has dwindled in real terms (Bawden & Palmer 1984, 185–6). Central government programs targeted for reduction or demise include many of the cash and in-kind anti-poverty, housing, and community development programs initiated or vastly expanded in the 1960s. Various inter-governmental grants programs have been trimmed by more than a third since the late 1970s (Peterson 1984, 228). All in all, the Congressional Budget Office estimated that between 1982 and 1985, over $57 billion were cut from federal programs aimed at the poor (Wolch & Akita 1988). Reductions have been implemented through several means, including direct benefit and service reductions, increased stringency of eligibility requirements, and contracting-out services to nongovernmental organizations.

How has the slack created by this public sector retrenchment been taken up? Four ways have been widely acknowledged. First, many cuts have simply been absorbed, resulting in declining real income and wellbeing for some segments of the population, particularly the disadvantaged (Danzinger & Feaster 1985, Wolch & Akita 1988). Second, service recipients themselves have borne the burden of cuts, because service providers have raised fees and charges to offset collective revenue shortfalls (so-called "throw-back" policies). Third, some functions have been "privatized," i.e., assumed by private for-profit and voluntary (not-for-profit) sectors. Privatization has been prompted by the discovery of market potential in some services, the availability of government contracts, and the development of public/private partnership arrangements. Contracting and partnerships are not new phenomena, but have been steadily increasing since the 1960s (Kramer 1986). Fourth, charities and voluntary organizations have expanded their fund-raising efforts, and reoriented their spending in order to help fill service delivery gaps (Salamon *et al.* 1986).

These dynamics are in part determined outside of local institutions (e.g., tax laws, bureaucratic regulations, etc.). But metropolitan regions are the sites where the impacts of economic and welfare state restructuring are expressed and felt by groups and individuals. This localization of effects is exacerbated when a primary vehicle for restructuring is decentralization of responsibility for social provisions and "knotty problems" from higher to lower tiers of government. Decentralization is designed to increase central government legitimacy and popular support (by "giving government back to the people"), and to solve problems that central government institutions have been unable to resolve (Cockburn 1977). Reagan's "new federalism" is an excellent example of how the politically treacherous course of welfare cutbacks is negotiated by passing problems to lower-order branches of the state apparatus. As state and local governments try to ameliorate or pass these dilemmas on to other institutions and agents, the metropolitan region becomes the locale in which struggles over rights, obligations, entitlements, and state penetration in daily life are played out.

My concern in this chapter is with the shift of public service responsibilities to the voluntary sector, and the resulting formation of a "shadow state" apparatus. In section 1, the conservative, liberal/left and pragmatic rationales for this transfer of responsibility are discussed, to demonstrate the variety of arguments which have given impetus to increased voluntarism. The broad appeal of a voluntary sector strategy foreshadows the emergence of the shadow state, in the form of voluntary organizations with collective responsibilities, which operate outside traditional democratic controls, yet are strongly affected by state resources and constraints.

Then, in section 2, the voluntary sector in a specific metropolitan locale, Los Angeles, is evaluated. This analysis indicates how a national policy of shifting collective service responsibility from the state to voluntary organizations can be expressed and intensified at the level of urban locale, as it intersects with local-level dynamics.[1] The study details: (1) the extent of local voluntary sector activity, and hence penetration into daily concerns; (2) patterns of community access to voluntary services and opportunities to participate in voluntary groups; (3) degree of voluntary sector reliance on government funding, state regulatory provisions, and their own entrepreneurial activities, which either constrain political action or influence service delivery patterns; and (4) level of client and membership responsibility for service cost burdens. These findings imply broad transformations in public and voluntary sector service delivery patterns, and the emergence of a shadow state apparatus. Section 3 details these transformations, which will affect what, where, and to whom services are provided, who pays the service provision tab, and how popular control will (or will not) be exerted in the realm of service provision. In conclusion, section 4 argues that while the shadow state has progressive potential, it could also lead to weakened citizen entitlements to welfare program benefits, and an increasing statization of everyday life.

Making sense of voluntarism

Voluntarism is the current fashion among welfare state analysts and politicians. But careful reading of the sector's promoters reveals that the reasons for the new emphasis on voluntarism are not everywhere the same. A basic contrast is between an advocacy of voluntarism based on: (1) ideological conviction (right and left); and (2) pragmatic political and budgetary considerations.

The right

Ideological rationales for the shift toward voluntarism come from the political right, as articulated by politicians such as Reagan and Thatcher, and by neoconservative theoreticians, including public choice analysts. The right claims that voluntary groups are the cornerstones of freedom and democracy, to be encouraged in preference to expansion of an unresponsive state increasingly beyond popular control. Nisbet (1962), for example, early on argued that voluntary organizations and government are inherently in conflict, and blamed the development and expansion of a monolithic state for mounting social problems and alienation. American Enterprise Institute authors Berger & Neuhaus (1977) have expanded on this theme, suggesting that with religious institutions, neighborhoods, and the family, voluntary organizations are "mediating structures" necessary to forestall state monopoly over service provision and the destruction of private initiative and responsibility. They call for public policy to protect and foster mediating structures by not damaging them, and empowering them through provision of resources.

Public choice variants state the benefits of voluntarism in economistic terms; Weisbrod (1977) argues that voluntary service provision serves to increase individual consumer utility and choice, since those left unsatisfied by government can club together to provide desired services. Other authors stress the importance of voluntary organizations in promoting self-sufficiency and individual initiative, and thus the capitalist system. This is in stark contrast to the government sector which is most often seen as squelching the individual. As Pines (1982) argues, voluntary action is essential to the maintenance of traditional American values. The types of voluntarism he champions are conservative, directed at promulgating the virtues of free enterprise; against abortion; stricter criminal sentencing; and deregulation. Thus voluntarism is necessary to counter views which are hostile to capitalism, economic growth, hard work, and discipline.

199

Key politicians have adopted this approach as well. In Ronald Reagan's view:

> We've let Government take away many things we once considered were really ours to do voluntarily out of the goodness of our hearts and a sense of community pride and neighborliness (quoted in *New York Times*, September 27, 1981).

Such a view harkens back to a "mythical Golden Age of voluntary-sector purity" uncontaminated by government funding (Salamon 1987, 108). It has become official doctrine; the President's Task Force on Private Initiatives, for example, extolled the virtues of voluntary action, as well as highlighting its practical advantages.

In the context of trenchant and persistent attacks on the welfare state, Margaret Thatcher articulates a less idealized but essentially similar perspective:

> I believe that the volunteer movement is at the heart of all our social welfare provision. That the statutory services are the supportive ones, underpinning where necessary, filling the gaps and helping the helpers (quoted in Ungerson 1985, 214).

The British Conservative Party Manifesto in 1979 also stressed voluntarism:

> In the community, we must do more to help people to help themselves and the family to look after their own. We must also encourage the voluntary movement and self-help groups acting in partnership with statutory services (quoted in Sugden 1984, 70).

As Webb & Wistow comment about the Thatcher government, "an endorsement of the voluntary sector . . . was centre stage. No minister could refer to the personal social services (provided by the voluntary and/or informal sectors) without 'beatifying' these sources of nonstatutory provision" (1987, 92).

The left

The case for voluntarism is also heard from the political left. Liberal and left-wing politicians and social analysts have advocated grassroots participation, decentralization of power, alternative economic development strategies, and greater self-determination, especially for the disadvantaged. The concrete legal/institutional form for these efforts is most often the voluntary group. Boyte (1980), for instance, championed the "backyard revolution" by which local community groups were able to gain greater power over the fate of their neighborhood environments and institutions. Gladstone (1979) and others have argued that voluntary organizations can liberate people from the domination of professionalized service providers. Ken Livingstone, former leader of the now-defunct Greater London Council, targeted the voluntary sector as a means to increase democracy, promote involvement of ethnic minorities, women, gays and lesbians, and develop "noncapitalist" forms of production. Most recently, Dahrendorf has called for "a hundred if not a thousand local initiatives" by voluntary groups in order to transform capitalist society in the wake of increasingly bankrupt forms of economic and social organization (1987, 15).

Pragmatic rationales for voluntarism

More pragmatic assessments of the voluntary sector voiced by analysts of various political persuasions suggest that the development of voluntarism offers numerous practical advantages. The voluntary sector has in effect become a *shadow state*: that is, a para-state apparatus with collective service responsibilities previously shouldered by the public sector, administered outside traditional democratic politics, but yet controlled in both formal and informal ways by the state. The shadow state may be seen as a corporatist strategy, designed to create "partnerships" with components of civil society, and necessary to the continuing legitimacy of the state. Devolution to voluntary groups is also attractive as a solution to increasingly complex welfare state crises. Governments of many ideological stripes are beating a retreat from an increasingly unpopular platform of centralized state planning and management. They seek alternatives to a state monopoly over human services provision which turned out to be expensive, alienating to clients, and riddled with persistent delivery problems.

The pragmatic arguments for the shadow state that neoconservative (or neocorporatist) writers stress are essentially four-fold. First, voluntary service delivery is efficient. It is claimed to be cheaper than government, to stimulate competition, and to promote economies of scale (Savas 1982, Spann 1977). Second, voluntary provision allows the state to fund and oversee programs, but permits voluntary groups the discretion and flexibility necessary to be innovative and deal with local conditions. Third, expansion of voluntarism increases the general stock of welfare services (and hence presumably citizen wellbeing) without expanding the size of the state apparatus, and in some instances, allowing a reduction in state welfare provision. Fourth, funding a variety of voluntary groups is politically expedient, to insure the support of affected groups for new service programs and incorporate new constituencies into the political system.

Liberal and left arguments for the pragmatic merits of voluntarism, as proffered by Salamon (1987), Kramer (1987), Gladstone (1979) and Donnison (1984), include the efficiency argument, but also stress the potential of voluntarism for extending the spectrum of welfare services, the flexibility and innovative character of voluntary groups, and the political attractiveness of voluntarism in rallying support. Salamon and Kramer, for example, both note the long historical tradition of a "mixed economy of welfare," and stress that the pragmatic advantages of voluntary group service provision have led to effective "public-private partnerships" – a tradition which should be expanded upon.

Gladstone (1979) summarizes the rationale for adopting what he terms "gradual welfare pluralism," which would include a larger role for the voluntary sector in service provision: (1) adaptability of voluntary groups; (2) their cost-effectiveness; (3) the opportunity for enhanced participation; and (4) greater coordination of welfare services. But in addition, progressives have argued that voluntary groups can play an important role in fostering nontraditional (and noncapitalist) economic development. As Donnison (1984) suggests, voluntary groups involved in a wide variety of areas can assist local small business, promote worker and consumer cooperatives, and increase community development potential of an area or group through the activities of alternative arts groups, youth clubs, women's groups, and centers for the unemployed.

Summary

Neoconservative and liberal/left ideologies combined with pragmatic arguments for voluntarism have formed justifications for corporatism in general, and for various forms of privatization in particular. The Reagan and Thatcher governments, for example, have already used privatization within corporatist arrangements to accomplish the reorganization and retrenchment of the welfare state. Their general strategy has been to let the voluntary sector meet service shortfalls and solve emergent social problems; to privatize the delivery of state responsibilities for welfare support via contracts, fees, and vouchers for efficient voluntary organizations in order to cut costs and expand consumer choice; and to implement these strategies via a multitude of public-private partnerships. Liberal/left strategies for privatization, implemented largely at the local level, have entailed expanding (instead of cutting) human services through wide voluntary sector participation and power in welfare service production and policy-making. Policy and service provision directions have been established by formal and informal "joint planning" by voluntary sector and state.

Thus, despite ideological differences and varying emphases on its practical advantages, there is substantial support for the transfer of collective service responsibilities to the voluntary sector. As this consensus has been translated into public policy, the voluntary sector has expanded rapidly, signaling the emergence of a shadow state apparatus. Ultimately, the shadow state will privatize and reorganize the welfare state.

Voluntarism in the locale

The metropolitan locale is especially appropriate for detailed study of the shadow state, since it is within urban communities that the impacts of shadow state formation are experienced on an everyday basis. In this section a cross-sectional profile of the Los Angeles voluntary sector is developed which: (1) identifies the magnitude of the sector's political and economic resources, as measured by numbers and types of organizations, their financial power, and extent of sectoral job-control; (2) permits an assessment of voluntary sector resource distribution; (3) examines the spatial distribution of those resources in the region; and (4) indicates sources of financial support for voluntary group revenues and constraints associated with reliance on these various funding sources (particularly government funds, business revenues, membership dues, and client fees-for-service).[2]

Political and economic resources

Political and economic resources of the voluntary sector in Los Angeles derive from citizen participation and membership, charitable contributions, government funds, business activities, and job creation. More indirectly, groups can use their ties to public agencies and decision-makers to gain organizational objectives (Wolch 1987). There is little or no information on either voluntary group membership, the quality of leadership/membership participation, amount of volunteered labor, or intangible political resources such as institutional ties. But, other vital dimensions of the sector's resource base indicate the potential for voluntary sector impact on everyday life and for political influence. The most important of these indicators are the number of associations and their types, their organizational wealth, and their control over local jobs and income.

NUMBERS AND TYPES
The mere *existence* of a viable voluntary organization constitutes a potential voice in the political process, regardless of group size or additional capacities, although clearly each group voice will not have equal impact on decision-makers. In Los Angeles County, there are over 8,500 voluntary entities, ranging in size from small neighborhood associations to large trusts and foundations. The vast majority of organizations is small, as defined by total revenue received: almost half have annual revenues under $10,000, and about three-quarters receive less than $100,000 per year. The large organizations have supplementary resources which often allow them a specific type of voice in the political arena; for instance, public administrators often play dual roles as foundation directors, and group leadership may have social links to political elites (Harris & Klepper 1976, Commission on Foundation and Private Philanthropy 1970). Small groups, none the less, may use alternative avenues and generate significant change: for example, landmark legal decisions concerning urban land use have been instigated by well-organized neighborhood associations.

What types of groups comprise this sector? One distinction can be made on the basis of use of funds for grant-making, or for direct service provision. Donor groups, such as foundations, raise funds from various sources (family wealth, corporate gifts, and bequests), and make grants to other voluntary organizations. The latter are *service-providing* agencies which utilize funds from donor groups and other sources to produce services for the final consumer (e.g., health care, social services, or recreation services). No one type of group can be characterized as more or less politically active; donor groups such as the United Way or the ARCO Foundations are demonstrably involved in local public affairs in Los Angeles, but so are organizations which would be categorized as service-providing groups, such as the Boy Scouts, the University of Southern California, and the Salvation Army.

The donor/recipient distinction can create problems for analysis, for two reasons. First, although the definition of group types is clear enough in theory, in practice many groups perform both roles – that is, they give grants to other organizations, and provide services as well. Second, because funds from one organization's revenue pool are donated and become part of another organization's revenue, totals for the sector may reflect a partial double-counting. In the present case double-counting is not an issue; of interest is the aggregate financial resources of all agencies to effectuate political change and meet organizational objectives, rather than a total which reflects the sector's contribution to the economy alone. Thus, a donor may deploy a grant strategically, and the service-providing organization receiving that grant may in turn use it to secure political or institutional goals.

Since questions of service cost incidence are of major concern, segregating recipients from donors is useful. Organizations which use 10 per cent or less of their revenue pool to make grants or allocations to other voluntary agencies have been grouped together, and are labeled *service-providing recipients*. This is a restrictive criterion, since many groups giving more than 10 per cent still provide significant amounts of client services. Nevertheless, this partitioning allows for a clear picture of differences between the circumstances of donors and recipients who deliver services directly to users. The service-providing segment still comprises 3821 or approximately 45 per cent of the entire sample.

Another distinction can be made on the basis of functional purpose, defined here as involvement in one of seven categories: religion, education and research, international activities, health, general (multi-purpose) charity, the arts, and social welfare and community services. Among those reporting to the State of California for purposes of obtaining tax exemptions, voluntary groups engaged in social welfare and community service activities were by far the most numerous, followed by education and research groups, general charitable agencies, health agencies, arts and culture associations, religious-affiliated groups, and agencies involved in international relief and exchange efforts (Table 9.1).

Some of the most common types of organizations (200–700 groups per category) are general social welfare agencies, service clubs, music and performing arts groups, and groups promoting a specific purpose by educational means. Slightly less numerous are groups with a general educational purpose, veterans' organizations, civic leagues, organized youth activities, social service agencies, and ethnic/racial associations.

204

Table 9.1 Functional purposes of Los Angeles voluntary organizations.

I	Broad categories	TOTAL	8,560	(100%)
	Social welfare and community services		3,483	(40.69%)
	Education and research		1,947	(22.75%)
	General purpose charitable		1,087	(12.70%)
	Arts and culture		806	(9.42%)
	Health		791	(9.24%)
	Religion		303	(3.54%)
	International services		143	(1.67%)

II   Common organizational types (number of groups per type)

General social welfare agencies	670
Service clubs	739
Music and performing arts groups	438
Groups promoting a specific purpose by educational means	406
General educational groups	299
Veterans' organizations	282
Civic leagues	258
Organized youth activities	251
Scholarships	249
Social service agencies	231
Ethnic/racial associations	212
Cultural interest associations	198
Civic improvement	177
Alcoholics assistance	145
Handicapped assistance groups	133
Hospital auxiliary organizations	123
Child welfare agencies	122
Housing assistance	121
Aid to needy persons	120

Source: Geiger & Wolch (1986).

Other purpose categories are more than 100 groups strong, involved with alcoholics, children, the handicapped, the needy, housing, civic improvement, and specific cultural interests.

The functional purpose of a group is only a very broad indicator of the degree to which organizations will use their resources for specific political purposes, at any specific spatial scale or tier of government. However, some organizations are more likely to be active on the local political scene than others. Other types of groups, including international relief agencies, may have greater involvement in state and national policy-making spheres, or divide their efforts across several levels of government. The functional purpose profile also illustrates the mix of voluntary services in the region, provided in response to a complex of factors including donor preferences, client service demands and needs, business opportunities, and cash and in-kind resources available from local communities, state government, and the federal government. This mix is likely to have undergone change recently, government resource availability, targeting and incentives, corporate philanthropic activities, and client needs having shifted.

ORGANIZATIONAL WEALTH

Financial resources and capital assets permit agencies to become formally established, obtain knowledgeable and skilled staff, acquire and generate information about issues of concern, mobilize public and official opinion to support organizational ends, and pursue legal, legislative, and administrative routes to achieve their objectives. Los Angeles' voluntary sector controls $7.74 billion in assets, including $1.19 billion in land and property, and receives $3 billion in revenues. These organizations contribute $2.32 billion to the regional economy, in the course of their grant-making, administration, fund-raising, and service production activities. Over $566 million goes to employees in the form of salaries and wages. Viewed on a per capita basis, the sector raised $464 and held $1196 worth of assets per resident of incorporated jurisdictions in Los Angeles County.

The distribution of revenues and assets by organizational purpose indicates that not only are social welfare and community service organizations the most numerous, but they also collected the largest share of voluntary sector revenues, followed by general purpose charitable groups, educational and research organizations, and health agencies. The pattern of asset holdings is quite different, however. The arts and general charitable sectors each controls slightly less than one-third of voluntary assets, followed by social welfare and community service groups with less than one-fifth of the total. The remaining purpose groups trail far behind, each enjoying less than 10 per cent of the voluntary asset base (Table 9.2).

Consideration of resource rankings of purpose categories, based on numbers of groups, revenue, and assets of purpose categories leads to some interesting disparities. For instance, the Los Angeles arts subsector is first in terms of assets, but trailing in terms of revenue and numbers of

Table 9.2 Revenues and assets, by organizational purpose category.

Functional purpose	Revenues		Assets	
	(in millions; percent total)			
Social welfare and community services	836	(28%)	1,420	(18%)
Education and research	416	(14%)	573	(7%)
General purpose charitable	680	(23%)	2,198	(28%)
Arts and culture	322	(11%)	3,328	(30%)
Health	386	(13%)	672	(9%)
Religion	109	(4%)	249	(3%)
International activities	254	(8%)	296	(4%)

Source: Geiger & Wolch (1986). Dollar amounts rounded to nearest million.

groups. Social welfare and community service agencies, first in terms of aggregate revenue and numbers, are relatively asset-poor. And education and research, the second largest group in terms of numbers, comes in third and fifth, respectively, in terms of revenue and assets. Thus the mix of these political resources is such that no one functional subsector captures top rank.

Several distinct investment profiles emerge from the data. Overall, the Los Angeles voluntary sector invests almost half of its capital assets in securities; about 15 per cent in property (land, buildings, and equipment); and 14 per cent in savings and cash investments. This aggregate profile varies markedly across functional purpose groups, however. For example, although investments in securities form the largest single share of assets for all except social welfare and community services, that share ranges from between 27 per cent (education and research) to 70 per cent (the arts). Savings and cash reserves were more significant for international agencies (47 per cent of total assets) which require liquidity given the nature of their relief undertakings. Property was the favored form of investment for social welfare and community service groups (45 per cent), and also important for education and research organizations. This is not surprising, for two reasons. First, many of these organizations require a place-specific facility, and are likely to own their own building – a sound management strategy for organizations with highly uncertain revenue generation capabilities. Second, a portion of organizations in these subsectors tend to be asset-starved, and operate on the basis of current revenues; the average asset pool per education and research unit was only $294,000; per social welfare and community service group, $408,000. These figures stand in comparison with average asset pools of between nearly $1 to over $2 million for groups in other functional categories.

Expenditure patterns – what voluntary groups actually use their revenues and liquified capital assets for – indicate that organizations spend the largest share of resources (27.9%) on salaries and wages and related employee costs (benefits, payroll taxes), followed by miscellaneous expenditure (25.3%) and grants and allocations (23.9%).[3] This pattern varies by sector; the dominance of donor-type groups in some categories is apparent, most notably in the arts and general purpose charity groupings.

Is this organizational wealth of the Los Angeles voluntary sector significant in comparison with the magnitude of local public service provision? Voluntary sector expenditures per 1980 incorporated county resident were $358; the County of Los Angeles spent (in 1985) $579 per capita, while in 1984 aggregate municipal spending averaged $829 in expenditures per resident of incorporated jurisdictions in the county (Los Angeles County 1985, State of California 1985). Most federal and state funds are channeled through the county, and so the county figure captures a substantial portion of total public sector welfare spending. Average per capita voluntary sector spending in incorporated cities was almost 60 per cent of the average per capita county expenditure, and 43 per cent of the average per capita municipal expenditure on urban public services. In addition, a fifth of the cities in the county used volunteer labor in the provision of urban services, typically in the areas of public safety, community development, and recreation (State of California 1985).

These comparisons overstate the importance of the voluntary sector's *independent* service provision to the extent that county and city governments are funding voluntary groups. But the sector's role would be far greater were consideration restricted to public expenditures for similar functions (health and human services, arts, culture, education, and recreation), but excluding spending on municipal and county enterprises (harbors, airports, golf courses), and public utilities which are usually self-financing or profitable. None the less, the comparison indicates the significant size of the "shadow" cast by the voluntary sector, and suggests that like the state, voluntary organizations are affecting daily life routinely and in a myriad ways.

JOBS AND INCOME

The expenditures of voluntary organizations create jobs and income which constitute political resources. Such indirect resources arise because the state requires a smooth-functioning economy in order to maintain a satisfied electorate and its own legitimacy. Hence it must rely on voluntary organizations (as well as proprietary firms and its own agencies) to provide employment and to build the social infrastructure necessary to the ongoing production of goods and services (Lindblom 1977). Significantly, this type of resource does not need to be mobilized in response to specific political issues; rather it is a contextual factor that conditions the decision-making environment of the state, and stands as a barrier to excessive state control of group activities (Wolch 1987).

Job- and income-generation characteristics of the Los Angeles voluntary sector have been estimated for 1977, the latest year for which data are available. The analysis indicates that the region's voluntary associations employed 141,999 workers, or about 5 per cent of the area labor force, who earned $1.67 billion or 4 per cent of total regional earnings. Via multiplier effects, voluntary sector jobs and income were linked to an additional 151,948 jobs in other industrial sectors and $1.33 billion in indirect earnings. The most significant inter-industry linkages occurred with trade and service sectors. Voluntary organizations contributed approximately 10 per cent to the total job base in personal services, agriculture, utilities, and transportation, local government transit, and postal service sectors. It also provided between 5 and 7 per cent of the job base in such other major sectors as business services, real estate and insurance, and retail and wholesale trade. Together, direct and indirect employment linked to the voluntary sector comprised almost 300,000 jobs (9 per cent of the regional labor force), and $3 billion in earnings (7 per cent of the regional total).

In the Los Angeles region the voluntary sector's size (defined in terms of employment) relative to other major sectors is not insignificant. It is almost one-fifth the size of manufacturing; three-fourths the size of wholesale trade; one-fourth the size of retail trade; one-third the size of for-profit services; and almost one-half the size of the local government sector. These magnitudes reinforce the argument that the state is apt to be sensitive to the political priorities of the voluntary sector, just as it considers the impact of its policies on other major industrial sectors. This sensitivity may be actually pronounced since the voluntary sector is one of the fastest growing segments of the economy, expanding by about 30 per cent between 1977 and 1982 (Weitzman & Hodgkinson 1986; Independent Sector Inc. 1984).

Concentration of wealth

A central condition for widespread participation in public life is the existence of many groups each of which have some useful political resources, and can provide meaningful opportunities for citizen involvement in the democratic process. But analysis of the distribution of organizational wealth among Los Angeles voluntary groups varying by size (asset-class) and by functional purpose, indicates intense concentration of resources in large organizations. Seven per cent of all voluntary organizations in the region control 77.24 per cent of total voluntary sector revenues. This elite group is comprised of the 591 organizations with assets in excess of $1 million. In contrast, the group of smallest organizations (assets under $10,000) is almost eight times the size, but controls less than 3 per cent of all revenues. Across functional purpose groups, some variation emerges; the arts group, for example, is by far the most concentrated, with 3 per cent or 24 organizations (with $1 million or more in assets) controlling 84 per cent of revenues. The least concentrated is the general purpose charitable category, most of which are donor-type organizations. Thus there may be many opportunities for community participation, but resources for participation may be highly unequal.

Other research suggests that heightened competition among voluntary groups to secure gifts, grants, and public contracts, may work to the detriment of smaller less financially secure organizations. Large and well-established associations may be better able to secure government grants, enter into public-private partnerships, advertize services, revamp management to maximize efficiency in the face of public funding austerity, or conduct business activities in support of their mission (Kramer 1986). Governmental reliance on voluntary groups may thus inadvertently work in their favor, and simultaneously disadvantage small and/or newer groups, leading to a self-reinforcing cycle that intensifies resource concentration.

Spatial distribution of organizations and resources

Uneven availability of voluntary services has been noted in the past (Gilbert & Specht 1974, Kahn 1973). It is not unexpected given the nature of the sector and residential differentiation of metropolitan regions. However, the unevenness, inconsistency, and inequity in the distribution of voluntary resources may become increasingly detrimental as public services shrink (Kramer 1986). In Los Angeles, the geographical distribution of voluntary resources indicates differential development of this sector across metropolitan space, implying uneven access to services, and also to opportunities for participation and involvement in voluntary group activities. In the case of social welfare and community service agencies across cities, the pattern of voluntary organizations was shown to favor middle-class, inner-ring suburbs (Wolch and Geiger 1983). The most impoverished, heavily service-dependent inner-city zones enjoyed relatively few voluntary sector resources. The same was true of industrial enclaves, which have small populations, and affluent suburbs on the outskirts of the region, where residents purchase services in the marketplace.

More recent city-specific data indicate a spatial pattern that is not dissimilar. Considering numbers of organizations, revenues and assets per capita (as a proportion of regional totals), the city of Los Angeles clearly dominates the region. It has 4,349 organizations (over half the regional total), receiving two-thirds of total revenues, and controlling three-quarters of the voluntary sector asset pool. This is not particularly enlightening, since the city houses almost half the county population and is highly heterogeneous in terms of socio-economic and demographic characteristics. It is also the headquarters location for many region-serving institutions, particularly foundations with high asset bases, and the mailing address of groups providing services in other cities.

More interestingly, middle-to-high income inner-ring jurisdictions are voluntary service-rich, while the lowest income jurisdictions (along with industrial enclaves) are voluntary service-poor. Some of the contrasts are striking. For instance, 336 voluntary organizations in affluent Beverly Hills collected $2,787 in revenues and $7,953 in assets per capita (estimated 1984 population 33,421), while the six groups in Maywood (estimated 1984 population 23,337), one of the region's most impoverished cities, collected less than $5 in revenues and controlled less than $4 per capita in assets. It is not known how much of the revenue collected by voluntary organizations is spent on local residents in either location.

Analysis of the distribution of voluntary resources by functional purpose reveals that cities in the region have widely variable levels of voluntary services in each purpose category. Again, some variation is expected since (among other things) certain types of service facilities have larger catchment areas than others (e.g., arts centers), demands may vary, and access to resources is not uniform. But with the exception of social welfare and community services, no distinct spatial pattern emerges for individual purpose categories. For the social welfare and community service group (the largest segment of the sample) middle-to-high income inner-ring jurisdictions are voluntary service-rich, whereas the lowest income jurisdictions (along with industrial enclaves) tend to be service-poor.[4]

Constraints on voluntary action and service delivery

The extent to which voluntary organizations can utilize their various resources to achieve political objectives or fulfill service missions cannot be directly estimated. Constraints on political action and service provision are often subtle and intangible. This is particularly true in the case of legal sanctions on specific types of voluntary group political activity (i.e., electioneering and lobbying) and Internal Revenue Service compliance procedures (Wolch 1987). However, the evidence for Los Angeles provides some clues concerning the constraints on voluntary action and factors affecting service delivery patterns. This analysis focuses on the extent of voluntary sector dependence on government funding, regulatory provisions, business-linked income, and client fees.

Reliance on public grants and contracts, and regulatory provisions, limit the extent to which voluntary groups can advocate social change, since excessive activism can lead to curtailment of public support or loss of tax-example status. Moreover, "vendorism" brings with it bureaucratic oversight, monitoring, and evaluation, as well as fiscal and management complications which are especially troublesome for small voluntary organizations (Kramer 1986). Clearly, some degree of oversight is justified to protect the public and maintain accountability, but the question of degree is open to continual redefinition. Entrepreneurial activities and reliance on client fees indicate the response of some voluntary groups to fiscal retrenchment, as well as ongoing management trends in voluntary organizations which place greater emphasis on administrative efficiency and fund-raising strategies (Chisman 1986, Salamon *et al.* 1986). The development of these revenue sources has helped to replace lost public funding and make voluntary organizations more competitive. However, with the rising numbers of for-profit service firms, the results may include increasing commercialization of service outputs, and potentially regressive changes in the cost burden of service delivery.

GROUP RELIANCE ON PUBLIC FUNDS

In the aggregate, voluntary organizations in Los Angeles receive $361 million per year in revenue from government (at all levels), or 12.03% of total revenue. This degree of reliance on public support is below the latest figures for the nation; in 1980, federal, state and local governments provided more than one-quarter of the voluntary sector's revenue, a proportion that had been steadily increasing since the early 1970s (Independent Sector Inc. 1984, 34). And in a major Urban Institute study of 16 local voluntary sectors across the country, government funding was found to account for almost 40 per cent of total revenue, the largest single source of funds (Salamon *et al.* 1986).

Among Los Angeles organizations of varying functional purpose, the majority receive *some* public monies, but the extent of utilization ranges from almost negligible in the international services and arts categories, to approximately 16 per cent of total revenues amongst education and research units. The average amount received from government sources also varies by purpose group: from about $5,000 per organization in the international area, to $78,359 amongst health agencies. Many of the most common types of groups received little public support, however; for example organized youth activities, music groups, veterans' associations, civic leagues, and general social welfare associations received only around 5 per cent of their revenues from public sources.

A different picture emerges when organizations are grouped according to size (asset-size class), and when only *collective* revenues (gifts, grants, and contributions), and service-providing groups are considered. Large organizations ($1 million or more in assets) receive *much less* public funding in proportion to their revenue base than do smaller groups. The exceptions are in the health and general purpose charity areas, where the reverse holds (although general purpose charities do not receive a great deal of public support in any event). Small groups (under $10,000 in assets) involved in the arts gain more than 10 per cent of their revenues from government; large groups only 1.3%. In education, the comparable figures are 25.1% and 12.9%; and for social welfare and community services, 28.9% and 15.2%. This pattern implies that smaller groups (which tend to be newer, more flexible, and socially innovative) are the most vulnerable to state-imposed constraints effectuated via funding relationships, reinforcing other findings which suggest that the survival of these groups has been threatened by retrenchment (Kramer 1986).

The significance of government funding for specific sectors is highlighted when the government share of collective revenues (gifts, grants, and contributions) is considered. This type of revenue represents all funds that can be raised beyond self-generated income (fees, dues, sales, investments, or other business activities) from individual, corporate and foundation donations, and from government grants. Again, the government share of collective revenue remains negligible for international and arts subsectors, and is under 5 per cent for religious groups. But public monies account for a major share of collective revenues in other sectors: 13.7% for general purpose charities, 16.5% for health agencies, 18.4% for education units, and 45.9% for social welfare and community service groups.

The reliance of strictly service-providing groups (as opposed to donors) on government funding is of particular concern from the perspective of welfare state restructuring. For these are the groups that have shouldered most of the shift of responsibilities from the public sector. The primary question is: are these organizations more dependent upon public sector sources than the voluntary sector in general, thus making them (and their clients) vulnerable to further social spending reductions? The answer appears to be yes. From the subsample of groups engaged primarily in service provision (45% of total organizations and 66% of total revenues), government sources account for 17.4% of total revenues – 5 per cent more than the aggregate sample. The representation of service-providing groups varies considerably by purpose category; international agencies and general purpose charities are primarily donors (making 50–70 per cent of total expenditures on grants). Thus, not surprisingly, reliance on public funds varies dramatically by purpose group. Arts and international organizations still utilized a negligible amount of public funding, but for some areas most central to the "social safety net" public monies are much more vital. Service-providing social welfare and community service groups, for instance, received more than a fifth of their revenue from government sources.

Nevertheless, the reliance of Los Angeles voluntary organizations on government funding is still comparatively low. There are several plausible explanations for this. Some are related to the sample, which is more extensive than those used in other studies. Other reasons are linked to the nature of the region itself. For example, like other newer cities, Los Angeles grew up in an era of changing values and rapid mobility. It may not have developed the same public/voluntary partnership arrangements which characterized older cities with philanthropic and charitable sectors established prior to the 20th century expansion of public service responsibilities. More recently, following the precedent set by the city of Lakewood in the 1950s, most of the newer cities in the region contract with the County of Los Angeles for many collective services. In fact, all the county's cities contract for the provision of at least one major urban service (law enforcement, sanitation, health) with the county (Los Angeles County 1985). Lastly, governmental fiscal restraints and the presence of a strong for-profit service sector in the region may be leading to greater competition for public contracts between for-profit and nonprofit organizations.

REGULATORY PROVISIONS

Government regulatory provisions such as tax exemptions, reduced mailing rates, and tax deductibility of individual and corporate charitable contributions, are tax expenditures made by the public sector in support of voluntary activities. These tax expenditures amount to indirect income that is essential for many voluntary organizations, rendering them "fundable." This implicit income is significant: for example, in a recent study of tax policy and the national voluntary arts sector, Feld *et al.* (1983) found that the amount of direct public sector funding was less than half the magnitude of tax expenditures or revenues foregone due to favorable tax code provisions. Such indirect support may make voluntary groups cautious and conservative, to avoid risking the loss of tax-exempt status.

No attempt has been made to estimate the total monetary value of regulatory provisions for voluntary organizations in Los Angeles. Such an analysis is extremely complex and has never been accomplished for an individual metropolitan region. However, three types of indirect public income can be estimated: local property tax exemptions; federal tax expenditures for charitable contribution deductibility allowances (for individuals, corporations, bequests, foundations, and trusts); and non-profit postal subsidy.[5] State-level subsidies are omitted from this analysis. Assuming that property owned by the sample of voluntary organizations is located within the region, revenues foregone by local governments in Los Angeles due to their grant of property tax relief amount to more than

$13 million; this is an underestimate since universities, hospitals, and churches are not included. This results in an average local government subsidy to sample organizations of more than $1500 per year, with a range of $611 for education and research units, to $2671 for religious-affiliated organizations.

Federal tax expenditures for the charitable contribution deduction exceeded $9 billion for the nation in 1982 (US Office of Management and Budget 1984). Multiplying the US average per capita share by the 1982 population of Los Angeles County results in about $315 million for the region (US Department of Commerce 1984). A proportional allocation of the nation's total mailing subsidy to voluntary organizations implies a $27 million share for Los Angeles. These three estimates total $355 million of indirect income from government, suggesting the value of regulatory provisions for Los Angeles voluntary organizations. Benefits are apt to be disproportionately enjoyed by large organizations and communities in the metropolitan region which have many voluntary resources.

USE OF PROGRAM FEES AND BUSINESS-RELATED REVENUES

Given a relatively low dependence on government contributions, voluntary organizations in Los Angeles are much more reliant on program service fees paid by users, and on business-linked activities such as investment income, interest and dividends, sales of memberships and goods, and rental income. Program service fees represent a full fifth of aggregate voluntary sector revenues, but were most important in religion (26.9%); education and research (25.8%); health (25.7%); and social welfare and community services (32.4%). These rates, in conjunction with the public funding pattern, imply an extensive "throw-back" of service cost burden on client groups. To the extent that previously subsidized middle-income service users now pay for services, new fee structures may be welcomed. It remains unclear, however, whether fee levels have restricted access to services for the poor and disadvantaged.

Business-linked activities are also crucial, accounting for more than one-third of total revenue and ranging from a low of 11.8% for international agencies, to almost 60 per cent in the arts category. Even traditional institutions in the social welfare area received more than a fifth of their revenues from business-type activities. This emphasis on entrepreneurial activity is mirrored in a startling rise in for-profit "spin-offs" used to "cross-subsidize" voluntary service efforts faced with declining public support and increasing competition from for-profit service delivery firms (Chisman 1986, Kramer 1986, Salamon et al. 1986).

Again, differences in dependence on business-related revenues are evident across size classes. Large organizations tend to be able to raise this sort of funding much more readily than small and/or new groups (Heuchan 1986). For example, small religious-affiliated agencies raise 5.6% of their income in this way, while large organizations raise more than one-third of their total. Also striking are size-related differences in the social welfare and community service field: small organizations raise 11.3% of their income from business activities, while large agencies bring in almost one-quarter of their total revenues from business receipts.

Program service fees are even more important for the solely service-providing organizations, while business-linked earnings play a smaller role. Overall, almost 30 per cent of this group's revenues come from fees, compared to about 20 per cent for the entire sample. Similarly, business-linked revenues constitute about 10 per cent less for service-providers than for the aggregate sector. Thus expressly service-providing voluntary organizations not only are more dependent on the state, but their clients foot more of the service-delivery bill.

Summary

In summary, the Los Angeles analysis suggests that the structure of voluntarism – and hence the nature of service provision – has been transformed as a result of the changing macro and local contexts in which organizations are situated. In Los Angeles, the sector commands important and influential political and economic resources. Moreover, the distribution of resources between and within sectors is highly uneven. An extremely small number of groups wield the vast majority of vital voluntary sector resources, while thousands of groups subsist on minimal footing. Inequality is also manifest geographically, since some local jurisdictions have far greater voluntary resources upon which they are able to draw for services augmentation, public sector substitutions, and political action. Often these voluntary resource-rich jurisdictions are not those in which social problems and needs are most pressing; instead, more affluent communities with more total public and private resources may boast the richest panoply of voluntary groups. The analysis of group resources in Los Angeles also hints at a changing orientation of the voluntary sector. Voluntary groups in the metropolitan area, particularly large ones, are becoming more entrepreneurial, raising more resources through business activities and client fees. Government funds and program service fees remain vital for smaller groups and service-providing recipient organizations, however.

The rise of the shadow state

The Los Angeles analysis suggests a variety of speculative, subtle conclusions about how the role of voluntarism is transforming service delivery and the nature of the local welfare state. First, conventional assumptions about the benefits of voluntarism remain highly questionable. Given the extensive tax expenditures on voluntarism, the *net* efficiency gains from voluntary sector versus public production need to be assessed before any firm conclusions regarding relative efficiency can be made. Moreover, the concentration of resources in large organizations, and the geographical disparities in voluntary resources suggest that the expansion of consumer choice may be limited by the oligopolistic behavior of large organizations, and by uneven access of metropolitan residents to service organizations. Resource concentration and uneven spatial development also imply that opportunities to participate in a democratic political process are constrained because so many groups have marginal political resources to command, and because many communities simply do not have viable voluntary organizations.

Second, the patterns and incidence of collective services (in Los Angeles and perhaps elsewhere) may change as a consequence of altered patterns of local government spending (Lowe 1986, Geiger 1986). County and municipal priorities may therefore be reordered due to the changing political resources of voluntary associations, as public spending cuts and reorganization of the sector proceed. Some voluntary groups will gain, others will lose, and the mix of resources within each organization may shift. Together, this shifting distribution and pattern of voluntary resources will affect the abilities of groups to pressure local officials and to generate public support for alternative fiscal and service provision arrangements. The net result will be changes in the pattern of government as well as voluntary sector outputs.

Third, patterns of service inequities across metropolitan regions may be changed, as local state resource commitments shift and the voluntary sector expands unevenly. These changes are apt to be regressive, since smaller organizations and groups in areas without ready access to affluent donors will probably suffer disproportionately from any further public spending reductions. They are likely to be less able to maintain their service levels by increasing user fees and raising levels of business activity than larger, entrenched and traditional institutions which have greater reserves and ability to absorb changing financial circumstances. Ironically, the activities of such groups are often targeted to disadvantaged populations, providing legal aid, advocacy, job training, housing assistance, and social services (Salamon *et al.* 1986). They tend to be most responsive to emergent social problems and predominate in areas of highest social need.

Fourth, the incidence of service cost burden may shift, and the nature of the voluntary services provided by the sector may be altered. The large shares of revenue derived from user fees, particularly among service-providing groups, indicates ongoing displacement of responsibility for supporting services on to clients themselves. The extensive reliance on business-related revenues suggests that voluntary groups are increasingly involved in business ventures and investments in order to keep solvent the service delivery side of their operation. In turn, this may bring some types of groups into competition with for-profit firms, prompting changes in voluntary group service organization, pricing, and program mix. These changes could result in "goal-deflection," or changes in orientation, service mission, and clientele (Kramer 1986).

Fifth, and perhaps most fundamentally, the relative expansion of the voluntary sector will affect future patterns of institutional interaction between the sector and government. Some argue that the heightened reliance on the voluntary sector could represent the re-emergence of an older, pre-welfare state pattern of state/voluntary sector powers and roles, but the new state and voluntary sector interdependencies are not likely to mirror earlier forms. Instead, the rise of the voluntary sector signals the emergence of a shadow state apparatus: a sector with an increasing share of formerly public welfare responsibilities, and increased political resources with which to affect public policy. It may be (a) less accountable to the public, (b) outside key formal democratic controls, and (c) remain circumscribed by government. Its influence is less direct, more difficult to detect, and harder to change since channels for public response to service gaps or inadequacies are more diffuse and fragmented.

Conclusion

Society may face profound dangers with the emergence of the shadow state. A voluntary sector strategy may be accompanied by purposive erosion in basic welfare entitlements. As long as voluntary organizations (and other institutions such as the family) can be seen to meet the shortfalls in services and benefits created by entitlement reductions, these reductions may appear justifiable and politically attractive to many. But state funding which keeps a shadow state operating at acceptable levels can be reduced or discontinued with much less difficulty and negative publicity than cuts in state agencies and directly provided programs. Such flexibility is, after all, one of its main attractions to government contractors – along with lower wages and fewer benefits which make voluntary sector workers more attractive than protected and better paid public employees. The only defense against encroachments on entitlements and expanded use of poorly paid voluntary agency workers is to use the sector's political resources to maintain the flow of state funds and (over time) to enhance voluntary sector remuneration and working conditions. But as costs creep up, so too does state resistance.

The statization of life inherent in the shadow state is also worrisome. The term "statization" has been used to refer to the increasing penetration of state control in matters of everyday life. The long-standing dilemma implicit in this trend is, of course, that of defining the appropriate limits of state intervention. It is somewhat paradoxical that, in a time of apparent retreat from state intervention, the shadow state analysis should point to an extended and increasingly diversified pattern of state intervention via voluntary groups. Viewed pessimistically, the shadow state may be interpreted as a deliberate strategy of social control on the part of the state. That is, far from orchestrating a neoconservative roll-back of the welfare state, current policies are in fact designed to intensify the grip of the state. Hence, even though the formal edifice of the state apparatus is being held in check or even scaled down, state control of social life is permeating ever more deeply into the practices of voluntary associations, local communities, the home, and personal life. We have not had sufficient experience of the shadow state to form definitive judgments on the potential threats posed by statization. However, those threats seem to be at least as substantial as the commonly-voiced rhetoric on the "economic efficiency" and "political realism" associated with voluntarism. The shadow state may yet have the potential to cast a heavy blanket of social control over our lives, and as such, its extension is something to be carefully monitored. Statization via voluntarism is to be strongly resisted if the only beneficiary is the power of the state apparatus.

Despite these dangers, voluntary groups have been at the cutting edge of progressive social change and pressure for local autonomy. Their efforts have altered popular consciousness and expectations of the state. For example, through the impetus of Great Society funding and popular organizing efforts, new voluntary groups dedicated to improving urban conditions met with some success (Castells 1983). Voluntary organizations in London have been vital to filling welfare service funding shortfalls, and have often served to mobilize resistance to regressive central government policies, including abolition of the Greater London Council (Wolch 1987). Voluntary mental health groups in the US have organized to battle for better community-based mental health care, and against state intitiatives promoting resinstitutionalization of the mentally disabled. In California, they have succeeded in translating some of their concerns into legislation and funding programs (Dear & Wolch 1987). So while the behavioral latitude of many voluntary groups is constrained by structural political economic forces, the determination and persistence of individuals pursuing voluntary initiatives can also alter those constraints.

Thus the rise of the voluntary sector could also signal a very different configuration for the political economic system more in line with the classical pluralist vision. Instead of transformation into a shadow state apparatus which performs certain functions of a shrinking welfare state, the increased size, variety, and resources of the voluntary sector could hold the potential for expanding popular influence on the state and for allowing individuals greater self-determination. In short, expanded voluntarism may allow for progressive social change in the metropolis.

Notes

1 Los Angeles is a particularly appropriate locale for study, since the region provides a preeminent example of urban restructuring. Moreover, as the largest metropolitan center in the first tax-revolt state, public sector retrenchment and privatization have proceeded rapidly in many instances. But is Los Angeles a typical case or a harbinger? There are insufficient urban-level data on the voluntary sector to make definitive claims, with recent studies tending to substantiate contrasts in the pattern of reliance on voluntarily provided services (Salamon *et al.* 1986). But existing evidence, however, suggests that Los Angeles may be typical of newer Sunbelt cities with respect to the development of their voluntary sectors (Wolch 1987). In these cities, responsibilities for collective service provision are being transferred from public agencies to the voluntary sector in a diffuse fragmented way, oftentimes with enhanced development of commercialized services. In older Northeastern cities, a firmly entrenched and well-endowed voluntary sector appears to be influencing the reorganization of the local welfare system (Wolpert & Reiner 1985). The basic

patterns of voluntary sector/state relationships are thus apt to be differentially expressed in various metropolitan locales.

2 The analysis is based on a sample of 8,560 voluntary organizations located in 80 of the incorporated cities of Los Angeles county. This information is based on the filings required by the Internal Revenue Services for exempt organizations on Form 990. Additional material was collected from Form CT-2 which is required by the State of California and is maintained by the Registry of Charitable Trusts. The data obtained from these sources is from 1983. In addition, a variety of other data sources is utilized in analyses of economic resources and spatial representation of the voluntary sector, as reported in Wolch & Geiger (1983, 1986) and Geiger & Wolch (1986).

3 The miscellaneous category is large because many organizations simply fail to itemize expenses as required, and lump many smaller expense categories in the residual box.

4 This is the case even if we discount for the presence of region-serving organizations in voluntary service-rich cities, and the fact that some of these cities have a concentration of legal and financial services which serve as mailing addresses for donor organizations.

5 Methods for estimating tax expenditures are as follows. Property tax subsidy was determined by multiplication of total value of property (land, building and equipment) listed on the Registry of Charitable Trusts file, by the average property tax rate paid per $100 of assessed value in the County of Los Angeles in fiscal year 1983 to 1984 (1.123%; Deukmejian 1985). The federal nonprofit mailing subsidy for Los Angeles is simply the region's per capita proportional share of the 1984 national total ($801 million; Alliance of Nonprofit Mailers 1984, US Department of Commerce 1982).

References

Alliance of Nonprofit Mailers 1984. *Alliance Report, September 28, 1984.* Washington DC: Alliance of Nonprofit Mailers.

Bawden, D. L. (ed.) 1984. *The social contract revisited.* Washington DC: Urban Institute Press.

Bawden, D. L. & J. Palmer 1984. Social policy: challenging the welfare state. In *The Reagan record*, J. Palmer & I. Sawhill (eds.), 177–216. Washington DC: Urban Institute Press.

Berger, P. L. & R. J. Neuhaus 1977. *To empower people: the role of mediating structures in public policy.* Washington DC: American Enterprise Institute.

Boyte, H. 1980. *The backyard revolution: understanding the new citizen movement.* Philadelphia: Temple University Press.

Butler, S. M. 1982. *Philanthropy in America: the need for action.* Washington DC: Heritage Foundation.

Castells, M. 1983. *The city and the grassroots.* Berkeley: University of California Press.

Chisman, F. 1986. Privatization as a social issue. In *Philanthropy, voluntary action, and the public good*, Independent Sector Inc. & United Way Institute. Washington DC: Independent Sector Inc.

Cockburn, C. 1977. *The local state.* London: Pluto Press.

Commission on Foundations & Private Philanthropy 1970. *Foundations, giving and public policy.* Chicago: University of Chicago Press.

Dahrendorf, R. 1987. Why we can't afford an underclass. *New Statesman* **113**, 12–15.

Danziger, S. & D. Feaster 1985. Income transfers and poverty in the 1980s. In *American domestic priorities*, J. Quigley & D. Rubinfeld (eds.), 89–117. Berkeley: University of California Press.

Dear, M. J. & J. R. Wolch 1987. *Landscapes of despair: from deinstitutionalization to homelessness.* Oxford: Polity Press.

Deukmejian, G. 1985. *Economic report of the governor.* Sacramento, CA: State of California.

Donnison, D. 1984. The progressive potential of privatization. In *Privatization and the welfare state*, J. LeGrand & R. Robinson (eds.). London: Allen & Unwin.

Feld, A. L., M. O'Hare & J. M. D. Schuster 1983. *Patrons despite themselves: taxpayers and arts policy.* New York: New York University Press.

Geiger, R. K. 1986. *Institutional interdependence and voluntary organizations.* PhD dissertation. University of Southern California, Los Angeles, CA: School of Urban and Regional Planning.

Geiger, R. K. & J. Wolch 1986. A shadow state? Voluntarism in metropolitan Los Angeles. *Society & Space* **43**, 351–66.

Gilbert, N. 1983. *Capitalism and the welfare state.* New Haven: Yale University Press.

Gilbert, N. & H. Specht 1974. *Dimensions of social welfare policy.* Englewood Cliffs, NJ: Prentice-Hall.

Gladstone, F. 1979. *Voluntary action in a changing world.* London: Bedford Square Press.

Harris, J. & A. Klepper 1976. *Corporate philanthropic public service activities.* New York: Conference Board.

Heuchan, L. M. 1986. A survey of nonprofit charitable organizations. In *Philanthropy, voluntary action, and the public good*, Independent Sector Inc. & United Way Institute. Washington DC: Independent Sector Inc.

Independent Sector Inc. 1984. *Dimensions of the independent sector.* Washington DC: Independent Sector Inc.

Kahn, A. 1973. *Social policy and social services.* New York: Random House.

Kramer, R. 1986. The future of voluntary organizations in social welfare. In *Philanthropy, voluntary action, and the public good*, Independent Sector Inc. & United Way Institute. Washington DC: Independent Sector Inc.

Kramer, R. 1987. Voluntary agencies and personal social services. In *The nonprofit sector: a research handbook*, W. W. Powell (ed.), 240–57. New Haven: Yale University Press.

Lindblom, C. E. 1977. *Politics and markets.* New York: Basic Books.

Los Angeles County 1985. *Contract Services Record.* Los Angeles County, Los Angeles Ca.: Chief Administrator's Office.

Lowe, B. 1986. *Metropolitan area supply of community-based elderly services.* PhD dissertation. University of Southern California, Los Angeles, CA: School of Urban & Regional Planning.

Mishra, R. 1984. *The welfare state in crisis.* Brighton: Harvester Press.

Nisbet, R. A. 1962. *Community and power.* New York: Galaxy.

Palmer, J. & I. Sawhill (eds.) 1984. *The Reagan record.* Washington DC: Urban Institute Press.

Peterson, G. 1984. Federalism and the states: an experiment in decentralization. In *The Reagan record*, J. Palmer & I. Sawhill (eds.). Washington DC: Urban Institute Press.

Pines, B. Y. 1982. *Back to basics: the traditionalist movement that is sweeping grassroots America.* New York: Morrow.

Salamon, L. 1987. Partners in public service: the scope and theory of government-nonprofit relations. In *The nonprofit sector: a research handbook*, W. W. Powell (ed.), 99–117. New Haven: Yale University Press.

Salamon, L., J. C. Musselwhite Jr. & C. J. de Vita 1986. Partners in public service: government and the nonprofit sector in the welfare state. In *Philanthropy, voluntary action, and the public good*, Independent Sector Inc. & United Way Institute. Washington DC: Independent Sector Inc.

Savas, E. S. 1982. *Privatizing the public sector*. New York: Chatham House Press.

Soja, E., R. Morales & G. Wolf 1983. Urban restructuring: an analysis of social and spatial change in Los Angeles. *Economic Geography* **59**, 195–227.

Spann, R.M. 1977. Public vs. private provision of government services. In *Budgets and bureaucrats: the sources of government growth*, T. Borcherding (ed.), 71–89. Durham NC: Duke University Press.

State of California 1985. *Annual report of financial transactions concerning cities of California: fiscal year 1983–4*. Sacramento, CA: Office of State Controller.

Sugden, R. 1984. Voluntary organizations and the welfare state. In *Privatization and the welfare state*, J. LeGrand & R. Robinson (eds.), 70–94. London: Allen & Unwin.

Ungerson, C. 1985. *Women and social policy: a reader*. Harmondsworth: Penguin.

US Department of Commerce 1982. *Projections of the population of the United States 1982–2050*. Current Population Reports. Series P-25. Washington DC: US Bureau of the Census.

US Department of Commerce 1984. *Local population estimates*. Current Population Reports. Series P-26, vol. 4. Washington DC: US Bureau of the Census.

US Office of Management & Budget 1984. *Special analyses, budget of the United State Government, FY 1984*. Washington DC: Executive Office of the President.

Webb, A. & G. Wistow 1987. *Social work, social care, and social services since Seebohm*. London: Longman.

Weisbrod, B. 1977. *The voluntary nonprofit sector: an economic analysis*. Lexington, Mass.: D. C. Heath.

Weitzman, M. & V. Hodgkinson 1986. Measuring the size, scope and dimensions of the independent sector: a progress report. In *Philanthropy, voluntary action, and the public good*, Independent Sector Inc. & United Way Institute. Washington DC: Independent Sector Inc.

Wolch, J. R. 1987. *The shadow state*. University of Southern California, Los Angeles, CA: Planning Institute Monograph.

Wolch, J. R. & A. Akita 1988. The federal response to homelessness and its implications for American cities. *Urban Geography* (forthcoming).

Wolch, J. R. & R. K. Geiger 1983. The urban distribution of voluntary resources: an exploratory analysis. *Environment & Planning A* **15**, 1067–82.

Wolch, J. R. & R. K. Geiger 1986. Urban restructuring and the not-for-profit sector. *Economic Geography* **62**, 3–18.

Wolpert, J. & T. Reiner 1985. The not-for-profit sector in stable and growing regions. *Urban Affairs Quarterly* **20**, 487–510.

10

Social reproduction and housing alternatives: cooperative housing in post-war Canada

VERA CHOUINARD

The rise of "new right" politics and state programs, in an era of international restructuring in advanced capitalism, has helped to highlight the importance of understanding how social relations and practices are reproduced through struggles over service provision. But despite a growing international literature on how industrial restructuring is altering peoples' lives in particular places (*see*, e.g., Massey 1984, Cooke 1983, 1985, Scott & Storper 1986, Bluestone & Harrison 1982), relatively little is known about changing experiences of the state and other sites of social reproduction beyond the point of production. How have cutbacks, restraint and procedural changes in various social programs influenced organizing around the state in different places? To what extent have such transformations been contested independently of workplace issues (e.g., technological change and job loss)? Has this limited class–based political struggles and if so, why? Answers to questions like these will contribute not only to our understanding of contemporary capitalism, but also to practical strategies for tackling such pressing issues as homelessness and "shelter poverty" (Stone 1983).

This chapter examines how post-war struggles for cooperative housing alternatives have helped to reproduce specific social relations and practices within Canadian capitalism. A Marxist theory of social reproduction through struggles over state formation (section 1) is used to help to explain the development of cooperative housing programs in post-war Canada (section 2). I argue that these programs have had contradictory implications for the reproduction of social relations and practices: creating opportunities for relatively decommodified housing alternatives, but also contributing to marginalization of collective relations of housing provision within the state (section 3).

Social reproduction and housing policies

Recent debates have stressed the need for further analyses of how housing provision influences political practices. Marxist and Weberian analysts have challenged abstract, logical assumptions that housing tenures determine mass attitudes and practices (Harvey 1978, Castells 1979, Saunders 1984). They advocate further empirical study of the impacts of housing use and tenure on political practices (like voting), in order to establish the precise role of housing in social change. Research results to date lend support to this view, by documenting variations in the impacts of housing tenure on social attitudes and practices (Gray 1976, Rose 1979, Agnew 1981, Pratt 1986a,b, Saunders 1984, 1986).

It is also clear that housing is a very significant site of struggle over social reproduction in capitalist societies. One of the stakes in current conflicts over housing and social service policies is whether or not some members of the "lumpenproletariat" will be reproduced at all, as the homeless populations in advanced nations remind us. For people able to consume private or state-assisted shelter, housing is not only regarded as a basic necessity, but also requires considerable investment of finances and time. The provision and use of housing is therefore central to the reproduction of labor.

Housing may also have symbolic meaning for individuals and groups: advertisements for new Canadian housing developments often rely on selling particular lifestyles, for example. To the extent that people identify with these messages, it is likely that housing experiences will influence political attitudes and practices. And the rules governing access to and use of housing are of critical importance in shaping the politics of place. As Chouinard's (1988) analysis of cooperative housing in Toronto indicates, struggles for and against non-profit assisted housing involve conflict over the extent to which private market rights should determine access to shelter.

How then can we conceptualize the process of reproduction involved in struggles over housing provision? Social reproduction in capitalism involves the re-creation and/or transformation of specific forms of capital, labor, inter-class and intra-class relations (Dickinson & Russell 1986). The material sites of this process include relations between capital and labor in the workplace, intra-class relations of cooperation and authority in the workplace and residence community, relations in and against the state at and beyond the point of production, and relations between family, economy, and state.

Given this broad definition, struggles over housing provision contribute to social reproduction in the following ways. They influence the extent to which people are integrated into dominant class relations of housing production and exchange, by determining whether access to housing will be solely on the basis of individual market rights to commodity consumption (cf. non-profit/public housing alternatives). Struggles over housing provision also help to determine possible relations of housing exchange and use in places. The availability of public housing, for example, influences the extent to which commodity relations between owners, renters, and landlords will prevail in a neighborhood, as opposed to more "mixed" relations including those between housing clients and the state. And these struggles also influence possible relations in and against the state with respect to housing provision. The mix of housing alternatives in a place influences whether people will relate to the state through individualized legal relations like property-owner/renter or legal ward of the state, versus collectivities like non-profit housing boards.

As I have argued elsewhere, struggles over housing provision and policies are thus part of a contradictory process of class and state formation (Chouinard 1986, 1987, 1988). The capitalist state is defined as the dominant political apparatus involved in regulating antagonistic class relations and practices. Capitalist and working-class groups contest the administrative and legal characteristics of the state, as part of the struggle to ensure conditions favoring their own reproduction. Economic assistance programs that place priority on job creation are likely to favor reproduction of the working class for instance.

These struggles contribute to changes in the institutional arrangements, programs, legislation, obligations, and rights which constitute the state as an apparatus of collective political intervention in production, exchange, and reproduction. Struggles over state formation thus influence class capacities to contest dominant social relations and practices. The implementation of back-to-work legislation limits working-class capacities to use strike action to contest conditions in the workplace. State defense of the existing distribution of the means of production, through refusals to broaden legal definitions of property rights for instance, can limit the capacity of class and sub-class groups to reduce their dependence on wage-labor or welfare programs. Long-standing Native Canadian land claims are a dramatic example of such limits (Matthews & Morrow 1982, Little Bear *et al.* 1984).

Thus the characteristics of state programs and the manner of program implementation can limit or enhance class capacities to struggle for conditions favoring reproduction. Canada's housing programs have favored capitalist production of housing, through characteristics like generous tax incentives, direct funding of commodified housing, and aggressive advertizing of private ownership and rental programs. Program implementation influences peoples' experiences of political subjection to the state, and thus their capacities for class and sub-class struggles. If procedures treat different class and sub-class groups as abstract, equal subjects of the state, through universal, market-based rules of competition for funding for instance, then members of subordinate classes are likely to be disadvantaged by lack of control of the means of production (e.g., money, machinery, expertise). Canadian non-profit and cooperative housing producers, for example, have found it difficult to satisfy requirements for minimization of project costs (since they lack capitalist advantages like access to land assemblies) (Chouinard 1986).

Class and sub-class capacities to contest housing provision will also be influenced by experiences of housing and development issues within localities. According to cooperative housing activists in Toronto, it has been more difficult to develop projects in suburban communities, which lack a history of resistance to housing redevelopment and low income displacement, than in central city neighborhoods where people have had direct experiences of such changes (information provided to author by the Labour Council Development Foundation and Cooperative Housing Federation; see also Chouinard 1986, 1988).

Thus the housing programs of capitalist states can be conceptualized as an important terrain of inter- and intra-class conflict over class and state formation. I now want to use this theory to help to explain post-war changes in Canadian cooperative housing policies and procedures.

Continuing cooperative housing in post-war Canada

Analysis of the post-war development of Canadian cooperative housing programs reveals three distinct phases in class and state formation. First, a period in which housing programs virtually excluded non-profit and cooperative housing groups from decommodified housing provision. This was followed, in the late 1960s and early 1970s, by a permanent federal funding program, drawing coop groups into competition for state assistance. Since the late 1970s, a new phase of "privatization" has emerged. Private loans have replaced direct loans, and through a series of measures, the state has encouraged the "recommodification" of relations in cooperative housing delivery and use.

In what follows, I use the theory outlined above (section 1) to help to explain these changing terrains of class conflict and class formation within the Canadian state.

State formation and the politics of exclusion

Prior to the 1970s, cooperative housing was provided almost exclusively through building coops (concentrated in Nova Scotia and Quebec). Unlike continuing cooperatives, in which housing is collectively owned and managed on a long-term basis, building cooperatives favor commodified relations of housing exchange and use since units are owned on an individual, market basis after construction. As late as 1973, there were fewer than 20 continuing housing cooperatives in existence outside Quebec. This was despite struggles by labor, student, church, and other housing activists for increased government support, and expansion of the cooperative housing movement beyond Atlantic Canada (Cooperative Housing Foundation 1978, Laidlaw 1977, Macpherson 1983, Chouinard 1986, Akpan *et al.*, n.d.).

This absence of "decommodified" cooperative housing can be explained in part by the fact that a broadly based, national cooperative housing movement did not emerge in Canada until the 1960s. It was then that the Canadian Labour Congress (CLC) and the Cooperative Union of Canada (CUC) formed a National Labour Cooperative Committee (NLCC) to encourage liaison between the labor and cooperative movements on issues like housing. In March 1968, the CLC, CUC and Canadian Union of Students established the Co-operative Housing Foundation (CHF) "to promote Canada-wide initiative, sponsorship and co-ordination of various types of cooperative housing" (cited in CHF 1978; Chouinard 1986).

But limited development of continuing cooperatives also reflected political resistance to decommodified housing alternatives. Although cooperatives were eligible for funding under federal limited dividend programs in effect from 1944 to 1964 and in 1969, and under a small non-profit program from 1964 to 1968, repeated requests for low interest loans by the CUC and its supporters were rejected by officials of the federal housing agency (Canada Mortgage and Housing Corporation or CMHC). CMHC and housing ministry officials argued that collective ownership and management conflicted with the intent of existing legislation to provide assisted rental accommodation to those unable to purchase housing. They also insisted that subsidies for such housing would destabilize the housing credit system and economy generally, and that collective homeownership might encourage negative attitudes toward conventional private housing provision and consumption (Dennis & Fish 1972, 250–1). Laidlaw's (1977, 43–4) account of Willow Park, developed from 1960 to 1964, illustrates how limited federal assistance and resistance from financial institutions and municipal government helped to impede development of Canada's first continuing coop:

Almost nobody, especially in the housing establishment, understood what they were about; most sources of financing would not touch anything so unconventional; housing officials were skeptical because it was outside their ken and experience. The antipathy towards cooperative action for housing can be judged from the fact that only one Winnipeg alderman out of eighteen voted for a co-op proposal presented to city council. After months of dogged effort on the part of organizers, city officials finally consented to lease land in an area that could hardly be considered prime location, and with misgivings Central Mortgage and Housing Corporation made a loan commitment, with certain strings attached, at the regular interest rate. Though Willow Park Cooperative was incorporated in 1961, they had to wait until 1963 before the land was serviced. Construction started in 1964 and the first residents took possession in August 1965. . . . The project encountered a soft housing market and was handicapped by vacancies for about a year, but the Cooperative Credit Society Manitoba, with the backing of Federated Cooperatives and other Western co-operative groups, carried Willow Park through a difficult financial beginning.

Initial exclusion of cooperative housing from Canada's assisted housing programs reflected a process of state formation geared toward subsidizing the mass production and consumption of commodified housing. During this century, housing assistance programs have been aimed at stabilizing the housing credit system, subsidizing private housing production and consumption, and (since World War 2) using housing as a countercyclical tool of economic management. These policy priorities were developed through close consultation between state officials and capitalists, and gave rise to programs which emphasize tax and direct subsidies for investment in mortgages, rental housing, and individual homeownership (Bacher 1980, 1985, Patterson 1972, Dowler 1983, Chouinard 1986, Belec 1984). These programs helped to sustain an intensive regime of accumulation in post-war Canada, but were not "predetermined" by the needs of capital. Ineffective struggles by non-profit housing activists, and the relative prosperity of the immediate post-war years severely limited possibilities for decommodified housing provision (Chouinard 1986).

Under the new programs of the 1970s, exclusion gave way to marginalization of coop groups in struggles for federal funding, and in conflicts over regulations governing assisted housing provision.

Class formation and the development of cooperative housing programs

By the late 1960s, CMHC had begun to fund continuing cooperative housing projects on an experimental basis, through its part five research program. This was a response to nation-wide political pressure for community-based housing and redevelopment projects, and to concerns to find alternatives to public housing in an era of escalating subsidy costs (Chouinard 1986). But experimental assistance was implemented in ways that reproduced the marginal position of continuing cooperatives as assisted housing producers. A 1973 study of 56 community-based housing and development projects reported that almost all groups found CMHC head office staff unresponsive when approached for assistance. CMHC was reluctant to evaluate and fund projects, and groups reported that other levels of government were generally unsupportive and sometimes even hostile toward their project proposals. Lack of government support, low project budgets, and lack of technical assistance funds limited project development (Happy 1973, 31–47).

Resistance to decommodified assisted housing was also evident during negotiations between federal officials and the coop sector in the late 1960s and early 1970s. In 1969, CMHC's president informed the housing minister that the corporation remained reluctant to fund continuing cooperative housing, despite growing demand, because collective owner-ship was inconsistent with conventional, assisted private rental housing:

> The co-ops . . . while relying on the technicality that the occupant is a tenant promote entirely the concept of ownership rather than tenancy and then have indeed proposed that the tenants provide the equities required to finance their units. We feel this to be inconsistent with the basic concept of the legislation as it is currently written (memo from the president, CMHC to the federal Housing Minister, September 26, 1969, CMHC file 112–1–2, vol. 8).

As late as 1972, CMHC's president questioned whether experimental assistance to continuing cooperatives was justified by public demand and community support (memo to MSUA from CMHC president, March 21, 1972, CMHC file 112–1–2, vol. 9). Reports from cooperative housing groups in British Columbia during the early 1970s indicated CMHC and other government agencies lacked understanding of the special problems faced by groups providing cooperative housing. Officials were reluctant, for example, to recognize the need for increased financial and technical assistance to compensate for inexperience in housing provision (report for CMHC, Capital financing for cooperative housing, Westminster County Cooperative Building Society, 1972, CMHC file 112–1–2, vol. 9). The delivery of experimental assistance thus contributed to continued marginalization of cooperative housing in Canada.

In 1973, amendments to the National Housing Act (NHA) introduced Canada's first permanent funding program for continuing cooperatives. The product of lengthy negotiations between the Cooperative Housing Foundation (CHF) and CMHC, this program was characterized by direct, low interest loans (8%) and joint federal-provincial rent subsidies (modeled after the public housing program). From 1973 to 1978 7,529 cooperative housing units were financed under this program (Chouinard 1986, Table 3.3). But federal spending patterns continued to marginalize cooperative housing producers. Federal expenditures on loans and subsidies for housing cooperatives as a proportion of total spending on housing assistance varied from a low of 1.6% in 1974 to a high of 3.2% in 1975 and 1978 under the 34.18 program (Chouinard 1987, Table 3.3). According to the Cooperative Housing Foundation, these funding levels fell far short of demand (Chouinard 1987, Table 3.2).

The manner in which activists contested program characteristics and implementation also contributed to marginalization of coop housing alternatives. Opportunities for development of a self-conscious, collective sense of working-class opposition to commodified housing have been limited by ambiguities and tensions in the political philosophy and practices of the cooperative housing movement. The CHF characterizes the Canadian movement as one directed toward developing a society based on neither capitalist nor socialist social relations: "co-operation is more than just a new method of solving problems. It involves a new way of looking at society. Neither capitalistic nor socialistic in nature, it can truly be named the third sector in our national economy" (Cooperative Housing Foundation 1982, 1). Alexander Laidlaw (1977, 10), an internationally recognized Canadian activist, stressed consumer opposition to profiteering in land and housing as the subjective underpinnings of the cooperative housing movement:

This new idea in housing is based on the belief that no individual has a right to unearned increment in land and shelter, and that no one can extract a profit which someone else does not have to pay. Moreover, housing co-ops represent the ultimate in consumer control: the opportunity for citizens to participate in deciding the kind of housing, the type of neighbourhood and the quality of life to which they aspire.

Organizations like the CUC, Canadian Labour Congress and New

Democratic Party have echoed these commitments to advancing the rights of "ordinary people" and "consumers" to affordable housing through the provision of coop housing (Chouinard 1986). But this collective sense of opposition to "capitalistic" and "profit-oriented" relations in housing provision, and advocacy of increased consumer/community control of housing, has yet to be explicitly linked to working-class struggles for increased control of production and the state. Is the new, cooperative "third sector" society to remain an island of "alternative relations" within capitalism, or is the movement's agenda to develop a society based on popular control, and if so, on what relations will this be based?

In the absence of a clear, class-based agenda for social change, struggles over the cooperative housing program came to focus on immediate details of program administration. After the introduction of a "permanent" funding program in 1973, conflict between activists and government officials shifted away from the issue of whether or not continuing cooperatives were legitimate subjects of housing assistance, and towards details of program administration. Although coop groups were able to win some concessions on procedures which discouraged housing provision (e.g., a relaxation of 80–50 percent presale rules prior to loan commitment), they were unable effectively to challenge the general manner of project regulation by the state and thus marginalization of coop housing groups as assisted housing producers. In particular, activists were unsuccessful in challenging government insistence that coop housing groups compete with private housing producers for assistance,· producing housing comparable in cost and quality despite disparities in financial and technical resources (letter from President Teron to David Peters, Ontario Habitat Foundation, CMHC file 112–1–41, vol. 10).

Difficulties in mobilizing cooperative members helped to limit action on common issues like marginal funding and competitive regulation. Leaders in the cooperative housing movement and the CHF report that most members are concerned with immediate consumption issues (e.g., personal housing charges), rather than issues affecting the cooperative housing sector generally. This emphasis on housing results and consumption benefits helps to explain why policy struggles tended to focus on details of program administration rather than relations in housing delivery (e.g., whether or not coops should be treated like any other private housing producer) (Chouinard 1986, 1987). And, since these regulatory relations within the state went uncontested, officials were able to constitute coop activists as abstract, apparently equal assisted housing producers, without claim to special concessions from the state.

CMHC insisted, for example, that the Cooperative Housing Foundation and technical resource groups (regional groups facilitating coop development) rely on fees-for-service rather than government assistance. This policy was consistent with CMHC's position that no housing producer should use federal assistance for political activities, but also meant that technical resource groups were placed in competition for unit allocations by CMHC (since revenues depend on services provided to projects in the planning and development stages). Moreover, when CMHC announced cutbacks in funding to the Cooperative Housing Foundation, leaders were able to contest only the timing of (rather than rationale for) funding withdrawal, since they had not effectively challenged the assumption that all national/regional organizations of assisted housing producers should be independent of state funding (Chouinard 1986, 1987).

Winning the battle and losing the war: "recommodifying" cooperative housing

Reproduction of an abstract, marginal political position within the Canadian state had important implications for struggles over redesign of the cooperative housing program in the late 1970s. As early as 1976, coop groups were protesting CMHC efforts to revise operating agreements with projects; revisions which would have required that 90 per cent of residents be low income (defined as earning less than 4 times rent charges). This move reflected the adoption of spending restraint by the Canadian state after 1975, and related efforts to target housing assistance more "efficiently" to lower income people (CMHC 1983a,b). At the same time, political pressure to channel more federal assistance to the private sector grew as crises in the housing sector and Canadian economy deepened (Chouinard 1986).

Despite nation-wide struggles for input to the new program, it was not until April 1978 that the Cooperative Housing Foundation won concessions from CMHC. These were: a delay in the introduction of the revised program (until 1979), reduction of the minimum required low income households to 15 per cent of units, and a decision not to disentangle cooperative housing assistance to the provinces (the fate of other non-profit housing programs). These were clearly "last minute" break-throughs. For as late as March 1978, the Cooperative Housing Foundation was complaining of exclusion from "secretive" CMHC reviews of its non-profit and coop housing programs, and voicing concerns that coop funding might be discontinued altogether:

> we know that CMHC is doing an evaluation of their social housing programs, but their deliberations seem to be very secretive. There has been absolutely no consultation with the cooperative housing sector on the evaluation and there has been no hint of the contents of the evaluation. . . . Local CMHC offices are refusing to give start-up funds or mortgage commitments, postponing decisions on anything and everything until "after the evaluation" (letter from G. Haddrell, CHF, March 21, 1978, CMHC file 112–1–41, vol. 10).

The new federal assistance program was in place by January 1979 and represented a "victory" for cooperative housing activists in terms of immediate goals like continuation of funding, central rather than fragmented state administration (which groups believed would help prevent political fragmentation), and some details of program implementation (more flexible tendering procedures, for example) (Cooperative

Housing Federation 1978, 24–5; Chouinard 1986). It is also worth noting that more units (12,312) of cooperative housing were produced under the 56.1 program from 1979 to 1983 than had been produced under the earlier program from 1973 to 1978 (7,529 units) (Chouinard 1986); lending some support to claims that the new system of subsidized private mortgage loans was more "efficient" than the previous direct lending program.

But the general regulatory procedures included in the new program favored the "recommodification" of cooperative housing delivery, thus threatening class capacities to reproduce alternative, collective relations in housing provision and use. As NDP housing critic Bob Rae pointed out in the House of Commons, the new privatized loan system forced groups to rely on financial institutions to support less conventional forms of housing provision, and offered no political means of ensuring that such support would continue or increase (Canada: House of Commons Debates, 1979, vol. 4, 3867). The new system of housing charges (based on market rather than operating costs) helps to discourage collective management of housing projects, since it removes economic incentives to reduce general operating costs through collective participation in project maintenance (Goldblatt 1978, Chouinard 1986). Recent case studies suggest that this change in housing charges, in conjunction with federal withdrawal from interest subsidies as incomes rise within projects, is likely to decrease the capacity of coops to retain moderate income households. This is because the individual economic advantages of choosing cooperative housing over private market alternatives are diminished (Chouinard 1986). ,

Recent struggles over Canadian cooperative housing policy have thus been unable to prevent the introduction of regulations that encourage more commodified, individualized relations in housing provision and consumption. Implementation criteria favoring equal treatment of assisted housing producers irrespective of class, market or political positions (within the state), and the absence of effective challenges to these abstract forms of subjection to state authority, have made it difficult to resist these trends. For example, one of the more intriguing (and still poorly documented) symptoms of privatization has been direct competition for funding between coop groups and private developers within non-profit programs. By late 1976, CMHC had begun to receive inquiries from finance institutions and development corporations interested in producing "turn-key" housing projects. These projects are produced by the private sector and then "turned over" to non-profit/coop housing groups for management. CMHC's social housing division director indicated that there were no objections to such projects (providing existing program criteria were met), despite previous official expectations that housing coops would be sponsored by community-based rather than for-profit groups (letter to CMHC, Kitchener, from Social Housing Division, CMHC, December 24, 1976, CMHC file 112–1–41, vol. 7).

Coop activists indicate that competition from private turn-key projects remains an important issue, a claim underscored by a 1984 conflict between several Toronto coops and CMHC over the housing agency's transfer of funding from coop projects to projects sponsored by a private developer (Chouinard 1986). Yet struggles against this and other aspects of the "recommodification" of cooperative housing provision have been limited to specific cases like that in Toronto, or individual protests by leading activists. Regional or national opposition to privatized state regulation of coop housing has yet to emerge. Such opposition would require a political shift toward contests over relations in the delivery of state assistance to coop housing groups.

Cooperation or cooptation? Prospects for decommodified housing alternatives

Post-war efforts to develop cooperative housing have thus had contradictory implications for provision of decommodified housing alternatives. In an absolute sense, opportunities for reproduction of collectively owned and managed assisted housing have increased. By September 1983, there were 1,049 housing cooperatives in Canada (402 of these in the development or planning stages), representing 41,628 housing units. Over 27,000 of these units were concentrated in Ontario and Quebec (Cooperative Housing Foundation, September 1983). Current estimates (to March 1987) place the number of Canadian housing cooperatives at 1,140, with 88 projects under development (representing a total of 46,652 units) (information provided to author from Cooperative Housing Foundation, research division, June 1987).

While these gains are unprecedented in Canada, cooperative housing still represents a tiny fraction of the country's housing stock (of the order of 0.003% based on 1981 data; author's calculations based on CMHC 1982, and Cooperative Housing Foundation September 1983). And public and non-profit assisted housing have constituted the bulk of non-profit housing alternatives (Chouinard 1986).

It is also clear that recent struggles over cooperative housing policies and procedures have not effectively checked reproduction of more individualized and commodified relations in housing provision and consumption (e.g., shifts to private loans and market charges). Cooperative housing advocates have so far won the "battle" to retain some state support for coop housing provision, but have lost the "class war" to develop state programs that increase capacities to deliver coop housing in collective, decommodified forms. In this sense, the development of an effective terrain of conflict over coop housing within the state marked the beginning of political incorporation into relations that reproduced the marginal position of coop activists in and against the state.

What then are the prospects for decommodified housing alternatives in Canada? During negotiations between CMHC and the Cooperative Housing Foundation in the summer of 1986, federal officials reported the first absolute cutbacks in funding for cooperative housing projects. In view of the limited support available for housing cooperatives at both the federal and provincial levels of the state, absolute funding reductions are a serious setback indeed (CMHC 1983b, Chouinard 1986). According to the

232

Cooperative Housing Foundation, these cuts have reduced the number of units planned for development from 1986 to 1987 from 5,000 to 3,230 units (or by over 1700; information provided to author from Cooperative Housing Foundation, research division, June 1987). Unless the cooperative housing movement can effectively mobilize a broadly-based opposition, such reductions seem likely to continue under a Conservative government with a stated commitment to "privatization."

But effective mobilization will require changes in political consciousness and practices. As the preceding analysis indicates, recent struggles have tended to be incorporated and contained within doiminant forms of assisted housing regulation. The challenge for the cooperative housing movement lies in shifting the terrain of class and state formation towards concrete, class-based issues in housing provision, and away from the competitive and abstract modes of regulation developed to date. Insistence that governments recognize the marginalized position of coop housing producers in assisted housing provision through special support programs, and political advocacy which explicitly links decommodified housing to working peoples' struggles to exercise greater control over their lives, would be considerable advances in this direction. Such actions are unlikely to overcome the state's concentration of assistance toward commodity production and consumption (so long as capitalism persists). But they will contribute to the formation of more collective class capacities to reproduce decommodified housing alternatives in and against the state.

Conclusions

This chapter has argued that housing policies constitute a significant, contested site of social reproduction in capitalist societies. Analysis of recent struggles over Canadian cooperative housing programs has revealed a contradictory process of class and state formation. Although opportunities for the provision of decommodified housing have increased with federal funding programs, class capacities to contest marginalization and "recommodification" of cooperative housing delivery have been limited by state regulation and political practices.

These results suggest several ways to strengthen our analyses of struggles over service provision in advanced capitalism. Further theoretical and empirical work is needed to account for contradictions between objective class capacities to reproduce services like decommodified housing, and subjective identification with "non-" or "quasi-" class objectives like increased consumer control. Theories of class formation in places can contribute to this task, by helping to specify how local social relations and practices shape social reproduction (e.g., Katznelson 1981, Cooke 1983, Cox & Mair 1987, Chouinard 1988).

Equally important are studies of geographic variations in class formation between places. Are there some localities, perhaps those experiencing severe deindustrialization, in which struggles for coop housing have been closely linked to struggles for increased working-class control over production and the state? How have subjective working-class capacities to resist state programs which "recommodify" coop housing been influenced by objective links to labor, church, and student movements in different places? Answers to questions like these will contribute to our understanding of the role of contingent conditions in shaping class conflict and formation within localities. More importantly, they encourage close examination of links between experiences of the state and service provision, and political mobilization beyond local territories around issues like the delivery of cooperative housing.

Further work on the process of political incorporation within the capitalist state is also needed. This analysis has indicated some of the ways in which state regulations and political practices combined to reproduce coop housing activists as marginalized, "non-class" subjects of cooperative housing programs; subjects with limited capacities to contest recommodified housing delivery or reductions in funding. But the links between this process of class formation in and against the state, and the constitution of state subjects within localities requires further specification. Greater theoretical emphasis on the range of political relations involved in the reproduction of state subjects would contribute to this task. Studies of variations in experiences of state programs between localities, and how these encourage, displace, transform or spatially "contain" class conflicts, will also play a vital role in advancing our understanding of political incorporation and resistance under capitalism.

Perhaps the most challenging implication of these findings is that political life and culture must be given a central place in analyses of struggles in and against the capitalist state. If we hope to develop "non-economistic" explanations for continued privatization and/or cut backs in housing assistance programs, in an era when millions are homeless and "shelter poor" (Dear 1987, Stone 1983, Harloe 1981), it is important to examine how advanced capitalist cultures help to shape the politics of housing. We need to know, for example, how individualized values (such as definitions of social status in terms of effective competition for consumption commodities) have contributed to difficulties in mobilizing support for decommodified housing alternatives. How are these cultural norms reproduced or resisted in struggles within the state, workplace, and residence community?

Current debates offer a number of perspectives for such analyses of state development. This author agrees with Levine *et al.* (1987) that one of the more promising strategies is to develop better Marxist explanations of the "micro-foundations" of peoples' experiences and practices within capitalism. This requires further elaboration of "intermediate" theoretical categories which specify causal processes of change in the politics and culture of places (e.g., constitution of class capacities, state formation, and localities), as well as so-called "locality studies" (Wright 1985, Cooke 1983, 1987, Smith 1987, Chouinard 1988, Katznelson 1981). A potential advantage of this approach is that it encourages more rigorous analyses of the politics and culture of capitalism, without neglecting the role of "macro" forces in limiting the possibilities for social change in places (e.g., economic limits to decommodified service provision under an intensive regime of accumulation; see Chouinard & Fincher 1987; Chouinard 1986). We can look forward to exciting advances in research on state and class formation, as analysts draw on Marxist and other critical social theories to understand the uneven possibilities for social change in capitalist societies.

References

Akpan, U., E. Kirolos, R. Moras, A. Moshinsky, A. Ruggero & H. Steyer (n.d.). Cooperation in the city. Paper for Social Science 486. Toronto: Cooperative Housing Federation.

Agnew, J. A. 1981. Homeownership and the capitalist social order. In *Urbanization and urban planning in capitalist society*, M. Dear & A. J. Scott (eds.). London and New York: Methuen.

Bacher, J. C. 1980. Keeping to the private market: the evolution of Canadian Hamilton, Canada: Department of History, McMaster University.

Bacher, J.C. 1985. Keeping to the private market: the evolution of Canadian housing policy, 1900–1949. Unpublished PhD dissertation. Hamilton, Canada: Department of History, McMaster University.

Belec, J. 1984. Origins of state housing policy in Canada: the case of the Central Mortgage Bank. *The Canadian Geographer* **28**, 377–82.

Bluestone, B. & B. Harrison 1982. *The deindustrialization of America*. New York: Basic Books.

Canada: House of Commons Debates (1979) "Official Report", 4th session, 30th parliament, vol. 2. Ottawa: Queen's Printer.

Castells, M. 1979. *The urban question*. Cambridge, Mass.: MIT Press.

Chouinard, V. 1986. State formation and housing policies: assisted housing programmes and cooperative housing in postwar Canada. PhD dissertation. Hamilton, Canada: Department of Geography, McMaster University.

Chouinard, V. 1987. Class formation, conflict and housing policies. Paper presented to the annual meeting of the American Association of Geographers, April 21–7, 1987, Portland, Oregon and submitted to the *International Journal of Urban and Regional Research*.

Chouinard, V. 1988. Explaining local experiences of state formation: the case of cooperative housing in Toronto. *Environment and Planning D: Society and Space* (forthcoming).

CMHC (Canada Mortgage and Housing Corporation), files on cooperative housing policy: 112–1–41, vols. 7, 8, 9, 10.

CMHC 1982. *Canadian Housing Statistics*. Ottawa, Canada: Mortgage and Housing Corporation.

CMHC 1983a. Social housing evaluation. Draft report, Program Evaluation Division, April 1983. Ottawa: Canada Mortgage and Housing Corporation.

CMHC 1983b. *Section 56.1 Non-Profit and Cooperative Housing Program Evaluation*. Program Evaluation Division, November 1983. Ottawa: Canada Mortgage and Housing Corporation.

Cooke, P. 1983. *Theories of planning and spatial development*. London: Hutchinson.

Cooke, P. 1985. Class practices as regional markers: a contribution to labour geography. In *Social relations and spatial structures*, D. Gregory & J. Urry (eds.). London: Macmillan.

Cooke, P. 1987. Clinical inference and geographic theory. *Antipode* **19**, 69–78.

Cooperative Housing Federation 1978. *Annual report*. Toronto: Cooperative Housing Federation.

Cooperative Housing Foundation 1978. The first ten years of the Cooperative Housing Foundation. Research memo. Ottawa: CHF.

Cooperative Housing Foundation 1982. Introduction to cooperative housing. Ottawa: CHF.

Cooperative Housing Foundation Sept. 1983. Directory of housing co-operatives. Ottawa: CHF.

Cox, K. & A. Mair 1987. The new spatial politics. Draft paper available from authors, Geography Department, Ohio State University.

Dear, M. 1987. The year of living dangerously: Canadian cities face the International Year of Shelter for the Homeless, 1987. Paper presented to the annual meeting of the Canadian Association of Geographers, May 25–31, 1987, Hamilton, Canada.

Dennis, M. & S. Fish 1972. *Programs in search of a policy: low income housing in Canada*. Toronto: Hakkert.

Dickinson, J. & B. Russell 1986. Introduction: the structure of reproduction in capitalist society. In *Family, economy and state: the social reproduction process under capitalism*, J. Dickinson & B. Russell (eds.). Toronto: Garamond Press.

Dowler, R. G. 1983. Housing related tax expenditures: an overview and evaluation. Major report no. 22, Centre for Urban and Community Studies, Toronto: University of Toronto.

Goldblatt, M. 1978. CMHC: deathblow to co-ops?. *City Magazine* **3**, 11–16.

Gray, F. 1976. The management of local authority housing. In *Housing and class in Britain*, 75–86. London: Conference of Socialist Economists.

Happy, J. R. 1973. Funding community organizations: a review of the experience of Central Mortgage and Housing Corporation. Ottawa: Canada Mortgage and Housing Corporation.

Harloe, M. 1981. The recommodification of housing. In *City, class and capital*, M. Harloe & E. Lebas (eds.). London: Edward Arnold.

Harvey, D. 1978. Labour, capital and class struggle around the built environment in advanced capitalist societies. In *Urbanization and conflict in market societies*, K. R. Cox (ed.). Chicago: Maaroufa Press.

Katznelson, I. 1982. *City trenches*. Chicago: University of Chicago Press.

Laidlaw, A. 1977. *Housing you can afford*. Toronto: Green Tree.

Levine, A., E. Sober & E. O. Wright 1987. Marxism and methodological individualism. *New Left Review* **162**, 67–84.

Little Bear, L., M. Boldt & J. A. Long (eds.) 1984. *Pathways to self-determination*. Toronto: University of Toronto Press.

Macpherson, I. 1983. *Building and protecting the cooperative movement: a brief history of the Cooperative Union of Canada*. Ottawa: Cooperative Union of Canada.

Massey, D. 1984. *Spatial divisions of labour: social structures and the geography of production*. London and Basingstoke: Macmillan.

Matthews, G. J. & R. Morrow (eds.) 1982. *Canada and the world: an atlas resource*, 13–14. Toronto: Prentice-Hall.

Patterson, J. O. R. 1972. First phase policy determinants and outputs in relation to housing and urban affairs in Canada. Draft report, part 1, available from author. Toronto: Social Planning Council of Metropolitan Toronto.

Pratt, G. 1986a. Housing-consumption sectors and political response in urban Canada. *Environment and Planning D: Society and Space* **4**, 165–82.

Pratt, G. 1986b. Against reductionism: the relations of consumption as a mode of social structuration. *International Journal of Urban and Regional Research* **10**, 377–400.

Rose, D. 1980. Toward a re-evaluation of the political significance of homeownership in Britain. In *Housing, construction and the state*. Political Economy of Housing Workshop: Conference of Socialist Economists. Nottingham: Russell Press.

Saunders, P. 1984. Beyond housing classes: the sociological significance of private property rights and means of consumption. *International Journal of Urban and Regional Research* **8**, 202–27.

Saunders, P. 1986. Comment on Dunleavy and Preteceille. *Environment and Planning D: Society and Space* **4**, 155–63.

Scott, A. J. & M. Storper (eds.) 1986. *Production, work, territory: the geographical anatomy of industrial capitalism*. Boston: Allen & Unwin.

Smith, N. 1987. Dangers of the empirical turn: some comments on the CURS initiative. *Antipode* **19**, 59–68.

Stone, M. E. 1983. Housing and the Economic crisis: an analysis and emergency program. In *America's housing crisis*, C. Hartman (ed.). Boston: Routledge & Kegan Paul.

Wright, E. O. 1985. *Classes*. London: Verso.

11

Privatization and dependency on the local welfare state

GLENDA LAWS

Introduction

Local communities are bearing the brunt of the restructuring of the Canadian welfare state (Fincher & Ruddick 1983, SPCMT 1985, Riches 1986). Similarly, in the United States urban governments and local community groups are forced to deal with the growing problems of homelessness and poverty (Katz 1986, Dear & Wolch 1987). The British experience is comparable (see various issues of *Critical Social Policy*). The inner cores of North American cities have long been recognized as the point at which transient groups find refuge and support services. But no longer is it the transient, skid-row population identified by the Chicago school sociological models which dominate this part of the city. Now, families, young people, and single parents are joining the poor of the inner city. The concentration of such people and the day-to-day problems with which they are confronted are symptomatic of much more widespread processes, a central element of which is the restructuring of the social policies of the modern welfare state.

Newspapers around the world now regularly report on "privatization," an explicit move by governments to deal with the problems of balancing budgets and reducing national debts. Most often these reports deal with the sale of so-called "hard services" (airlines, railways, utility companies) to the private sector, but social services are also undergoing the same process. Both by design and default private organizations are being asked to take greater responsibility for the provision of a range of social services (SPCMT 1984).

Any process, such as privatization, which is occurring internationally is not likely to manifest itself evenly in all places and at all times. Local acceptance or resistance of a central state policy will modify its implementation and potential, and actual, outcomes. How, then, does such a policy affect local areas? And what are the consequences of privatizing social services for the consumers of these services? These are the questions which form the focus of this chapter. To understand the local consequences of changes in state policy we must first understand the context in which these changes are occurring. In the next section the recent restructuring of the welfare state is shown to be occurring internationally, the most significant indicator of this being the reductions in the rates of growth of social service expenditures. One policy outcome of these shifts in public expenditure patterns has been the promotion of private sector involvement. After reviewing arguments for and against this process, I argue that one of the fundamental errors in this literature is the tendency to treat privatization as a generic process, and thus to overlook the variations in origins and outcomes of three different forms of privatization. These are non-profit, commercial, and domestic forms of private service provision.

The chapter then considers the question of the local state, and how privatization might affect the levels of dependency upon the local welfare state, and hence the uneven development of the state. After all, it is at the local level that the state is most accessible to the demands of social service consumers and their advocates. The degree of local activity in response to a central state initiative is argued to have several potential outcomes. However, local pressures on the welfare state should not be seen as merely reactive. Local pressure groups might, in fact initiate changes that will affect central policy.

In the fourth section a case study is presented to illustrate the arguments developed in the earlier sections. The privatization of residential care facilities in Hamilton illustrates the ways in which privatization affects, and is affected by, the development of the local welfare state. The consequences of this process for Hamilton's psychiatrically-disabled population are considered. The case study notes that the local welfare state has perpetuated the dependency status of this group. The concluding section points to research questions and methodological issues that are raised by the analysis in this chapter.

Restructuring the welfare state: the case of privatization

Since the Great Depression there has been a phase in the development of the capitalist state that has generally been referred to as the rise of the welfare state. Several features characterize this period. First, there has been a "commitment," on the part of western states, to provide some level of social security for their constituents. This has resulted from the demands made upon the state by competing interest groups which seek a redistribution of society's resources, and some compensation for the problems associated with economic growth and decline. The degree to which this social minima is either satisfactory or adequate is questionable, but none the less the stated intention is there. Second, this commitment has led to the evolution of a state apparatus that deals almost exclusively with welfare issues. Third, the growth of the welfare state since the close of World War 2 has seen a massive injection of public money into social expenditures, defined by the OECD (1985) to include expenditures on health, education, pensions, other income maintenance programs, and programs related to the general welfare of the population. While this growth began in the years following the Depression, it accelerated after

Table 11.1 Social expenditure as a proportion of total government expenditure, selected OECD countries, 1960–80 (percentages).

	1960	1965	1970	1975	1980
Canada	37.95	43.97	52.26	56.95	57.63
USA	36.35	41.89	47.51	57.95	59.50
UK	38.88	44.40	48.90	51.04	54.09
Australia	43.11	41.60	44.32	59.55	62.39
Germany	60.90	61.08	63.15	70.92	68.74
Finland	54.51	54.98	63.88	65.57	67.95
Ireland	34.08	37.62	45.04	54.76	54.00

Source: OECD, 1985: Annex C.

the war and increased most rapidly after the early 1960s. In Canada, for example, in 1961 some 37 per cent of government expenditures were allocated to social services; by 1977 this figure was almost 59 per cent. Similar trends can be found in other countries including the United States, the United Kingdom, and Australia (Table 11.1).

Another indication of the evolution of the welfare state over the last two decades is given in Table 11.2. As a proportion of gross domestic product social expenditures doubled in the countries listed between 1960 and 1981. Close inspection of Table 11.2, however, reveals that most of this growth occurred prior to 1975. Since then the annual growth rates for social expenditures have slowed quite dramatically. Figure 11.1 shows that compared to the growth rate of gross domestic product and the growth rate of all public expenditure, the annual rates of increase in social expenditure have dropped the most, from a high of around 11 per cent in 1975 to 4 per cent in 1971. There has been some recovery, but between 1975 and 1981 the average annual growth rate for social expenditures has been 4.8% compared with 8.4% for the 15 years prior to 1975 (Table 11.2).

This change in expenditure patterns is indicative of the *restructuring* of the welfare state. Restructuring refers to the rationalization and reorganization of the state's activities that occurs as demands are made upon society's resources via the state apparatus. There are two broad sources of pressure for the state to restructure. First, there is pressure internal to the state, the most obvious example being the fiscal crisis experienced by most western economies (O'Connor 1973). Problems of balancing deficits have caused many states to initiate methods of decreasing expenditures and increasing revenues. Taxes can be increased; crown corporations may be sold to the private sector; and public spending can be reduced. Also at work to restructure the state is a range of external pressures. For instance, the business community will periodically call upon the state to funnel its resources into activities that will maintain and promote a profitable business climate. (Trade tariffs, for example, represent a state policy which is enthusiastically greeted by many manufacturers.) In addition, the working class looks to the state to provide it with support of various

240

Table 11.2 Social expenditures compared with Gross Domestic Product (GDP), 7 major OECD countries, 1960–81.

	Social expenditure as % of GDP			Annual growth rate of real GDP (%)		Annual growth rate of deflated social expenditure (%)	
	1960	1975	1981	1960–75	1975–81	1960–75	1975–81
Canada	12.1	21.8	21.5	5.1	3.3	9.3	3.1
France	13.4	23.9	29.5	5.0	2.8	7.3	6.2
Germany	20.5	32.6	31.5	3.8	3.0	7.0	2.4
Italy	16.8	26.0	29.1	4.6	3.2	7.7	5.1
Japan	8.0	14.2	17.5	8.6	4.7	12.8	8.4
UK	13.9	22.5	23.7	2.6	1.0	5.9	1.8
USA	10.9	20.8	20.8	3.4	3.2	8.0	3.2
Average	13.7	23.1	24.8	4.7	3.0	8.3	4.3
OECD Average	13.1		25.6	4.6	2.6	8.4	4.8

Source: OECD, 1985.

Figure 11.1 Annual growth rate of Gross Domestic Product (GDP), public expenditure and social expenditure in the OECD area, 1960–81. Source: OECD, 1985, Charts 1, 2.

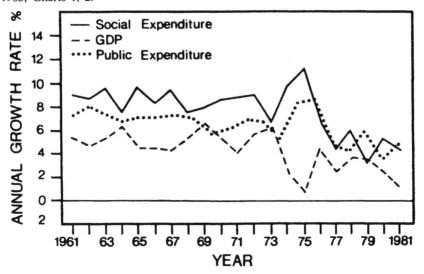

forms including the provision of health care and income-maintenance programs. While these separate sources of calls for restructuring exist, it is important to note that (in any historical instance) such distinct categories may not be observed. On certain occasions, for example, the interests of business and labor may coincide. Representatives of these groups might, therefore, unite in making their demands regarding the form that restructuring should take, if it is necessary, for instance, for the state to provide large subsidies to an ailing industry to save jobs.

Accompanying the programs of fiscal restraint introduced by many states in the last decade have been attempts to promote the use of the private sector in the provision of welfare services. *Privatization*, the term coined to describe this process, involves the shifting of responsibility for the provision of "public" goods from the state to the private sector. The extent to which privatization is a recent phenomenon differs between jurisdictions. Clearly, the United States has had a much longer history of involvement by the private sector than, say, the United Kingdom. In Canada, prior to the inception of the welfare state, it was the private sector (the church, families, and philanthropists) that was mainly responsible for welfare services. The literature therefore often refers to "reprivatization," the *return* of the responsibility for service provision to the private sector (Le Grand & Robinson 1984).

Those on the political right are the strongest advocates of privatization. The Thatcher Government in Britain has used the privatization of housing, for example, as one means of achieving its conservative policy goals (Crook 1986). The arguments for privatization are grounded on the principles of efficiency, equality, and individualism (Le Grand & Robinson 1984, Walker 1984). According to this viewpoint, "greater competition among providers generates more powerful incentives for reducing costs of production" (Pruger & Miller 1973, 22). Competition, therefore, is to be promoted since it will result in costs "substantially below that of the public sector" (Fisk *et al.* 1978, 2). State-provided services, in contrast, are inefficient because they undermine the incentive to work, and allocate resources inefficiently so that productivity is reduced (Le Grand & Robinson 1984; Mishra 1984; George & Wilding 1976).

Privatization, according to this view, might also be able to rectify the inequalities created by the welfare state. While the welfare state purportedly provides mechanisms which seek to minimize social inequalities, these critics would argue that this is not the case. Universal social services, available to everyone regardless of their income or wealth, are inegalitarian, inflationary, and costly (Block 1983). Universal services which are more often used by the well-off suggest that the welfare state has failed to achieve any real measure of equality (Le Grand 1982).

The erosion of individual liberty is the third dimension along which the welfare state is attacked. Block (1983, 26) claims, for instance, that universal social services are "costly in terms of the use of coercive taxation." Further, individuals should be free to choose among a variety of producers. To restrict this choice, by way of state-provided services, is an infringement upon an individual's liberties. The state should be subservient to the interests of the individual, who has the right to self-determination. Private, competitive provision of social services is therefore to be preferred.

Such criticisms of the welfare state lay behind monetarist policies aimed at undoing "the 'mischief' done to western economies by Keynesianism and the liberal democracy during the last thirty years" (Mishra 1984, 43). Emphasis is placed on controlling the money supply and balancing the budget. To do so government deficits must be cut, meaning a reduction in state activity. It is assumed that if there is a demand for a (social) service the market will respond accordingly. Monetarist policies therefore have an explicit theoretical rationale behind their promotion of private sector activity.

It is not, however, only the neoconservatives who criticize the welfare state. Critiques of the state, coming from the political left, also see it as inefficient and ineffective in that it has not achieved any significant redistribution of power or wealth. Indeed, evidence suggests that, in Canada at least, there is a greater and greater concentration of wealth in the hands of a very few people (Ross 1986). Offe (1984, 154) further criticizes the welfare state as being repressive, and responsible for "a false ('ideological') understanding of social and political reality within the working class." This is because welfare state policies often focus on the individual's responsibility for their current plight rather than on the macro-forces which create social problems. The welfare state therefore creates a dependency status among the working class, and is thus unlikely to promote an egalitarian society.

Attempts to increase consumer control (as opposed to state control) over social services amount to calls for privatization, but for very different reasons to those argued by the neoconservatives. State social services are seen to be contradictory. While they do provide assistance to certain groups, they simultaneously "control" these people by promoting a dependency relationship between the producer (the state) and the consumer. If privatization took the form of a process that would overcome this contradiction between the control element and that aspect of these services that truly improved the lot of the consumer, then it would be a process that would receive the support of organized labor, and other community groups.

Regardless of who may be arguing for privatization, there are (as with many state policies) critics arguing against this shift in policy. There is some evidence that the private sector may be able to provide services at a lower cost (e.g., CCHS 1985), but this also raises questions about the quality of care which can be provided at such reduced rates. Often the cost savings can be found by the lower labor costs paid in the private sector, both commercial and voluntary. Privatization may therefore involve the employment of less qualified staff. When the staff are appropriately qualified, privatization may then be creating a pool of underpaid jobs. The question of quality of care is often ignored in discussions of the efficiency of the private sector, since the focus is generally on economic efficiency. But, as Walker (1984) points out, economic efficiency is not necessarily commensurable with social equity.

Arguments that privatization will promote greater individual liberty have also come under close scrutiny. Liberty can be defined in terms of the absence of coercion (George & Wilding 1976, 22–4). The welfare state can therefore be seen as limiting freedom (Block 1983). This is partly due to the fact that social services create a dependency status among their clients (Galper 1975, Illich *et al.* 1977). However, liberty can also be defined as having the freedom to choose. In this sense, welfare state policies can be seen as increasing freedom, since they are aimed at increasing opportunities for individuals, especially low-income people (Le Grand & Robinson 1984). While there is some debate over whether or not the welfare state has increased or decreased the opportunities and choices available to an individual, there seems to be no clear evidence to say that a private system will necessarily improve the situation.

The local welfare state, service dependency, and privatization

The local state

The restructuring of the welfare state in general, and privatization more specifically, are processes that will impact differentially on localities and their populations. This is because the form and substance of the state's welfare policies are the products of social conflicts over the distribution of society's resources. That is to say, state policies must be understood not as the inevitable by-products of the capitalist social formation, but as the result of the intersection of the actions of human agents and the structures of contemporary capitalist society. One outcome of this interaction between structures and agents is the *uneven development* of the welfare state over space. The emphasis by some analysts on the structural determinants of this uneven development has recently been superseded by those who seek to accommodate the role of struggles by people as they react to, and seek to modify, structural constraints (e.g., Cooke 1983, Chouinard & Fincher 1985, Urry 1981).

The spatially uneven development of the state has given rise to the notion of *local states*. These are the smallest units within a territorially defined and hierarchically arranged state apparatus and, according to authors such as Corrigan (1979) and Cockburn (1977), local states are the arenas of local demands for concessions from the state. Thus, we would not expect to see the state develop in the same fashion in all places, since different localities will have different interests to be advanced. This has profound implications for the forms of social services that are found in various locations. The degree of politicization of particular communities and neighborhoods, for example, can significantly alter the patterns of social service delivery that will be available to local residents. The most conspicuous example of this in the geographic literature is the case of the location of group homes. Certain neighborhoods have been quite successful in having these facilities excluded, while other less politically-vociferous neighborhoods have been relatively saturated by these facilities (Dear & Taylor 1982, Dear & Laws 1986, Joseph & Hall 1985). Seley (1983) provides other examples of the ways in which local action can promote or prevent the location of particular facilities in certain local areas, and thus influence the development of the physical infrastructure of the local state apparatus.

The argument which can be developed from such cases is that local agents are able to modify policies of the central welfare state. Distinctive local welfare initiatives may therefore develop. There are three possible outcomes of these initiatives, and these can be illustrated with reference again to the group homes example. Throughout the early 1970s the City of Toronto had to deal with repeated conflicts over the location of group homes within the City boundaries. In 1978 the City passed a bylaw which permitted these homes to locate in all residential neighborhoods within its jurisdiction. This was an innovative move at the time and resulted from the lobbying of professionals and advocates who felt that residents of these homes should not be discriminated against when it came to siting their housing. Further, the bylaw acted as an important catalyst in the refining of Provincial policy on this matter. In fact, the Province of Ontario used the City of Toronto bylaw as a model for its own policy, which was announced later in the same year. So, the first outcome of the local development of the welfare state is the possibility of policy initiatives "filtering up" instead of being "imposed" from above. Secondly, other localities might choose to reinforce the central state's recently announced policy by adopting it within their own jurisdictions. The actions of local agents might therefore actively reinforce central policies, and thus the local development of the welfare state might be modeled closely upon central policies. (This will, of course, vary with the degree of autonomy operating between different levels within the state.) Finally, agents in any locality might resist central state policy recommendations. This did occur in the Toronto group home example as suburban municipalities sought to strengthen their own autonomy in determining local land-use legislation, and thus contested the power of the central state to intervene in this area. Clearly, then, we cannot underestimate the role of local actors in determining the form and function of the local welfare state (Dear & Laws 1986).

Characteristics of service-dependency

The common characteristics shared by all members of local service-dependent populations is their dependence upon the state as their major source of income. State-provided income-maintenance programs include pensions of different types, unemployment benefits, and welfare payments. Also, there are a range of programs which subsidize the cost of purchasing certain items which may otherwise be inaccessible. This financial dependence upon the state leads to other forms of dependence, one of which is dependence upon the immediate locality.

Locale takes on a special, and much restricted, meaning for individuals who do not have at their disposal the necessary resources for venturing beyond their immediate physical environment. The elderly, the mentally ill, and the retarded all rely on proximate services because of the limited transport services available to them. (Institutionalized persons are even more dependent upon their immediate surroundings.) Even when subsidized services are available, there are often requirements that the recipient is resident of a particular area. Thus, dependency upon a particular neighborhood, or locality, is created by state policies. Furthermore, this has partly been responsible for the concentrations of service-dependent people and the services they utilize in inner-city areas, referred to as the public city (Dear 1980) or the zone of dependence (Dear & Wolch 1987). Writing of the concentration of chronically mentally ill people in these locations, Mechanic (1979) suggests that these "colonies" of patients find the inner city a most appropriate, if not desirable, location because of its access to transport, social support, and the greater tolerance of local residents.

A third form of dependence is that upon the domestic sphere. Increasingly, households are being asked to accept greater responsibility for the welfare of their members (see Jones 1983, Guberman 1986). As the state makes changes to welfare policy (e.g., cutbacks, privatization), women's unpaid domestic labor can be tapped to meet household needs within the domestic sphere and within community organizations (Finch 1984, Armstrong 1984). Writing on recent trends in Quebec, Guberman (1986, 220) notes:

> The dismantling of the social welfare state means shifting, to the private sphere of the family, responsibilities that in recent years have been assumed by the state. The Quebec state is thus proposing to alter significantly the nature and work of families, and more specifically the work of women in families, not only as it pertains to minor children, but also with regard to the ill, the elderly, the physically, mentally, and socially disabled, and the new category of youth, defined as those between 18 and 30 years old.

One final form of dependence is that of dependence upon *professionals*. Dear & Wolch (1987), following Illich *et al.* (1977), write that the welfare apparatus tends to "produce disabling effects in a population as a prerequisite for receiving care. It can be argued that the apparatus does not 'cure' mental illness; that it 'produces' illness in clients and their social networks (especially the family); and that it encourages long-term dependency in those who enter the system" (see also Gaylin *et al.* 1981). While Dear & Wolch refer to the mentally ill, this argument can be extended to policies that affect other groups, including the elderly and those convalescing after some acute illness (e.g., Carlson 1975). Criticisms of the role of professionals in creating this dependency among special-needs groups often points to the fact that professionals block access to alternative forms of care; that professionals "mystify" the nature of the dependency-inducing problem; and they create an inaccessible and incomprehensible technology around their field of "expertise." In Wilding's (1982, 116) words: "The collective impact of these processes is to disable people, to destroy both their confidence and their capacity to act autonomously and independently. It is the end product of dependence."

This section has argued that welfare policies and the implementation of these policies can produce and reproduce patterns of dependency among special-needs groups. However, I do not want to imply that state policies only have negative or insidious outcomes for clients of the welfare apparatus. Service-dependency, as I argued earlier, has its roots in the inability of particular groups to participate in the wage-labor market. Social services do indeed provide a means of caring for these people. And as client groups and their advocates lobby for changes in the form and substance of social policy real gains can be achieved.

Privatization

Just as the local welfare state develops unevenly, so too there will be local variations in the ways in which the policy of privatization is implemented in, and received by, local communities. A serious flaw with theoretical debates for and against privatization is that they too often treat it as if it were a generic process. But, as I have shown, there are a variety of calls for restructuring in general and privatization in particular. Privatization will, therefore, not take one form. This point can be illustrated by discussing the variety of forms of privatization currently observable in Ontario. These are (1) the non-profit sector; (2) the commercial sector; and (3) domestic forms of service provision.

Included in the *non-profit* sector are both cooperative and voluntary service agencies, the legal mandate of which prevents them from making a profit. Although the degree of cooperative involvement in Ontario's social service delivery is not strong, there are cooperative arrangements in the areas of children's day-care, housing, and food distribution. In contrast, voluntary agencies are a well-developed component of Ontario's social service system with a long history of involvement (Splane 1965). Prior to the development of an institutionalized welfare state it was the voluntary sector that cared for the elderly, orphaned children, and the indigent. In the latter part of the 19th century some government grants were made to philanthropic and church organizations. This relationship with the state has strengthened significantly, and voluntary agencies now feel the pressure of government restraint programs (Tucker *et al.* 1984, SPCMT 1981). None the less they play an important and growing role in many fields including day-care, care for the elderly, community information services, and foodbanks. Both voluntary and cooperative agencies offer services at relatively low costs, or often free, and so provide valuable assistance when state alternatives are unavailable or inadequate.

For-profit services are becoming a more visible component of Ontario's welfare system. Proprietary agencies have organized themselves into powerful business associations (e.g., the Ontario Long Term Residential Care Association, associations of nursing home and day-care operators, etc.) which have exerted pressure on the government to increase their share of the market. For example, between 1973 and 1983 commercially-operated nursing homes increased the number of beds they provide for the elderly by 30 per cent. During the same period the number of publicly-provided beds increased by only 7.4% (SPCMT 1984). About half the licensed spaces offered by children's day-care operators are found in commercial facilities. In both cases Ontario is witnessing the entry of large US multinationals, evidence of the increasingly corporate nature of commercial privatization.

Asking households to accept greater responsibility for the care of their members is another form that privatization is taking in Ontario. This *domestic* form includes calling on self-medication treatments and outpatient services which leave individuals or their families responsible for care for most of the time. The Ontario government subsidizes the services of visiting homemakers, who go into homes of the elderly or someone convalescing after an illness, to help with their rehabilitation or daily activities. Another example of the ways in which the state promotes this domestication of care is the financial assistance provided by the Ontario government for the purchase of mechanical devices which will allow physically-handicapped children to stay in their homes.

The local welfare state plays an important role in the lives of the service dependent who, more than any other group, are dependent upon local resources. The restructuring of the local welfare apparatus by way of the privatization of service provision might be expected to affect the lives of these people. It is to this issue that this chapter now turns.

The privatization of residential care facilities in Hamilton, Ontario

Hamilton's residential care sector

What, then, are the implications of privatization for the development of the local welfare state and for local service-dependent populations? Can it reduce levels of dependency upon the local welfare apparatus and thus be thought of as a progressive policy? Or does it necessarily imply an erosion of needed services and thus a regressive move for the service-dependent? To consider these questions I draw upon the example of the deinstitutionalization of mentally ill people in Hamilton, Ontario, and the subsequent privatization (especially commercialization) of residential care for this group. Hamilton is the site of a large public psychiatric hospital. Like other such institutions across North America, it has experienced a significant reduction in its inpatient population. In 1969 Hamilton Psychiatric Hospital had a maximum bed capacity of 1,451; by 1980, this had been reduced to 502. It is known that most discharged patients tend to remain in the City of Hamilton, regardless of their address prior to admission to the hospital (Dear *et al.* 1980). Thus, it is logical to assume that there will be a significant demand for community-based accommodation for this group in Hamilton.

For the most part this demand has been met by the private sector, including both voluntary and commercial agencies. This in itself is a direct outcome of provincial policy implemented during the 1960s and 1970s

Table 11.3 Number of adult residential care facilities in Hamilton–Wentworth, 1986.

Type	Number	%
Lodging homes	1141	27.5
Nursing homes	1344	32.4
Homes for special care	144	3.5
Total commercial sector	2629	63.4
Homes for the aged	1132	27.3
Group homes	384	9.3
Total non-commercial	1516	36.6
Total	4147	100.0

Source: Author's survey, 1986.

(Laws 1987). As the Province shifted the locus of care for service-dependent groups from institutions to community facilities, it passed legislation which provided for the licensing of homes which would fulfill the demand for community housing. Such legislation amounted to an invitation to interested private individuals and corporations to participate in the delivery of care for those in need. Some legislation was directed only toward non-profit organizations; other acts included any for-profit groups which could meet government requirements.

The effects of this on Hamilton's residential care sector is telling. Table 11.3 shows the structure of this sector in 1986. In all, some 4,147 beds are available within the Region of Hamilton-Wentworth, 63 per cent of which are provided in the commercially-operated Nursing Homes, lodging homes and Homes for Special Care.[1] Group homes, operated by the voluntary sector, provide less than 10 per cent of the beds available in community-based accommodation. The remaining 27 per cent is provided in Homes for the Aged. Approximately 60 per cent of the region's beds are in Nursing Homes or Homes for the Aged, and are thus designated for the elderly. This leaves 1,670 beds to serve all other special needs groups, including the psychiatrically disabled, the mentally retarded, and substance abusers and former offenders.

Data reporting on discharges from Hamilton Psychiatric Hospital for the last 6 months of 1986 show that, after "private homes," it is commercially-operated homes which receive most clients (Table 11.4). The proprietary sector provides homes in the form of lodging homes, Homes for Special Care and Nursing Homes. A survey conducted by the Regional Department of Public Health in 1983 revealed that almost 56 per cent of lodging home residents had had some history of psychiatric illness and/or had been diagnosed as mentally retarded. Clearly, the local lodging home sector is an important source of accommodation for this group.

A final point to note in this overview of the residential care sector is the pattern of growth of the privately-operated homes. A survey of the

Table 11.4 Hamilton psychiatric hospital; discharge by type of accommodation, July–November 1986.

	No. of patients	%
Private home	218	60.4
Private room	6	1.7
Lodging home	58	16.1
Hostel	20	5.5
HSC[1]	2	0.5
Co-op home	1	0.3
Group home	13	3.6
Nursing home	2	0.5
HFA[2]	1	0.3
Correct. instit.[3]	2	0.5
Comsoc facil.[4]	3	0.8
Other	35	9.7
Total	361	100.0

Notes:
[1] Home for Special Care (operated by the Ontario Ministry of Health).
[2] Home for the Aged (operated by the Ontario Ministry of Community and Social Services).
[3] Correctional Institution (operated by the Ontario Ministry of Correctional Services).
[4] Various community facilities operated by the Ministry of Community and Social Services.

voluntarily-operated facilities found that almost 80 per cent of the homes operating in 1986 had opened since 1970 when the Province began actively promoting community-based care. Only 25 per cent (or 6) of the homes, however, have started operations since 1981. Compare this to the fact that the number of commercial lodging homes granted licenses to provide accommodation for special needs groups has increased from 34 in 1981 to 72 in 1986 (Laws 1987). This evidence suggests that the commercial form of privatization is being promoted in Hamilton, while the non-profit sector grows almost imperceptibly.

Privatization and the development of the local welfare state

The privatization of residential care facilities in Hamilton affords an excellent opportunity to illustrate the reciprocal relationships at work between local conditions and state policy, and the subsequent development of the local welfare state. As has been observed in other locations, residents in Hamilton did not always respond positively to the establishment of special-needs housing in their neighborhoods. The Province had given private operators a mandate to establish alternatives to institutional care, and wherever possible these alternate homes were to be located in residential neighborhoods.

In 1977 residents of the area immediately east of Hamilton's downtown began to voice opposition to the fact that their neighborhood was accepting more than its fair share of these homes. With the help of their council representative they began to lobby City Hall, demanding that some action be taken to alleviate this "problem." Community debate over definitions of residential care facilities and the appropriate means of regulating their location ensued. Finally, in 1981, the City passed a bylaw which imposed a distance-separation factor in an effort to disperse the homes and prevent an undue concentration of facilities and their residents.

This exemplifies the ways in which the actions of local residents can create a particular local response to a central state policy. In doing so, a particular form of the local welfare state is created. The effectiveness of this local policy is, however, in question as Demopolis (1984) has shown that since 1981 lodging homes (one form of residential care in Hamilton) have continued to open in close proximity to the downtown. There are several possible explanations for this, and each relates to the desire of a commercial operator to maximize profits. First, the architecture in this part of the city is more readily converted into multi-resident housing than is the more peripheral suburban architecture. Second, there appears to be a greater demand for this kind of housing in the inner city. Here is where the cluster of needed services can be found. Whether the supply of housing has created this demand and the cluster of services, or whether the demand has created the supply is not clear. But certainly there now appears to be a degree of inertia in Hamilton's service ghetto. It is also in the inner city that public transit is most accessible, and this is an important consideration for those housing people who have no independent transport available to them.

The licensing of commercial lodging homes is a second example of the development of the local welfare state. Following the death of one resident in a Hamilton lodging home, the provincial coroner recommended that improvements be made to the regulation of those homes which serve special-needs groups. The City of Hamilton responded to this recommendation by passing a bylaw which would license lodging homes according to the population served. The designation "second level" refers to those homes which provide accommodation to service-dependent groups such as the mentally retarded and psychiatrically-disabled. These homes must provide 24-hour supervision and "assistance in the activities of daily living," as well as meeting stricter health and safety regulations than homes which do not accommodate people with special needs. While initially resisted by the lodging home operators, this new designation has resulted in some benefits for those operators who have been able to meet the new regulations. The primary benefit is that such operators are eligible for a contract with the Regional Department of Social Services which provides them with an assured source of income. This bylaw has proved attractive to other localities which have been having problems with lodging homes and is now being adopted by several other municipalities in Ontario, thus illustrating the ways in which local initiatives can influence the development of the local welfare state in other places.

Dependency upon the local welfare state

The development of an increasingly privatized system of residential care has implications for the status of local service-dependent groups. After clients of Hamilton Psychiatric Hospital are discharged to one of the community-based residences, for the most part, they will find themselves living in poverty and, as already noted, many will find their way into the commercially-operated lodging homes, Nursing Homes or Homes for Special Care. It is this commercialization of care that will receive particular attention in this section.

A person who is unable to find employment (a common problem among discharged psychiatric clients) is eligible to receive an income from Ontario's General Welfare Assistance or Family Benefits schemes. In Hamilton, the regional Department of Social Services administers these funds. Payments are minimal and Kearns (1987) found that, in his sample of 66 psychiatrically-disabled clients, the vast majority lived below official poverty lines. Such a situation allows little scope for the former inpatient to break away from a financially-dependent relationship with the state.

In Hamilton this is further accentuated by the contract arrangement that operates between the local state and the operators of lodging homes. In 1986 the Regional level of government would pay an operator a $CAN25.00 *per diem* to provide accommodation, meals, and some basic supervision. The cost of this is taken from the monies the client is entitled to receive under the General Welfare Assistance Act. After payment of this accommodation cost, each client is entitled to $CAN77.00 per month for personal expenses such as entertainment, clothing and the like. Similarly, residents of the provincially-regulated Homes for Special Care are financially dependent upon the state. In this case the Province pays the operator a *per diem* fee of $CAN20.88. The operator then charges all other costs (clothing, outings, cigarettes) to the Ministry of Health. Money never passes through the clients' hands, and thus money-management skills are never acquired. This clearly has implications for the long-term dependency status of any individual.

As noted above, such financial dependence severely constrains physical mobility. The local welfare state can therefore create an environmental dependence. It is not mandatory for Ontario's municipalities to enter into contract arrangements for providing accommodation, but when they do (as is the case in Hamilton), it provides an advantage to client groups who may otherwise have difficulties in negotiating the rental market. It is therefore in the interest of the ex-psychiatric inpatient to remain in Hamilton where such assistance is available. And when they do, they find that the monthly allowance of $77.00 does not provide enough money to travel any great distance for entertainment, work, or support services. This is clearly a contributing factor in the creation of service-dependent ghettos. As already noted, some commentators have pointed to the supportive nature of these inner-city colonies, which are characterized by greater tolerance of the bizarre behaviors, and which provide the services necessary for continued community tenure.

One service required for this is an outpatient clinic that can provide necessary medications to stabilize the behaviors of the psychiatrically disabled. The privatization of residential care has been greatly assisted by the introduction of psychotropic drugs which make the behaviors of the mentally ill more socially acceptable. However, such a move has reproduced those patterns of dependence which characterized the institutional model of care. Clients must be in touch with a professional to maintain access to their medications and other forms of required therapy. In Hamilton, again it is the inner city that provides the nucleus of professional service delivery. One hospital, St. Joseph's in the downtown core of the city, offers an emergency psychiatric unit to ensure the availability of professional assistance on a 24-hour basis. Private non-profit organizations such as the Canadian Mental Health Association and the Association for the Mentally Retarded also provide professional social workers and counselors, and again these are located downtown.

A recently announced provincial program will see these professional services extended into the homes of the mentally ill. Under Ministry of Health funding, social workers, and recreational and occupational therapists will be allowed to enter local lodging homes to provide the mentally ill with programs related to rehabilitation. As it stands, operators of lodging homes in Hamilton have been given the power to supervise the taking of medication within the home to ensure controlled behaviors. Although this discussion needs to be more fully developed (can we develop indicators of professional dependency and thus measure changes in its level?), it does raise the interesting question of whether dependence upon professionals has actually increased with the move to privately-operated community residences. In the hospital environment, bizarre behavior and the refusal to take medicines could be tolerated to a far greater extent than within the community. In the community there is far greater pressure to conform to professional advice; the alternative being a return to the institution.

Table 11.4 shows unequivocally that most discharged patients return to some form of domestic care in a private home. Depending on the nature of their problem, this implies that a domestic form of privatization is, in fact, occurring in the Hamilton case. As yet, however, research has not followed these people to ascertain how they are coping with community living. From the experience of the Hamilton lodging home residents it is difficult to make any conclusive comments about the ways in which dependence upon the domestic sector has been affected. Elliott (1987) and Kearns (1987) have found that even for those Hamilton clients not living in lodging homes, very few are in family situations. However, they found that most of their sample still had regular, if not frequent contact with families. For this group the main source of support is the local welfare state: it assists in finding accommodation; administers income-maintenance programs; subsidizes local transit costs; and provides professional services.

The local state in Hamilton creates a dependent relationship between its clients and the local welfare apparatus, even though the private sector is becoming increasingly involved in the delivery of care.

Conclusions

Geographers and planners are interested in the ways in which state policies impact upon the political, social, and built environments of our cities. Given the interests of this chapter, it is important to consider how privatization might impinge upon local communities, since we have established that it is in local places that people struggle over the forms of state policy, and it is at this level that we can observe the outcomes of state policy. The local state will in turn shape, and be shaped by, conditions in the local community. To conclude this chapter I will outline some of the research themes that are in need of further investigation.

One of the most fundamental questions is that relating to the identification of those unique features of a locality that translate a policy of privatization into practice. Social policy in Canada is largely written at the provincial level and yet we can observe an unevenness about its development across space. Characteristics at the local level must therefore play an important role in the implementation of state policy. It may be that local electoral politics influence the form and outcomes of privatization. A conservative municipality, for example, might be interested in promoting the private sector in order to reduce the level of government activity in the locality. A more radical local government, on the other hand, might promote privatization on the basis that it can increase opportunities for consumer participation. Or the degree of community and advocate activism might be the critical variable. In Toronto, for example, there have been attempts to provide cooperative housing for discharged psychiatric clients. This privatized form of accommodation gives more scope for consumer participation in decisions regarding their housing. One methodological implication of the argument that local conditions are crucial in understanding the policy process and outcomes is the need for comparative research between places.

The form of the urban built environment might also influence the policy and process of privatization. The need of private, commercial operators for large properties which can generate maximum profits has meant that private residential care facilities are concentrated in particular parts of urban areas, particularly close to downtown. Of course, large homes are not only to be found in the inner city, but the built environment is partitioned according to zoning bylaws and the real estate markets. High costs of real estate prohibit the conversion of some properties into residential care facilities. And when costs may be acceptable, zoning regulations often restrict the type of use which can be made of the building. The commercial nature of some of the privately-operated facilities, for example, often prevents their location in purely residential neighborhoods, and so they are further concentrated in the inner city which is generally characterized by less restrictive zoning practices.

Further research is also required to clarify whether or not privatization improves the position of various groups within the city. In terms of methodology this implies that a set of indicators that would allow us to assess the impacts of state policy on the lives of these people needs to be developed. For instance, has privatization increased, decreased or maintained the dependency status of populations such as the discharged psychiatric client? This might, for example, be measured by the degree to which they require state subsidies to access the privately-provided services. So, commercialization might mean that a poor working family requires some level of financial assistance to cover the costs of caring for an elderly relative. The development of a voluntary agency, on the other hand, may decrease this dependence because the family could pay the nominal fees from their earned income.

Note

1 Homes for Special Care refer to homes for the care of a person requiring nursing, residential or sheltered care and who is regarded as an inpatient of a provincial psychiatric hospital. They are licensed under the provincial Homes for Special Care Act. A lodging home is defined by the City of Hamilton as "a house primarily intended or used as a dwelling, where persons are harboured, received or lodged for hire by the week or more than a week, but not for any period less than a week and are accommodated without any separate kitchen, kitchenette or kitchen sink but excepting a hotel, private hospital, public and private home for the aged, children's home or boarding school." A Nursing Home is a home maintained and operated for people who require nursing care, or in which such care is provided for two or more unrelated people. These homes are licensed by the provincial Ministry of Health under the Nursing Homes Act.

References

Armstrong, P. 1984. *Labour pains: women's work in crisis.* Toronto: The Women's Educational Press.

Block, W. 1983. The case for selectivity. In *Canadian Social Work Review, 1983*, G. Drover (ed.). Ottawa: Canadian Association of Schools of Social Work.

Carlson, R. J. 1975. *The end of medicine.* London: Wiley.

CCHS 1985. *Homemaker services in Canada survey 1985.* Toronto: Canadian Council on Homemaker Services.

Chouinard, V. & R. Fincher 1985. Local terrains of conflict in the Canadian state. Paper presented at the Annual Meeting of the Association of American Geographers, Detroit, April.

Cockburn, C. 1977. *The local state.* London: Pluto.

Cooke, P. 1983. *Theories of planning and spatial development.* London: Hutchinson.

Corrigan, P. 1979. The local state: the struggle for democracy. *Marxism Today* **23**, 203–9.

Crook, A. D. H. 1986. Privatisation of housing and the impact of the Conservative government's initiatives on low-cost homeownership and private renting between 1979 and 1984 in England and Wales; 1. The privatisation policies. *Environment and Planning A* **18**, 639–59.

Dear, M. 1980. The public city. In *Residential mobility and public policy*, W. Clark & E. Moore (eds.), 219–41. Beverly Hills: Sage Publications.

Dear, M. J., L. Bayne, G. Boyd, E. Callaghan & E. Goldstein 1980. *Coping in the community: the needs of ex-psychiatric patients.* Hamilton, Ontario: A Mental Health/Hamilton Project Funded by Young Canada Works.

Dear, M. & S. M. Taylor 1982. *Not on our street.* London: Pion.

Dear, M. & J. Wolch 1987. *Landscapes of despair. From deinstitutionalisation to homelessness.* London: Polity Press.

Demopolis, C. 1984. *The development of the lodging home ghetto in Hamilton, Ontario.* Unpublished BA research paper. Hamilton, Ontario: McMaster University, Department of Geography.

Elliott, S. 1987. *The geography of mental health: housing ex-psychiatric patients.* Unpublished MA thesis. Hamilton, Ontario: McMaster University, Department of Geography.

Finch, J. 1984. Community care: developing non-sexist alternatives. *Critical Social Policy* **9**, 6–18.

Fincher, R. & S. Ruddick 1983. Transformation possibilities within the capitalist state: co-operative housing and decentralised health care in Quebec. *International Journal of Urban and Regional Research* **7**, 44–71.

Fisk, D., H. Kresling & T. Mullen 1978. *Private provision of public services: an overview.* Washington: The Urban Institute.

Galper, J. 1975. *The politics of social services.* Englewood Cliffs, NJ.: Prentice-Hall.

Gaylin, W., I. Glasser, S. Marcus & D. Rothman 1981. *Doing good: the limits of benevolence.* New York: Pantheon.

George, V. & P. Wilding 1976. *Ideology and social welfare.* London: Routledge & Kegan Paul.

Guberman, N. 1986. Who's at home to pick up the pieces? The effects of changing social policy on women in Quebec. *Canadian Social Work Review/Revue Canadienne de Service Social 1986*, 219–28.

Illich, I., I. Zola, J. McKnight, J. Caplan & H. Shaiken 1977. *Disabling professions.* London: Marion Boyers.

Jones, C. 1983. *State social work and the working class.* London and Basingstoke: Macmillan.

Joseph, A. & B. Hall 1985. The locational concentration of group homes in Toronto. *Professional Geographer* **37**, 143–55.

Katz, M. 1986. *In the shadow of the poorhouse. A social history of welfare in America.* New York: Basic Books.

Kearns, R. 1987. *In the shadow of illness: A social geography of the chronically mentally disabled in Hamilton.* McMaster University: PhD in progress, Department of Geography.

Laws, G. 1987. *Restructuring, privatisation and the local welfare state.* McMaster University: PhD thesis, Department of Geography.

Laws, G. & M. Dear 1988. Coping in the community: a review of factors influencing the lives of ex-psychiatric patients. In *Location and stigma: emerging trends in the study of mental health and mental illness*, C. Smith & J. Giggs (eds.). London: Allen & Unwin.

Le Grand, J. 1982. *The strategy of equality.* London: Allen & Unwin.

Le Grand, J. & R. Robinson (eds.) 1984. *Privatisation and the welfare state.* London: Allen & Unwin.

Mechanic, D. 1979. *Future issues in health care. Social policy and the rationing of medical services.* New York: The Free Press.

Mishra, R. 1984. *The welfare state in crisis: social thought and social change*. Brighton: Wheatsheaf Books.

O'Connor, J. 1973. *The fiscal crisis of the state*. New York: St. Martins Press.

OECD 1985. *Social expenditures 1960–1990: problems of growth and control*. Paris: OECD.

Offe, C. 1984. Some contradictions of the modern welfare state. In C. Offe, *Contradictions in the welfare state*, John Keane (ed.). Cambridge, Mass.: MIT Press.

Pruger, R. & L. Miller 1973. Competition and the public social services. *Public Welfare* **31**, 16–25.

Riches, G. 1986. *Foodbanks and the welfare crisis*. Ottawa: Canadian Council on Social Development.

Ross, D. 1986. *Background document on income security*. Ottawa: Canadian Council on Social Development.

Seley, J. 1983. *The politics of public facility planning*. Lexington, Mass.: D. C. Heath.

SPCMT 1981. *Voluntary sector at risk*. Toronto: Social Planning Council of Metropolitan Toronto.

SPCMT 1984. *Caring for profit. The commercialisation of human services in Ontario*. Toronto: Social Planning Council of Metropolitan Toronto.

SPCMT 1985. *Social Report No. 1*. Toronto: Social Planning Council of Metropolitan Toronto.

Splane, R. B. 1965. *Social welfare in Ontario 1791–1893. A study of public administration*. Toronto: University of Toronto Press.

Tucker, D., R. House, J. Singh & A. Meinhard 1984. *An ecological analysis of voluntary social service organisations*. Hamilton, Ontario: Program for the Study of Human Service Organisations, McMaster University.

Urry, J. 1981. Localities, regions and social class. *International Journal of Urban and Regional Research* **5**, 455–74.

Walker, A. 1984. The political economy of privatisation. In *Privatisation and the welfare state*, J. Le Grand & R. Robinson (eds.), 19–44. London: Allen & Unwin.

Wilding, P. 1982. *Professional power and social welfare*. London: Routledge & Kegan Paul.

12

"Enough to drive one mad": the organization of space in 19th-century lunatic asylums

CHRIS PHILO

Dead, long dead,
Long dead!
And my heart is a handful of dust,
And the wheels go over my head,
And my bones are shaken with pain,
For into a shallow grave they are thrust,
Only a yard beneath the street,
And the hoofs of the horses beat,
Beat into my scalp and my brain,
With never an end to the stream of passing feet,
Driving, hurrying, marrying, burying,
Clamour and rumble, and ringing and clatter,
And here beneath it is all as bad,
For I thought the dead had peace, but it is not so;
To have no peace in the grave, is that not sad?
But up and down and to and fro,
Ever about me the dead men go;
And then to hear a dead man chatter
Is enough to drive one mad.
(Tennyson 1855, reprinted in Ricks 1969: 1086–7)

The above lines comprise the "madhouse canto" or "mad-scene" from Alfred Tennyson's *Maud*, and they convey how the hero's delusion of dying and then being buried under a city pavement shapes his interpretation of the lunatic asylum to which he is consigned.[1] He finds himself surrounded by the "clamour and rumble" of a busy institution, assaulted on all sides by patients sobbing in distress, praying to their "own great selves," pretending to be statesmen or being probed by "vile physicians," and in the midst of this restless crowd he can find no peace and can conceal no secrets. The "dead men chattering" are hence nothing but other inmates, and it is their incessant "idiot gabble and babble" that he finds "enough to drive one mad" (Tennyson 1855, reprinted in Ricks 1969: 1087–8; see also the commentary in Bucknill 1855b). Reading the "madhouse canto" in *Maud* has a similar effect to viewing the famous "Bedlam" scene in Hogarth's *The Rake's Progress* then, for the impression conjured up is not one of a well-ordered institutional environment, but one of noisy disorder and the chaotic arrangement of people, activities, and spaces. And it is this impression of "disorganized space," coupled with the notion that inmates themselves probably perceived their surroundings to be disorganized, which must be kept in mind during the following discussion.

Panopticons, panopticism and the socio-spatial reproduction of madness

There can be little doubt that most human societies past and present have "endured" a number of groups – usually minorities, but not necessarily so – who offend against the sensibilities of a hegemonic and usually majority population on the grounds of, for example, age, sex, shape, looks, ethnicity, culture, religion, politics. In many respects these offending groups are actually *produced* by their host society, both in a direct sense – think of the bodily deformities that arise as a result of inter-breeding in small, closed societies; think of the physical and mental scars sustained by victims of war and industrial accidents – and in the more indirect sense of people being labeled as different, deviant, or dangerous and then treated accordingly. Moreover, once initiated, the many "mainstream" fears and prejudices regarding certain "outsider" groups often feed into concrete social practices through which distinctions between these "mainstream" and "outsider" peoples are *reproduced* and even rendered more acute. And these concrete practices commonly boast a *spatial* dimension, as when societies seek to exclude their "outsiders" from normal places of living and working, perhaps by banishing them into the wilderness or perhaps by separating them off into what Julian Wolpert (1976) calls "closed spaces."

A number of geographers have begun to consider this spatial dimension of how "mainstream-outsider" relations are produced and reproduced (see Philo 1987a), and in particular they have considered the way in which respectable suburban communities strive to "purify" their surroundings by expelling and excluding all obviously different and "polluting" categories of person (see Evans 1978, Sibley 1981, 1987). One such category of person is the sufferer from mental illness – or sufferer from "madness," "lunacy" or "insanity," to use older descriptions of a distressed and disturbed mental state[2] – and Michael Dear, another geographer, claims that mentally ill people are greatly affected by the exclusionary tactics pursued by many suburban communities:

> The mentally ill, like other minority social groups such as the poor, are restricted in their selection of residence, workplace and recrea-tional outlets. Their continued isolation, in asylums historically and now in ghettos, can be interpreted as part of a wider system of socio-spatial organisation which causes the separation of antagonistic

groups. . . . The community in opposition to mental health facilities employs two indirect sources of power to exclude: the power of socio-spatial exclusion; and the power of state authority as manifest through planning policy (Dear 1981, 494).

Through a collection of empirical studies Dear and his various co-workers secure this thesis, and in so doing indicate that many mental health "clients" – and notably those patients discharged in the wake of deinstitutionalization policies – are now becoming "ghettoized" in the decaying cores of North American cities, in large measure because the community mental health facilities upon which they depend – the group homes, day-centers, outpatient clinics, "drop-in" clinics and so on – are themselves condemned to inner-city sites where community opposition is limited and politically weak (for a summary of these studies see Dear & Taylor 1982; see also Wolpert & Wolpert 1974, Wolpert, Dear & Crawford 1975, Dear 1980, Dear & Wolch 1987). Leading from these findings Dear (1981, Dear & Clark 1984) works towards a theoretical perspective on the "social and spatial reproduction of the mentally ill" that treats the encounter between mental health professionals and their "clients" as a form of *class relationship* buttressed by, and in turn constitutive of, both the unequal *authority relations* whereby "clients" are forced into a dependent "sick role," and the unequal *distributive groupings* whereby "clients" end up trapped in low-status inner-city areas at some remove from the higher-status homes of their doctors, psychiatrists, and social workers.[3]

Considerably more could be said about this adventurous theoretical and empirical synthesis, but for the purposes of this chapter I wish simply to flag two aspects of Dear's research into the socio-spatial reproduction of madness which I believe to be in need of further elaboration.[4] The first concerns his brief excursion into the "social history of the asylum," which proceeds from the opening assertion that

> The history of mental health care is a study in isolation and exclusion. Every culture appears to have had its "madness", and to have found some method of isolating the mentally ill. These methods have sometimes been crude (as when the mad were forced to live beyond the confines of medieval towns), and often well-intentioned (as with the development of the asylum – a true place of refuge). In all situations, however, the response has been to isolate and to exclude (Dear & Taylor 1982: 37).

This assertion may appear quite unexceptional, but I would argue that the stress on *isolation and exclusion* is actually somewhat overplayed, and that problems attach to what is effectively a transference of conclusions regarding present-day community suspicions about mental illness back into the historical record. This is not to pretend that the historical lot of mad people has ever been a particularly happy one, of course, but it is to acknowledge that there have been many occasions when, far from being isolated and excluded, mad people have experienced *tolerance* and even active attempts to *include* them in everyday scenes of living and working. Consider, for instance, the welcoming attitude towards "lunatik lollers" urged by *Piers Plowman* in late medieval England (see Neaman 1975: 131–3), or the assistance given by the early modern English state through the Court of Wards and the parishes to families rich and poor who looked after their own mad kinsfolk (see MacDonald 1981: 7). Furthermore, few historians would dispute that the wholesale confinement of mad people in those specialist "closed spaces" known as public or state lunatic asylums, charitable lunatic hospitals, private madhouses and workhouse insane wards is really a very modern invention, and that in consequence it is only recently that the history of mental health care has become dominated by the themes of isolation and exclusion.[5]

But it is the second aspect of Dear's research which I wish to reflect upon in more detail, and this concerns his excursion into the realm of institutional plans and architectures. Here, too, his emphasis lies squarely on the themes of isolation and exclusion (or separation), and he sets out his views on this matter in typically forthright fashion:

Care and/or treatment of the mentally ill has always proceeded from fundamental principles of isolation and separation of individuals in space. In penal systems the most intense architectural manifestation of this principle was Bentham's "Panopticon", which arranged individuals in isolated cells on tiered circles about a central observation area (Dear & Clark 1984: 67; see also Dear 1981: 491).

He then goes on to suggest four principles that mental health professionals have always mobilized "to describe a functional analytic space and to allocate patients for treatment within that space," and these he describes as the principles of "enclosure," "partitioning," "functional sites," and "rank" (Dear 1981: 491, Dear & Clark 1984: 67–8). He is certainly justified in highlighting the organization of institutional space as an issue worthy of study, so I would argue, but I remain worried that he paints a picture of an institutional history full of asylums whose internal spaces have been neatly chopped up into the minute portions demanded by Bentham's "Panopticon" design. The basic objective of this chapter is hence to provide empirical evidence that the organization of space in many actual and proposed asylums has departed greatly from "Panopticon" principles, and in the process to demonstrate that the role of space in the institutional relationship between mad people and their keepers has involved rather more than merely the furtherance of those "divide and rule" strategies underlying Bentham's vision. Signaling these aims at this point is perhaps to "run ahead" of myself, though, for I must first say more about the two pivots upon which Dear's account of asylum plans and architectures depends: Jeremy Bentham's initial "Panopticon" design and Michel Foucault's reconstruction of a "panopticism" that arose throughout 19th-century Western Europe.

Bentham and the "Panopticon"

The "Panopticon" or "Inspection-house" was born in a series of "contrived" letters which Bentham wrote during 1787, and later presented to the public with "postscripts" in 1791. In these he speculated on the appropriateness of his design for penitentiaries, prisons, houses of correction, manufactories, hospitals, schools, and madhouses (reprinted in Bowring 1843, IV: 37–172; see the commentaries in Himmelfarb 1968, Evans 1971). His objective was to use a "simple idea in architecture" to achieve a "new mode of obtaining power of mind over mind" (Bowring 1843, IV: 39), and to be more precise his hope was to foster an organization of space that ensured the *possibility* of inmates being *continually* inspected:

> the more constantly the persons to be inspected are under the eyes of the persons who should inspect them, the more perfectly will the purpose of the establishment have been obtained. Ideal perfection . . . would require that each person should actually be in that predicament [the "predicament" of being continually inspected] during every instant of time. This being impossible, the next thing to be wished for is that at every instant, seeing reason to believe as much and not being able to satisfy himself to the contrary, he should *conceive* himself to be so (Bowring 1843, IV: 40).

What this "predicament" for inmates was supposed to promote, so Bentham obviously believed but did not articulate all that explicitly, was a fear of being caught in the pursuit of those "habits" which might initiate or exacerbate lethargy, criminal propensities, perversions, illnesses, and other "tendencies," and it was presumed that this fear and the accompanying fear of punishment would cause inmates to refrain from all such misdemeanors. The "Panopticon" design hence flowed directly from the desire to create an illusion of continual inspection (see Fig. 12.1), and the blueprint called for numerous single cells to be positioned on the radii of a circle, each of which was to face inwards through iron gratings towards a central "inspector's lodge" from which it would be possible to see clearly the actions of every inhabitant of every cell.[6] The essence of the design thus lay in rendering every inmate competely visible to the institution's inspectors.[7] In part, this visibility would be maximized through the spatial separation of inmates. Bentham also believed that this separation would inhibit the "contagion" of bad thoughts, bad behavior, and disease.[8]

Bentham's "Panopticon" was planned down to the smallest detail, then, and the whole establishment was to be run in a tightly ordered fashion hostile to any disruption of either spatial organization or temporal routine.[9] In addition, it was to be a very peaceful establishment, since the separation of inmates would supposedly promote "security against annoyance by bad smells, bad sights, noise, quarrels and scolding" (Bowring 1843, VIII: 372). One implication of this intended orderliness

Figure 12.1 Jeremy Bentham's "Panopticon". Source: J. Bowring (ed.) 1843. *The works of Jeremy Bentham* vol. 4, between pp. 172 and 173. London: Simpkin, Marshall.

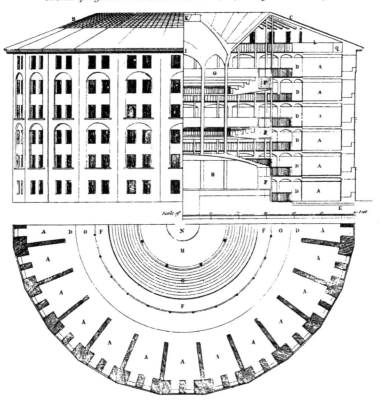

A General Idea of a PENITENTIARY PANOPTICON in an Improved, but as yet (Jan. 23d 1791) Unfinished State.
See Postscript References to Plan, Elevation, & Section (being Plate referred to as N° 2).

EXPLANATION.

Key	A	Cells
	B to C	Great annular skylight
	D	Cell galleries
	E	Entrance
	F	Inspection galleries
	G	Chapel galleries
	H	Inspector's lodge
	I	Dome of the chapel
	K	Skylight to dome

L	Storerooms with their galleries; immediately within the outer wall all round place for an annular Cistern Q
M	Floor of the chapel
N	Circular opening in dome (open except at church times) to light the Inspector's lodge
O	Annular wall from top to bottom for light, air, and separation

and peacefulness was that any institution constructed and operated according to proper "Panopticon" principles should never have experienced the disorganized "clamour and rumble" of inmates chattering, wandering about and acting out fantasies which confronts Tennyson's hero in *Maud*. And neither should it have descended into the turmoil and disorder depicted in myriad pictures of asylum interiors prepared in the tradition of Hogarth's "Bedlam" scene. This literary and artistic evidence is hardly conclusive by itself, of course, but it does mark an evocative first step in countering claims that the history of asylums has been full of "Panopticons."

Foucault and "panopticism"

In his remarkable text *Discipline and punish* (1979) Foucault portrays 19th-century Western Europe as shot through with disciplinary institutions that in one way or another resembled Bentham's precious brain-child. Indeed, springing from a detailed engagement with Bentham's own words (1979: 200–9), Foucault hacks an analytical swathe through a multitude of 19th-century social practices, and finally arrives at the revelation that a "whole series of institutions" coalesced to give "what one might call the carceral archipelago" (1979: 297). Along the way he pays particular attention to the organization of institutional space, and even observes that "in the first instance discipline proceeds from the distribution of individuals in space" (1979: 141). This observation leads him to discuss the "art of distributions," and in so doing to describe the techniques of "enclosure," "partitioning," "functional sites," and "rank" (1979: 141–9) which Dear suggests to be the four basic principles organizing the "functional analytic space" of asylums and other mental health facilities (see Dear 1981).[10]

But it is vital to realize that, although Foucault intends the term "panopticism" to cover a manipulation of institutional space akin to that urged by Bentham, he also intends it to embrace a whole lot more besides. In other words, he does not suppose "panopticism" to be conceptually reducible to the workings of the "Panopticon," and it occurs to me that *Discipline and punish* can be badly misunderstood if, as Felix Driver (1985b: 11) so aptly puts it, "Foucault's account of disciplinary power is identified too closely with the 'Panopticon'." There are perhaps two points to register here, and the first is that Foucault deploys the term "panopticism" to capture not only the role of institutional plans and architectures, but also the nature of many other "disciplinary techniques" (derived from the spheres of schooling, military training, accountancy, and so on), through which human subjects were converted into responsible "docile bodies" whose labors would serve to "strengthen social forces."[11] And the second point is that, even when he deals directly with institutional spaces, he describes buildings and programs which bore precious little *physical* resemblance to the "Panopticon," and he even fixes the "date of completion of the carceral system" in Western Europe at January 22, 1840 – the official opening date of a seemingly very "un-Panoptic" colony for young delinquents at Mettray in France (1979: 293):

264

In taking the Panopticon as the archetypal disciplinary institution, many of Foucault's critics have neglected his account of Mettray. Mettray has no walls of the kind that prisons do. It has no central observer and no individual cells. On the other hand, it is revered and widely emulated, even in England. Time and time again, those concerned with the treatment of delinquency refer to Mettray as their *example* (Driver 1985b: 10–11).

Instead of cells and an "inspector's lodge" Mettray had its spaces divided up into separate houses, each of which possessed its own workshop, dining hall, schoolroom and dormitory, and each of which was occupied by a "family" of about 40 boys who were supervised by two "elder brothers." Gone was the fear of continual visual inspection, but in its place the "apparently anti-institutional and natural weapon" of the *family* was used alongside a selection of related armaments – those of "cloister, prison, school, regiment" – as a yardstick for judgment and punishment compelling the delinquent children to become submissive, responsible, and versed in an "art of power relations" which they might themselves visit upon new arrivals (Foucault 1979: 293–5, Driver 1985b: 11–13).[12] The organization of space at Mettray hence differed greatly from that demanded by Bentham's "Panopticon," although Driver (1985b: 13) argues that "both were designed as spatial strategies in the production of socially useful individuals," and an appreciation of this subtlety written into the heart of Foucault's inquiry into "panopticism" marks a second important step away from envisaging a history of asylums full of "Panopticons." To take further steps, however, will necessitate a more sustained engagement with the empirical details upon which – in Britain at least – this particular history has turned.

"Panopticon" asylums designed, discussed, and discarded

In Letter Nineteen of his original series Bentham considered those "melancholy abodes appropriate to the reception of the insane," and in the process he proposed that erecting madhouses on the "inspection form" could have a "salutary influence" in relation to the object of such institutions:

> The powers of the insane, as well as those of the wicked, are capable of being directed either against their fellow creatures or against themselves. If in the latter case nothing less than perpetual chains should be availing, yet in all instances where only the former danger is to be apprehended, separate cells, exposed as in the case of prisons to inspection, would render the use of chains and other modes of corporal sufferance as unnecessary in this case as in any (Bowring 1843, IV: 60–1).

In his later writings Bentham returned to this theme when speculating that lunatics, as one amongst a number of particularly troublesome groups living in "Pauperland," probably warranted special institutional provisions in order both to improve their medical care and to rid ordinary industry-houses of a great annoyance (Bowring 1843, vol. 8, 394–5). Moreover, he claimed that his "central-inspection architecture" would "obviate whatever inconvenience might rise from the aggregation (of lunatics) if, instead of being but apparent, it were real,"[13] and he also implied that the expense per head of housing lunatics in "appropriate establishments" could be reduced below that incurred by parish authorities maintaining their insane charges in workhouses or private madhouses (Bowring 1843, VIII: 362).

And there can be little doubt that Bentham's "Panopticon" did exert an influence upon the plans and architectures discussed by early 19th-century asylum reformers, and it is clear that the "Panopticon" principle was familiar to those who tackled such matters on select committees and in official reports. For instance, when giving evidence to a select committee in 1815, Edward Wakefield did not feel it necessary to explain his description of the Guy's Hospital insane ward as being "built upon the panopticon system" (Report 1814–'15, First: 821), and in addition it was no coincidence that marked parallels existed between Bentham's "Panopticon" and the plans that James Bevans and William Stark drew up for asylums intended to serve respectively London and Glasgow (see Figs. 12.2 and 12.3). The historian Michael Donnelly has already spelled out these parallels elsewhere, in a chapter on the "architecture of confinement" (1983: 48, 58–67), but it does not require too much imagination to hear Bentham behind Bevans's assertion that "the whole and every part and every person, whether patient or keeper, is continually under the eye of the Governor or his assistant" (Report 1814–'15, Third: 970), or to hear Bentham behind Stark's assertion that "those who are inclined to disorder will be aware that an unseen eye is constantly following them and watching their conduct" (Report 1816, Third: 359).

Such explicitly "Panopticon" asylums were rarely built, though, in part, because they had to contend with the small purses possessed by county and borough authorities who might contemplate building a public lunatic asylum. In one case, to give an example, the Committee of Visitors appointed to oversee the opening of an asylum for Lincolnshire was forced to discard as impracticable and too expensive plans which

emulated the model prison, and [which] were rich in radiating galleries, high walls, peep-holes and other expressions of sombreness and security (Palmer 1854: 73).

But more importantly – and strangely unremarked upon in texts such as Donnelly's – the government's Commissioners in Lunacy established in 1845,[14] expressly rejected the "Panopticon" design because it was incompatible with certain requirements which they themselves had of asylum sites and lay-outs. As they stated in one of the "rules" appended to their *Further Report* of 1847:

Figure 12.2 James Bevans's proposed London Asylum. Source: *Report from Select Committee on the Better Regulation of Madhouses in England* (Parliamentary Reports, vol. 4, 1814–15), 1008.

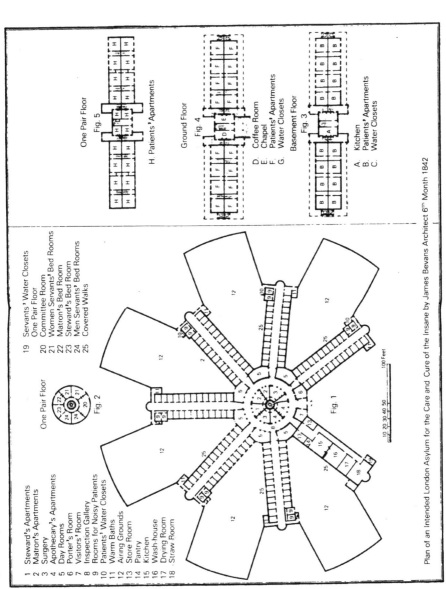

1   Steward's Apartments
2   Matron's Apartments
3   Surgery
4   Apothecary's Apartments
5   Day Rooms
6   Porter's Room
7   Visitors' Room
8   Inspection Gallery
9   Rooms for Noisy Patients
10 Patients' Water Closets
11 Warm Baths
12 Airing Grounds
13 Store Room
14 Pantry
15 Kitchen
16 Wash-house
17 Drying Room
18 Straw Room

19 Servants' Water Closets
    One Pair Floor
20 Committee Room
21 Women Servants' Bed Rooms
22 Matron's Bed Room
23 Steward's Bed Room
24 Men Servants' Bed Rooms
25 Covered Walks

One Pair Floor

Fig. 2

Fig. 1

10 20 30 40 50   100 Feet

One Pair Floor
Fig. 5

H. Patients' Apartments

Ground Floor
Fig. 4

D.   Coffee Room
E.   Chapel
F.   Patients' Apartments
G.   Water Closets

Basement Floor
Fig. 3

A.   Kitchen
B.   Patients' Apartments
C.   Water Closets

Plan of an Intended London Asylum for the Care and Cure of the Insane by James Bevans Architect 6th Month 1842

Figure 12.3 William Stark's proposed Glasgow Asylum (a preliminary plan). Soure: *Report from Select Committee on the Better Regulation of Madhouses in England* (Parliamentary Reports, vol. 4, 1816), 361.

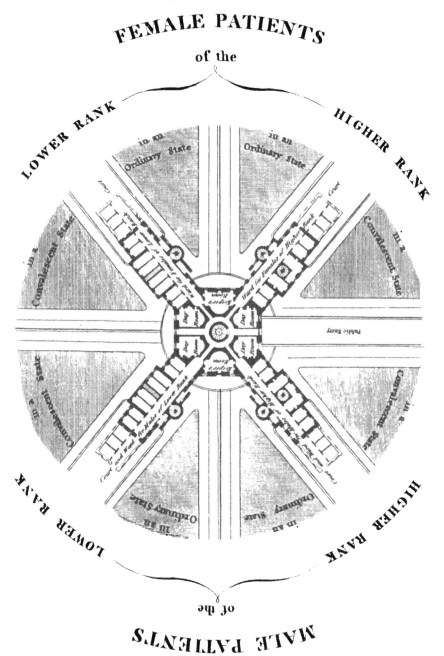

The general form of an asylum should be such as to afford an uninterrupted view of the country and the free access of the air and sun; and the several galleries and wards should, as far as possible, be so arranged that the medical officer and others may pass through all of them without retracing their steps. Plans upon what is called the radiating or windmill principle of design are unsuitable on account of the difficulties which they present in obtaining the advantages above alluded to (Commissioners in Lunacy 1847: 323–4).

There is no reason to suppose that the "radiating or windmill principle of design" criticized here was a reference to anything other than the "Panopticon" designs drawn up by Bevans, Stark, and other less prominent architects.

A further obstacle to the founding of "Panopticon" asylums was the popularity of an alternative model asylum inspired by the pleasing and mansion-like qualities of an institution such as William Tuke's York "Retreat" (see Digby 1985). The "Retreat" was a charitable lunatic hospital in which, alongside a host of other innovations, the hope was to envelop inmates in a nonprison-like "moral architecture"[15] that would press the lunatic's hitherto suppressed aesthetic sensibilities into the service of his or her own cure. It would be a fascinating exercise to compare the "Panopticon" with the "Retreat" – and it is perhaps surprising that a researcher like Donnelly (1983) does not say more about the tensions between these two institutional models – but to do so would be to voyage into waters that are not only troubled empirically, but which prompt a number of thorny theoretical questions: about the congruence of Foucault's study of Bentham and "panopticism" in *Discipline and punish* with his earlier study of Tuke and the "birth of the asylum" in *Madness and civilization* (1967);[16] about whether or not Bentham and Tuke were citizens of the same conceptual and discursive universe; about whether or not the popularity of the "Retreat" really amounted to a discarding of the "Panopticon."

These theoretical questions obviously tie in with the matter raised above – about the danger of identifying Foucault's "panopticism" too closely with Bentham's "Panopticon" – but it is beyond the scope of this chapter to tackle this treacherous theoretical terrain directly: instead, I propose to adopt a more oblique approach by narrowing the focus of my investigation to a selection of papers that surfaced in an organ called the *Asylum Journal* (which first appeared in 1853, became the *Asylum Journal of Mental Science* in 1855, and then became simply the *Journal of Mental Science* in 1859),[17] and which connected up to certain arguments put forward in their annual reports by the Commissioners in Lunacy. What this investigation reveals is the enormous variety of visions concerning asylum plans and architecture which could be sustained even within a community of "experts" who might have been expected to toe much the same line in this respect, and it also demonstrates that, while some "experts" extolled the virtues of asylums which took on board "Panopticon" principles (though even here proposals were rarely faithful to the letter of the

"Panopticon" design), many other "experts" favored asylum plans and architectures which departed markedly from Benthamite lines.

The spatial organization of asylums: views from the *Asylum Journal*

An understanding that featured in numerous contributions to the *Asylum Journal* was neatly voiced by J. T. Arlidge in his claim that

> the asylum is a machine through and by which the superintendent is to work out and develop his system of moral management; . . . in short, it is an instrument of cure, of the adequacy, utility and perfectness of which he must be the best judge (Arlidge 1858: 188–9).

And, writing in a similar vein, W. H. O. Sankey argued that success in the "long-continued siege" otherwise known as the treatment of insanity

> depends so greatly upon the regulation of the ordinary occupations and pursuits of the patient, and the facilities for these, upon the architectural arrangements of the building, that the form of the edifice and the disposition of its offices becomes of paramount importance to the medical attendant (Sankey 1856: 467).

The asylum was hence viewed as a machine – a tool, an instrument – which had to be properly constructed, maintained and operated if it was to assist the medical officers in their vital and onerous task of curing the insane. It was not supposed to be pretty, although this did not mean that it needed to be ugly, and it is revealing that on one occasion J. C. Bucknill (1855a: 4) explicitly objected to the niceties of "construction and arrangement" being stressed in some establishments over and above the pursuit of correct medical practices: as he exclaimed, "what a pity that so beautiful an appearance should have no brains." The attack here on the splendid stonework and landscaped drives of asylums imitating the York "Retreat" was but thinly veiled.

For Arlidge, Bucknill, and Sankey, therefore, the organization of institutional space had to be *functional*: it had to conform with a particular set of objectives which, rightly or wrongly, these "mad-doctors" believed to be invaluable in the medical assault on madness. Furthermore, it is evident that these doctors had inherited a Benthamite desire to manipulate spatial organization to maximize the "inspectability" of their charges. This desire certainly energized many of Sankey's proposals, which were the first on such matters to be published at any length in the journal, and here he emphasized – in a passage where "constant surveillance" replaced Stark's "unseen eye" – that

> it is essential in all intercourse with the patients that the attendant's conduct should be soothing. It must never be distrustful; but above

270

all, while a constant surveillance is necessary, it is important that he be not obtrusive or unnecessarily interfering. The day-rooms should be constructed, therefore, so as to afford the attendant a full view of his patients at a glance, without the necessity of his following them from place to place or dogging them like a spy (Sankey 1856: 470).

Remarks about the need to "see the whole" and to secure a "full view over the entire surface" were thus interspersed throughout Sankey's account as this progressed from a discussion of asylum wards in general – and note his complaint that isolating one ward from another brought "no advantages as far as the general government and discipline of the whole is concerned" (1856: 473) – to a more detailed assessment of day-rooms, dormitories, and galleries. With respect to the dormitories, he advocated the following spatial arrangements:

> if these dormitories were arranged somewhat in the form of a cross, of which the different branches or limbs of the cross were of unequal length, the nurses' apartment was stationed at the point of intersection in the cross, then – in the arrangement of dormitories thus formed – the larger rooms could be devoted to the quieter patients, and of these the quietist would be placed the furthest off. The smaller division could be given to the epileptic and the refractory, and those requiring the more frequent attendance of the nurse (Sankey 1856: 475).

And, as this quote illustrates, Sankey questioned the conventional wisdom that said asylums required numerous single sleeping-rooms rather than a few large dormitories, and in so doing he declared that a system where "attendants cannot see more than one or at most two patients at a time" was a severe impediment to the project of gaining "perfect surveillance" (1856: 474). The whole bundle of recommendations was highly reminiscent of the "Panopticon," then, but there was an intriguing "twist" in this connection because Sankey's striving for "constant surveillance" led him to favor *not* the dividing up of inmates into small groups and the associated division of an institution's floor-space into numerous small units, as was demanded by the "Panopticon," but the *aggregation* of patients into large, well-planned day-rooms and dormitories (Fig. 12.4).

This "twist" was also a feature of Arlidge's proposals for improving the "asylum as machine," by replacing the "ward and corridor" layout of many existing establishments with an alternative layout dependent upon having ground-floor day-rooms kept separate from upper-floor sleeping facilities. Arlidge objected to the usual arrangement of corridors and galleries flanked by small day-rooms, many of which would double as sleeping-rooms, chiefly because he reckoned this arrangement to be a hindrance to the "effectual superintendence and watching of patients" (Arlidge 1858: 191). And he was no happier with a related plan in which corridors contained large recesses *in lieu* of day-rooms, since this plan still

offer[ed] too many opportunities in its space and structure for those

Figure 12.4 Arlidge's proposals for improving the "Asylum as Machine." Source: J. T. Arlidge 1858. On the construction of public lunatic asylums. *Asylum Journal of Mental Science* **4**. Facing p. 194.

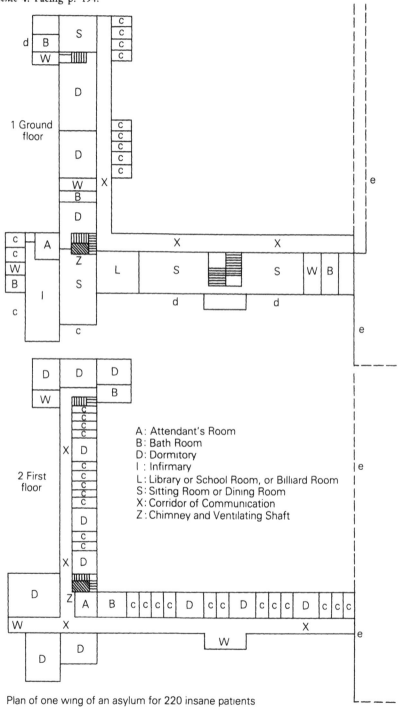

Plan of one wing of an asylum for 220 insane patients

who will to mope, and for the disorderly to annoy their neighbours without attracting the attention of attendants (Arlidge 1858: 191–2).

The solution, so Arlidge explained, was to provide large ground-floor day-rooms (a mixture of sitting-rooms, dining-rooms and recreation-rooms) where, alongside other advantages, "supervision becomes much more easy" as a result of there being "no corners or hiding-places" (1858: 198). Indeed, for both the medical superintendent and his underlings,

> the patients will be more immediately and constantly under their eye, than when distributed in a corridor and connected rooms; their requirements will be sooner perceived and more readily supplied; their peculiarities better detected and provided against, their insane tendencies more easily controlled and directed (Arlidge 1858: 198).

Arlidge reasoned in a similar fashion when considering the upper-floor sleeping facilities and, although he was less assertive than Sankey on this question, he evidently wished to see large dormitories in direct communication with the night-rooms of asylum attendants, given that this proximity constituted a "salutary check on conduct from the knowledge of the attendants being close at hand" (1858: 198–9). It should not be difficult to see that these recommendations from Arlidge and Sankey actually made a virtue out of herding asylum inmates *together* in both day-rooms and dormitories, the argument being that such aggregation would enhance inmate visibility to asylum staff, and it should also be obvious that this aggregation could lead very directly to the "clamour and rumble" which confronts Tennyson's hero in *Maud*.

Many other papers in the *Asylum Journal* contained lines or passages that imitated the visions of Arlidge and Sankey, but it should not be thought that every contributor agreed with their ideas. In one paper, for instance, Bucknill warned that the advocacy of large dormitories was obscuring the need for asylums to retain numerous *single* sleeping-rooms; he despaired that the controversy between advocates of the "big room" and supporters of the "little room" was as sharp as that between the "little endians" and the "big endians" on the island of Lilliput (Bucknill 1857b: 286), and he also worried that erecting many-storied asylums would bring with it serious drawbacks. In this latter respect he feared that those patients forced to "live aloft" would be "removed from the enjoyment of air and exercise," and he also suspected that a surveillance problem could arise from these inmates being visited less regularly by the asylum authorities (1857b: 286–8). As this brief gesture towards incompatibilities in the views of different "mad-doctors" should indicate, the arena of argument – or, as Foucault might say, the "field of discourse"[18] – was a tangled and embattled one. I would like now to consider some even more fiercely contested ground where several quite different blueprints for how to organize institutional space were struggling for supremacy. It is possible to identify three such blueprints from the pages of the *Asylum Journal*, each of which stood in a rather different relation to the concerns for separation and inspection enshrined in Bentham's "Panopticon": the first called for the "monster asylum," the second for a "detached block system," the third for a "cottage system."

Big asylums

One of the most remarkable papers to be found in the journal during the mid-19th century was that by Joseph Lalor, who boldly declared himself to be

> of opinion that asylums of large size are the best adapted for the curative and humane treatment of the insane (Lalor 1861: 105).

It is essential to grasp that his argument was not motivated solely by questions of economy, which was the principal reason for county and borough Visiting Magistrates permitting a mid-century explosion in asylum size,[19] for Lalor was unswerving in his belief that

> large asylums are not only the most economic, but what is of still greater importance, they can be made to present superior means of amusement, of employment and of instruction in much greater abundance, variety and perfection than small asylums (1861: 106).

His thesis involved several components, one of which stemmed from an equation of "moral treatment" for lunatics with techniques for educating the insane, and which thereby suggested that educational improvements leading to a "considerable diminution in the amount of individual teaching amongst the sane classes" should be paralleled by improvements allowing a reduced involvement in everyday "moral treatment" on the part of an asylum's resident superintendent (the "moral governor of old"). In other words, since it was becoming common to teach the sane in large numbers rather than in one-to-one situations, Lalor could see no reason why the insane should not be similarly treated (1861: 105–6). A second component of his thesis suggested that difficult or refractory patients might actually benefit from the example of "more orderly inmates," and that encounters between these different classes of patient would be more common in large establishments than in small ones where "separation" remained a key watch-word (1961: 109).[20]

A third – and perhaps the most telling – component of Lalor's thesis enlarged upon the sorts of claims registered by Arlidge and Sankey, and in so doing stressed the contribution that large asylums could make to the cause of "general supervision":

> Restraint and harsh treatment, so easy to practice and so difficult to detect in an asylum which is a complicated network of long and puzzling corridors, and of small or single cells, could not escape observation in those immense halls where large masses of patients in association would be exposed to the notice of each other, of

numerous attendants, and of every passing officer and visitor, assembled as the inmates would be on the most accessible floor and in the most central positions (1861: 110).

As this passage so vividly reveals, then, for Lalor there was a simple correlation between increasing asylum size, increasingly "rational" spatial arrangements and improvements in the "general supervision" of inmates, and seen in this light the strangeness is peeled from his assertion that

> association of large masses of insane people, far from producing the result dreaded by some of general turbulence and confusion, is found to be highly conducive to good order and quietude (1861: 108).

This was not the conclusion that the hero of Tennyson's *Maud* would have arrived at, however, since for him the aggregation of many insane patients led only to a "turbulence and confusion" that was plainly "enough to drive one mad."

Lalor's vision was not a popular one, it should immediately be noted, and statement after statement in the *Asylum Journal*, as well as in the annual reports of the Commissioners in Lunacy, decried as "madness" the escalating sizes of both existing and proposed asylums.[21] For example, despite his favoring of less warren-like establishments, Arlidge (1856: 368) spoke for most "mad-doctors" when complaining that "a large asylum can be no more honourable to a county than a large prison," and also when offering these somewhat sarcastic observations:

> As it is the era for great ships and big guns, so it is that for gigantic lunatic asylums, foremost among the promoters of which are the Visiting Justices of Middlesex. These gentlemen, in spite of the opposition of the Lunacy Commissioners and in antagonism to the decided opinions of all medical men conversant with the insane and their wants, have triumphed over all opposition, and at length succeeded in carrying out their pet project, the construction of two lunatic colonies most cunningly organised for impeding the treatment of the curable insane and for multiplying chronic lunatics (Arlidge 1860: 143–4).

Interestingly enough, Arlidge's remarks about the burgeoning "monster asylums" being impediments to treatment and liable to increase rather than decrease the numbers of "chronic lunatics" were not so far removed from the impression gained by Tennyson's hero. In a more typically sober analysis, meanwhile, the Commissioners in Lunacy declared that

> It has always been the opinion of this board that asylums beyond a certain size are objectionable. They forfeit the advantage which nothing can replace, whether in general management or the treatment of disease, of individual and responsible supervision (Commissioners in Lunacy 1857: 368–9).

Given this tide of official and "expert" opinion, it can easily be appreciated why Lalor's position remained a relatively lonely one.

Fragmented asylums

Faced with the need for existing asylums to expand and for future asylums to house more patients than hithertofor, most contributors to the *Asylum Journal* favored a "detached block system" that was carefully described in an anonymous paper claiming that the county of Surrey required a new public asylum to cater for 650 patients (Anon. 1861a). The author enumerated the advantages of what he referred to here as a "separate block system," and these included "simplicity of construction," "facility of natural ventilation," and "diminished risk of fire" (1861a: 600), but advantages of this specific and technical nature were reckoned to be small

> in comparison with that afforded by avoiding the evil of concentrating vast numbers of insane persons within a limited space, so that each patient to his great detriment becomes, as it were, surrounded by a thick atmosphere of insanity (Anon. 1861a: 600).

The basic principle of the detached or separate block plan was much as these names suggested, in that the objective was to achieve a "decentralization of buildings" while keeping these buildings within "due limits," and ensuring easy communication between the different blocks (1861a: 600). And it is not hard to see this principle at work in the text of the Surrey asylum paper, since here 6 large blocks were to be arranged around a selection of workshop yards and courts, walled airing courts, and patients' pleasure grounds (see Fig. 12.5).

This plan looked nothing like the "Panopticon," of course, and the departure from "Panopticon" principles was neatly symbolized by the way in which rooms occupied by medical officers ceased to be located where they could function as a central "inspector's lodge," and instead were to be sited on the periphery of an asylum's grounds. The architect of the plan in question thought that

> By distributing the residences of the officers, their influence will be at least as much felt throughout the asylum as if, in the more usual manner, they were all placed in the centre of the buildings (1861a: 600).

Having said this, though, the architect still followed the lead of Arlidge and Sankey in accepting that the spatial arrangements *within* individual blocks should "avoid internal galleries and passages" in favor of having large ground-floor day-rooms and large upper-story dormitories, both of which would "reduce waste and facilitate the easy observation of patients" (Anon. 1861a: 601).

The detached block ideal was received with equal enthusiasm in other journal papers, and in an anonymous review of a published text on the

Figure 12.5 An anonymous proposal for a new Surrey Asylum on the "Separate Block." Source: Anon. 1861. Description of a proposed new lunatic asylum for 650 patients on a separate block system for the County of Surrey. *Journal of Mental Science* **7**. Facing p. 602.

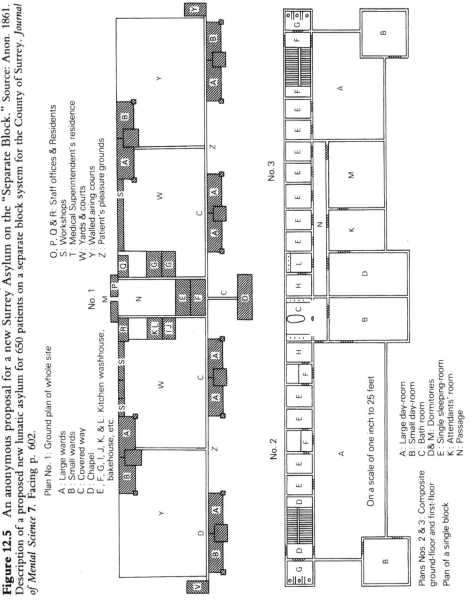

Plan No. 1: Ground plan of whole site

A: Large wards
B: Small wards
C: Covered way
D: Chapel
E, F, G, I, J, K, & L: Kitchen washhouse, bakehouse, etc.

O, P, Q & R: Staff offices & Residents
S: Workshops
T: Medical Superintendent's residence
W: Yards & courts
Y: Walled airing courts
Z: Patient's pleasure grounds

No. 1

No. 2

No. 3

On a scale of one inch to 25 feet

A: Large day-room
B: Small day-room
C: Bath room
D & M: Dormitories
E: Single sleeping-room
K: Attendants' room
N: Passage

Plans Nos. 2 & 3: Composite
ground-floor and first-floor
Plan of a single block

subject, Dr. Fairless's call for asylums comprising ten detached houses clustered about a central complex of refractory wards and an infirmary, it was claimed that

> It will not be needful for us to observe how much we prefer this system to the old model of asylum architecture, founded upon the long galleries and gloomy cells of desecrated monasteries, which were the first asylums (Anon. 1861b: 149).

Elsewhere Bucknill, who tended to have reservations about most drawing-board geometries, lent his guarded support to detached blocks so long as these did not become too "industrial" in character (1857a: 29–30). In addition, he was clearly pleased with the separate building that was erected at the asylum of which he was superintendent (the Devonshire Public County Lunatic Asylum near Exeter), since this was completed "with the best results, both in regard to economy of construction and efficiency of arrangement" (1858), and it might also be argued that the logical consequence of his dislike for asylums which expanded upwards into three- or four-story edifices was a preference for asylums which spread sideways into either additional wings or neighboring constructions.[22]

There can be little doubt that much of the impetus for using detached blocks emanated from the Commissioners in Lunacy, and it was as much to them as to the "mad-doctors" writing in the *Asylum Journal* – though subtle differences in the respective visions should not be overlooked – that responsibility for "breaking up" the built fabric of earlier asylum provisions must be allocated. In their *Tenth Annual Report*, for example, the Commissioners recommended that the best way to cope with the growing number of chronic or incurable patients in public asylums was

> not by making additions to the main structure, which must almost necessarily partake of the expensive character of the original building, but by erecting detached day-rooms and associated dormitories (1856: 524–5; see also Commissioners in Lunacy 1859: 537).

Furthermore, the Commissioners felt sure that

> the patients derive a direct benefit, in many ways, from residing in cheerful airy apartments detached from the main building and associated with officials engaged in conducting industrial pursuits (1856: 525, 1859: 537).

But there was something of an incoherence in this argument, for the Commissioners appear to have confused their advocacy of detached blocks for chronic cases (which would be less expensive to erect and operate than ordinary asylum accommodation, simply because chronic cases could seemingly derive no benefit from more elaborate provisions) with their advocacy of "separate residences" as a "probationary system for patients

278

who were convalescing" (1856: 525). This confusion raises an intriguing question about whether the Commissioners genuinely wished to have two different sorts of detached block on offer, one for incurables and one for convalescents, or whether they were really seeking to camouflage – or even to deny to themselves – a policy of "fobbing off" incurable patients with inferior provisions.[23]

No asylums

For some contributions to the *Asylum Journal* even the use of detached blocks looked too much like a punitive response to the problem of insanity, and for them the real needs of mad people called for a therapeutic regime that functioned outside the institution in a manner not unlike the "care in the community" ideal of more recent years. An extremely zealous ambassador on behalf of just such a non-institutional regime was the much-travelled J. Mundy, who peered beneath the attractive front presented by most asylums, whether public or private, to discover the real failings, confusions, brutalities, and miseries typical of everyday life in the majority of these establishments:

> Is not the systematic, unrestricted and indiscriminate sequestration of all the insane, which is still held as an established principle, a piece of barbaric routine that must be broken through, and the more so inasmuch as this proposed change has been proved, scientifically and practically, both possible and necessary? (Mundy 1861: 352–3).

And for Mundy (1861: 367) it was obvious that the shattering of this "barbaric routine" depended upon there being a massive diffusion of what he called the "patronal or family regime" away from the "one spot in the world" where it was already operated "on a large scale," in the Flemish district of Gheel. Here, in continuation of an ancient tradition, a whole village was involved in caring for an immigrant population of lunatics sent to Gheel from all over Belgium and beyond, and the practice was that almost every peasant family would receive into their home one or two lunatics for time periods decided upon by the colony's medical authorities. These lunatics hence traded the walls, bars, and gates of the asylum for the small cottages (some being brick-built; others being mere mud constructions) and the surrounding fields (where the lunatics labored alongside their hosts) that comprised the physical components of this remarkable "asylum without walls," and in making this trade the lunatics entered into a working family environment that would supposedly hasten their return to sanity quicker than would the prison-like qualities of the environment from which they had recently escaped.[24]

There are intriguing parallels here with Tuke's attempt to create a family atmosphere at the York "Retreat" (and also with the colony for young delinquents at Mettray (see above)), and much more could be said about why the *family* – and in particular why the family dominated by a strict but "fair" father figure – should have been seen as such a therapeutic influence on the insane and other "outsider" groups. However, it is notable that virtually everything written about Gheel in the *Asylum Journal*, even by relatively sympathetic commentators such as Dr. Coxe (1861) and Dr. J. Sibbald (1861), downplayed the "moral" aspects of the family regime in favor of evaluating the performance of Gheel against the medical and administrative criteria with which most of the journal's subscribers were familiar. And it is also clear that most commentators reckoned Britain's mid-century asylum system still to be performing exceedingly well – Sibald (1861: 31) remarked upon the "beautiful structures" for the insane which in England expressed "nearly the perfection of our present system" – and this meant that the thought of doing anything more than tinkering with existing provisions was not all that seriously entertained, except by a handful of enthusiasts such as Mundy. None the less, a measure of tinkering *was* contemplated, and it is interesting to learn that the example of Gheel was interpreted in some quarters as a justification for experimenting with detached blocks. Indeed, as Bucknill suggested in one address when voicing the need for British "experts" to inspect the merits of the Gheel colony for themselves:

> Although we may . . . not be persuaded that it is good to scatter lunatics over the face of the country, it is very possible that we may be convinced that it is better to scatter them over a large asylum area (Bucknill 1861: 313).

Furthermore, in an early issue of the *Asylum Journal* one writer (probably Bucknill 1855: 197) proposed that the authorities of public asylums should seek to cope with overcrowding, not by farming harmless chronic cases out to local workhouses, but by establishing "small colonies" of cottages or "small houses" which would be under the management of an attendant appointed by – and which would also need to be located at not too great a remove from – the "parent asylum" itself.

The cottages and fields of Gheel's insane colony bore little resemblance to the organization of institutional space usually advocated in the *Journal*, of course, but this is not to imply that spatial arrangements at the colony were entirely without a logic comprehensible to visitors. For instance, one visitor reported that

> patients are classified according to the very same principles which regulate classification in ordinary asylums. They are placed in groups or zones, which form the substitute for asylum wards. Thus, in the village and larger hamlets are placed the quiet, better behaved and more industrious of the patients, whilst the more noisy, dirty and least manageable are placed in the remoter hamlets and in the separate houses of the Winkelomsheide (Coxe 1861: 586).

The reasons behind this geography apparently included a desire to prevent the peace of more frequented places, of the village and the larger hamlets, from being disturbed by noisy, troublesome, and occasionally obscene patients, but it also arose because families living in these places tended to be "above the pecuniary necessity of receiving such patients", and so left them for the poorer families living in the remoter or less fertile districts (Coxe 1861: 586). Moreover, it is obvious from other observations dotted throughout Coxe's paper that this spatial patterning was reflected in the poorer accommodation, the greater incidence of fever, and the more common resort to mechanical restraint experienced by lunatics boarded out in the remoter or less fertile districts.[25] And, having unearthed these associations, Coxe arrived at the intriguing conclusion that

A visitor who should inspect merely the houses of the village, or of the larger hamlets, would be apt to form too high an opinion of the Gheel system of treatment; but, on the other hand, one who visited merely the remoter hamlets and the scattered cottages of the Winkelomsheide would infallibly fall into the opposite error. Herein, accordingly, may consist the explanation of the very different accounts which have been published of Gheel (1861: 586).

The geography of Gheel was hence one of small, isolated, and fragmented units into which the colony's authorities could only occasionally glance, and as such it could scarcely have been more different from a "Panopticon" asylum in which every nook and cranny was constantly visible to the "unseen eye" of the establishment medical staff. Unsurprisingly, then, a common criticism of the colony – and hence of the "cottage system" more generally – was that its spatial arrangements militated against the effective inspection and more active supervision of inmates. As Sibbald pointed out:

The most obvious defect of Gheel, as it at present exists, is the inadequate amount of medical superintendence. To inspect regularly and watch over the progress of 800 insane persons scattered over a district nine miles in diameter, is more than the present staff can overtake. . . . Consequently, two important evils must result; those cases which pass into forms requiring immediate attention and treatment are liable to be neglected, and instances of intentional or unintentional cruelty must pass undiscovered. The close attention which is necessary to obviate these occurrences even in a compact asylum establishes the unquestionable truth of this (Sibbald 1861: 49–50).

Coxe (1861: 589) noted that the difficulty of providing a sufficient level of superintendence was exacerbated by the medical director not being able to afford a horse, and Dr. Henry Stevens, a particularly damning critic of the "cottage system," repeatedly stressed that Gheel's "want of proper supervision" was responsible for countless patients being ill-treated, placed under mechanical restraint to prevent them from running away – and in the process being "fettered like the hedge-side vagrant donkey" – or turned into "semi-mechanical drudges" working ridiculously long hours on the land (1858: 433–5). It would appear that in the mind of a critic like Stevens the fact that Gheel was not run as some vast "Panopticon" asylum could explain all of the evils visited upon its inmates, and this fact undoubtedly blinded him to the success that Gheel could boast in fostering what Sibbald (1861: 43) saw as the "general contentment manifested by the insane."

Concluding comments: taking different institutions seriously

There can be little doubt that, even if the focus of inquiry is narrowed to a small group of officials and "experts" writing in quite specific publications such as the *Asylum Journal* or the annual reports of the Commissioners in Lunacy, the organization of space in asylums proposed – and subsequently in asylums modified or built from scratch – rarely followed the letter of Bentham's "Panopticon," even though inspection and the related task of supervision were often matters debated at great length. Indeed, even when the objective of improving inspection was given prominence, as in the conceptions of Arlidge, Lalor and Sankey, their proposed plans and architectures certainly did *not* proceed from Dear's "fundamental principles of isolation and separation". Rather, they depended upon deliberately *overcoming* isolation and separation in favor of bringing patients *together* in large open spaces where, instead of central areas being occupied by inspectors looking out at inmates on the periphery, central areas were to be filled with inmates under observation by medical officers and attendants who might be circling that periphery. Furthermore, many contemporary officials and "experts" did *not* regard inspection as such a key feature of therapeutic regimes, and for these commentators – and here I would mention Bucknill, the anonymous advocate of a new Surrey asylum, and also the Commissioners in Lunacy – questions of economy, ease of construction, accessibility of buildings, improving ventilation, reducing fire risk, and providing more homely surroundings for patients all came to the fore. But herein lay another curious "twist," for in seeking to tackle these additional questions, and in striving to dispel the "thick atmosphere of insanity" present in "monster asylums," the resulting plans and architectures began to necessitate the scattering of patients in detached blocks spread across an asylum's grounds. This meant that, while the aim of manipulating space to secure constant and minute inspection had disappeared at the scale of the asylum as a whole, the notion of dividing up the spaces of the asylum had actually returned to the drawing-board. And this was even more obviously the case for those writers – notably Mundy, but also to some extent Coxe and Sibbald – who were sympathetic to a "cottage system" that distributed patients across a landscape, and which substituted the "apparently anti-institutional and natural weapon" of the family for the indelibly institutional and imposed weapon of the "inspector's lodge." In short, then, the "Panopticon" demanded an equation of inspection with the separation of inmates and their spaces; the proposals of Arlidge, Lalor, and Sankey stressed the need for inspection, but sought to improve the visibility of inmates through aggregation rather than separation; the proposals of Bucknill, the Surrey asylum architect, and the Commissioners in Lunacy widened the range of objectives which could be achieved through the organization of space, and in so doing urged a scattering of inmates and their spaces for reasons which in no way reflected Bentham's equation of inspection and separation. Finally, in the most extreme version of the "detached block system" – namely, the "cottage system" – this Benthamite equation was lost altogether, and in the mind of an enthusiast like Mundy any hindrance to central inspection caused by the dispersion of patients in their far-flung cottages was more than offset by the therapeutic advantages of life in a normal family environment.[26]

The history of asylums has hence been far from full of "Panopticon" asylums, and neither has it been traversed by a timeless and universal equation of inspection with the "fundamental principles of isolation and separation," but this is not to suggest that a Foucauldian "panopticism" – as a subtle "calculus of power" acting on human materials to produce "docile bodies" – was not present in all manner of different institutional and non-institutional arrangements designed to shelter, restrain, and cure the 19th-century lunatic. Foucault did not identify "panopticism" too closely with the "Panopticon," despite occasional indications to the contrary, and neither should we. And neither did he regard "panopticism" as some neat and coherent *a priori* package that he would expect to find in all its neatness and coherence wherever he looked for it in the 19th century, and this is why he warned that

> There can be no question here of writing the history of the different disciplinary institutions, with all their individual differences. I simply intend to map on a series of examples some of the essential techniques that most easily spread from one to another (Foucault 1979: 139).

In other words, Foucault remained sensitive to *difference* and, by implication, to the uneven diffusion of different disciplinary techniques, and it is this sort of sensitivity that is needed when researchers examine the history of asylums. Asylums both actual and proposed often looked very different from one another, and they often took on board different techniques in different measures and sometimes even for very different immediate purposes, and it is vital that these many differences are not conceptually collapsed into the workings of one single institutional blueprint, whether this be the "Panopticon" or any other model institution for the insane. It is precisely this kind of maneuver that Foucault (1972: 9–10) objects to when seeking to replace *total history* – or "totalizing" accounts insensitive to differences fragmenting the unity of past happenings, periods, and places – with a *general history* sensitive to difference and to the way in which the fragments of past realities *do* relate to one another, but not in the sense of adding up to some tidy, articulated, and (grand) theorizable whole.[27] And seen in this light it is easier to appreciate why he speaks of differences between disciplinary institutions, but is still prepared to discuss a "panopticism" that infused into numerous 19th-century prisons, schools, hospitals, and asylums without ever dictating the precise arrangements, spatial or otherwise, that these were to display.

What this heightened sensitivity to differences between asylums also highlights, moreover, is that there was no single and straightforward relationship between 19th-century asylums and the socio-spatial reproduction of madness. It should be noted at once that, except in the case of proposals for cheap accommodations in which the chronically insane could live out their lives, the majority of arrangements proposed – from the prison-like "Panopticon" to the dispersed insane colony – were intended to reform or to cure the lunatic: to produce (or reproduce) sanity where previously there had been insanity. In the long term, however, the practical consequence of having a network of "closed spaces" devoted specifically to mad people was to produce and then continually to reproduce a population designated as different, deviant, and dangerous by "mainstream" society, and in the process to set in motion a "dynamic" leading fairly directly (as Dear indicates (see above)) to the present-day exclusion of mental health facilities from respectable suburban communities. In addition, it might be argued that the more short-term consequence of many institutional arrangements (many of which were intimately bound up with the organization of institutional space) was not to cure an inmate's mental unrest, but to reproduce this condition: to exacerbate it and in some cases maybe even to prompt madness where none existed before. This was the complaint voiced by Tennyson's hero when confronted with the "clamour and rumble" of overcrowded asylum wards, of course, but it could equally well be the complaint of a patient experiencing either the solitude and regimentation of life in a "Panopticon" asylum[28] or the hard work and shackles of life in a Gheel-like insane colony. The details of how 19th-century asylums could reproduce rather than settle an individual's mental unrest must have differed considerably from institution to institution, then, but it would certainly seem to be the case that each institution was in its own way a place "enough to drive one mad."

Notes

A version of this chapter was read at a conference on the History of Medical Geography held at the Wellcome Institute for the History of Medicine, 183 Euston Road, London, September 24–5, 1987, and I would like to thank participants at that conference for their encouraging comments and constructive criticism. I would also like to thank Derek Gregory, Roy Porter, and Jenny Robinson for their thoughts and assistance. A special note of thanks is owed to Felix Driver, however, since it has been through regular discussions with him that many of the themes explored here have emerged and acquired shape.

1 Tennyson's *Maud* is by no means the only work of 19th-century literature or poetry which I could have drawn upon here to illustrate the chaos of people, activities, and spaces in contemporary asylums, but I use it in preference to

other works because a commentary on its depiction of both madness and the career of a mad person appeared in the *Asylum Journal* (Bucknill 1855). This particular journal is an important source for my discussion elsewhere in the chapter, and so it seemed appropriate to follow its example in this respect as well.

2   Much controversy reigns over the question of whether mental distress and disturbance really amounts to an *illness* little different from any physical ailment, or whether labels such as "mental illness" are simply used as an instrument of *social control* through which "mainstream" society pathologizes and polices people whose only crime is to be slightly out of the ordinary.

3   In developing this theoretical perspective Dear draws heavily on the distinction that Giddens (1973) introduces between *proximate* factors, which include authority relations and distributive groupings, and *mediate* factors of class "structuration," and he also draws upon the work of geographers such as Hägerstrand, Harvey, Peet and Soja.

4   I provide a detailed inquiry into the cross-cutting theoretical and empirical endeavours of Dear and his various co-workers in the second chapter of my PhD thesis, which is currently in preparation (Philo 1988b).

5   This is a central theme of Foucault's magisterial *Madness and civilization* (1967), in which it is argued that the incarceration of madness in asylums, coupled with the tendency to cloak madness in the technical vocabularies of "mental illness," amounts to a *silencing* of the "other" that is Madness, and to the triumph of the "merciless language of non-madness" that is Reason. Seen in this light isolation and exclusion appear as relatively recent phenomena: as developments quite alien to earlier ages when a more open dialogue (buttressed by everyday encounters) was sustained between Madness and Reason.

6   In addition, the use of those "well-known and most effectual contrivances for seeing without being seen" (Bowring 1843, IV: 44) – in fact an elaborate system of blinds – would allow inspectors to look out on their charges, but would not allow these charges to see back into the "inspector's lodge."

7   Bentham also argued that the inspectors should themselves be open to inspection, for the "Panopticon's" doors were to be flung open to the public – "the great open committee of the tribunal of the world" (Bowring 1843, IV: 46) – who would be allowed to file in and out of the "inspector's lodge" at will.

8   As Bentham stated at one point, separating inmates would allow "1. Preservation of health from infection. 2. Preservation of morals from corruption. 3. Preservation of decency. 4. Prevention of unsatisfiable desires" (Bowring 1843, vol. 8, 372). But it should also be noted that he did countenance a measure of "appropriate aggregation" alongside separation, and this he would permit in the cases of "1. Matrimonial society. 2. Family society. 3. Nursing attendance. 4. Medical attendance. 5. Moral superintendence. 6. Instruction and direction of labour. 7. Intercommunity of work and labour" (Bowring 1843, vol. 8, 372).

9   The activities of inmates were to follow a strict chronological format through every minute of the day and night, and note that Foucault (1979: 149–62) discusses the temporal "control of activity" – the tyranny of time-tables; of marching rhythms; of drill – through which "docile bodies" might be produced.

10   Numerous commentaries and critiques have been spawned by *Discipline and punish*, but a contribution that I have found particularly useful is Driver 1985a.

11   Foucault dissects a subtle "calculus of power" that emerged in Western Europe

during the years pivoting about 1800, and which served the purpose – which was not conceived of by anybody in particular – of producing and reproducing a population weighed down with notions of responsibility, duty, self-restraint, and so on. In this way people became "docile bodies" without having to be terrorized into obedience by the spectacle of gruesome executions and tortures – those bloody manifestations of monarchical power – and Foucault hence describes "panopticism" as "the general principle of a new political anatomy whose object and end are not the relations of sovereignty (in which execution and torture were so deeply implicated) but the relations of discipline" (1979: 208).

12 Note the emphasis here upon *reproducing* the "art of power relations."

13 With this passage Bentham signals the way in which, while an "appropriate establishment" *would* bring large numbers of lunatics together in one place, the "Panopticon" principles of inspection and separation would prevent any *real* contact occurring between patients. Note too that Bentham planned to distinguish between the curable and the incurable insane, as well as between the "susceptible and unsusceptible of employment" and the "dangerous and innoxious" (Bowring 1843, VIII: 395).

14 The Commissioners in Lunacy were first established as a "national" body with jurisdiction (powers of licensing, rights of visitation, duties to criticize and propose improvements) over virtually all English and Welsh receptacles for the insane by the Lunatics Act of 1845 (8 & 9 Vict., c.100).

15 The term "moral architecture" is associated not so much with Tuke as with a North American "mad-doctor" called Kirkbride, who used institutional environments in a therapeutic fashion akin to Tuke (see Tomes 1981).

16 In the latter text Foucault suggests that Tuke replaced physical or mechanical restraint of lunatics with the "stifling anguish" of a "self-restraint" which insisted upon lunatics playing a crucial role in purging themselves of thoughts and actions unacceptable to "reasonable" society (1967: 241–55; see also Doerner 1981: 78–84; cf. Digby 1985). Seen in this light Tuke's objectives do not look so very different from Bentham's, and note that Bentham himself declared that the application of the "inspection principle" in madhouses would "render the use of chains and other modes of corporal sufferance unnecessary" (Bowring 1843, IV: 61).

17 I discuss this journal, and use its contents to probe the question of asylum siting in mid-19th century England, in Philo 1987b. My study here, as in Philo 1987b, is based upon a close reading of the journal's first 40 editions, as published between November 1853 and January 1862.

18 In this chapter and elsewhere (Philo 1987b) my analysis springs from a Foucauldian understanding of how oral and written "discourses" are constructed, fought over, and productive of tangible institutional develop- ments. I provide a more sustained theoretical account of this Foucauldian position in Philo (1988a).

19 In 1819, when there were six public county lunatic asylums, the average number of residents per establishment was 74.50: in 1844, when there were 21 asylums, the average number of residents had risen to 304.71; by 1860, when there were 37 asylums, the average number had risen still further to 456.43. This increase in numbers was due in part to a widening of the definitional "net," and in part to the accumulation of chronic cases who, notwithstanding the therapeutic optimism of many early reformers, could not be restored to reason.

20 This line of reasoning was foreshadowed in Tyerman (1857), where an

argument was pursued about the "moral, social and curative results likely to accrue" from difficult patients witnessing the "examples of their tranquil associates": hence, "some patients formerly prone to fill their pockets with food, and to gnaw their rations instead of using their knife and fork, have rectified these habits" (1857: 114–15).

21 In the *Asylum Journal* (see Anon. 1858, Arlidge 1856, 1860, Bucknill 1857a, 1859, Conolly 1859). With respect to the Commissioners in Lunacy, see their *Annual Report* 1857. In most cases the criticism of "monster asylums" in principle shaded into an attack on the massive enlargements sanctioned by the Visiting Magistrates of the two public county asylums serving Middlesex, Colney Hatch and Hanwell.

22 Bucknill (1857b: 287) detected a locational control in this respect: "in densely populated towns where land is sold by the foot and by the inch, it becomes a matter of economy to erect a lofty many-storied building; but in the country districts where asylums are placed, the value of building space is without weight." In Bucknill's view, asylums *should* be sited in "country districts" (see Philo 1987b), and so the need to build upwards should never have arisen.

23 Most of the officials and "experts" were finding it hard to accept that many cases really could not be returned to "normal," and there was a widespread reluctance to let go of the therapeutic optimism that had carried through the reforms of the early 19th century.

24 Sibbald (1861: 51) explicitly compared the family environment encouraged at Gheel with the "part palace, part barrack, part prison" experienced by most asylum inmates, and he stressed the therapeutic value of the former to *certain* types of mental patient: "to be boarded in a cottage with a family of his own rank in life, joining in the family meals, to watch the amusements or the employments of the children, to take part in the various incidents of home life, with the probability of having healthy affections and emotions excited by the social relations between himself and his new friends, and to mix little with any who would be likely to encourage feelings of discontent, would afford to such cases all the advantages which the most benevolent philanthropy could devise."

25 See Coxe 1861: 585 for remarks on accommodation; 1861: 587–8 for remarks on illness; 1861: 587 for remarks on restraint. Note that the Winkelomsheide was a "miserable sandy heath" (1861: 582) from which it was barely possible to eke a living through agricultural pursuits.

26 This summary perhaps overstates the differences between many of the officials and "experts" concerned: in practice there was really a "gradation" of views from support for a "Panopticon" asylum to support for a dispersed insane colony, and there were certainly many overlaps between – and often only subtle disagreements separating – the views of most commentators.

27 Foucault is hence very suspicious of intellectual exercises that reduce all of the complexity of history and society to the workings of some foundational "spirit," "principle," "mechanism" or whatever: as he says, "nothing is fundamental; that is what is interesting in the analysis of society" (1982: 18). This means that his words dovetail neatly with – and in many ways have been a crucial influence in the construction of – a *post-modernist* critique of the Enlightenment search for coherence, transcendental truths, and an enduring rationality (see Gregory 1987, 1988). But Foucault does not suppose that this line of argument leaves researchers with nothing to do but contemplate differences and celebrate fragmentation; rather, he urges the study of how differences connect up in particular historical and geographical contexts, and

also the study of the very real human conflicts, miseries, creations and happinesses that are revealed once the quest for all-consuming and ordered "depth" explanations is abandoned.

28 It should be noted that concern was commonly expressed in the 19th century on the question of whether or not solitary confinement was likely to disturb the mental health of prisoners, and it was even the case that Bentham modified his initial "Panopticon" plan – where single-person cells had been called for – to allow for two-, three- or four-person cells, in part because this modification made the plan more economically viable, but in part because he feared that complete isolation might interfere with the mental wellbeing of inmates (see Bowring 1843, vol. 4: 71–6). Moreover, when Field (1858) put forward proposals for a "separate system of imprisonment" which combined residence in single cells with regular visits from prison staff and occasional activities in common with other prisoners, he had to argue that his favored system would not be harmful to the sanity of inmates.

References

Anon. 1855. The accumulation of chronic lunatics in asylums: question of further accommodation. *Asylum Journal* 1, 193–8.

Anon. 1858. The new lunatic hospital at Nottinghamshire: laying the foundation stone. *Asylum Journal of Mental Science* 4, 296–304.

Anon. 1861a. Description of a proposed new lunatic asylum for 650 patients on the separate block system, for the County of Surrey. *Journal of Mental Science* 7, 600–8.

Anon. 1861b. Review of Fairless's "Suggestions concerning the construction of asylums for the insane." *Journal of Mental Science* 7, 149–50.

Arlidge, J. T. 1856. The Middlesex County Lunatic Asylums and their reports for 1855 and 1866. *Asylum Journal of Mental Science* 2, 354–79.

Arlidge, J. T. 1858. On the construction of public lunatic asylums. *Asylum Journal of Mental Science* 4, 188–204.

Arlidge, J. T. 1860. Thirteenth Report of the Commissioners in Lunacy. *Journal of Mental Science* 6, 141–56.

Bowring, J. (ed.) 1843. *The works of Jeremy Bentham.* Simpkin, Marshall & Co.: London (see vol. IV: 37–172 for "Panopticon, or the Inspection-House"; vol. VIII: 358–461 for "Tracts on Poor Laws and Pauper Management").

Bucknill, J. C. 1855a. Ninth Report of the Commissioners in Lunacy. *Asylum Journal of Mental Science* 2, 1–16.

Bucknill, J. C. 1855b. Tennyson's "Maud" and other poems. *Asylum Journal of Mental Science* 2. 95–104.

Bucknill, J. C. 1857a. Tenth Report of the Commissioners in Lunacy. *Asylum Journal of Mental Science* 3, 19–30.

Bucknill, J. C. 1857b. Asylum architecture and arrangements. *Asylum Journal of Mental Science* 3, 285–96.

Bucknill, J. C. 1858. Description of the new house at the Devon County Lunatic Asylum, with remarks upon the seaside residence for the insane which was for a time established at Exmouth. *Asylum Journal of Mental Science* 4, 317–28.

Bucknill, J. C. 1861. Valedictory address at the Annual meeting. *Journal of Mental Science* 7, 310–18.

Commissioners in Lunacy 1847. *Further Report* (Reports from Commissioners, vol. XXXII, 1847–'48).

Commissioners in Lunacy 1856. *Tenth Annual Report* (Reports from Commissioners, vol. XVIII, 1856).

Commissioners in Lunacy 1857. *Eleventh Annual Report* (Reports from Commissioners, in British Parliamentary Papers, vol. XVI, 1857, Sess. 2).

Commissioners in Lunacy 1859. *Thirteenth Annual Report* (Reports from Commissioners, vol. XIV, 1859, Sess. 2).

Conolly, J. 1859. President's address at the Annual Meeting. *Journal of Mental Science* 5, 71–8.

Coxe, Dr. 1861. Lunacy in France and at Gheel. *Journal of Mental Science* 7, 560–89.

Dear, M. J. 1980. The public city. In *Residential mobility and public policy*, W. A. V. Clark & E. G. Moore (eds.), 219–41. Beverly Hills: Sage Publications.

Dear, M. J. 1981. Social and spatial reproduction of the mentally ill. In *Urbanization and urban planning in capitalist society*. M. Dear & A. J. Scott (eds.), 481–97. London: Methuen.

Dear, M. J. & G. L. Clark 1984. *State apparatus: structures and language of legitimacy.* Boston: Allen & Unwin.

Dear, M. J. & S. M. Taylor 1982. *Not on our street: community attitudes to mental health care.* London: Pion.

Dear, M. J. & J. Wolch 1987. *Landscapes of despair: from deinstitutionalization to homelessness.* Oxford: Polity Press.

Digby, A. 1985. *Madness, morality and medicine: a study of the York Retreat, 1796–1914.* Cambridge: Cambridge University Press.

Doerner, K. 1981. *Madmen and the bourgeoisie: a social history of insanity and psychiatry.* Oxford: Basil Blackwell.

Donnelly, M. 1983. *Managing the mind: a study of medical psychology in early-nineteenth century Britain.* London: Tavistock.

Driver, F. 1985a. Power, space and the body: a critical assessment of Foucault's "Discipline and Punish". *Environment and Planning D: Society and Space* 3, 425–46.

Driver, F. 1985b. Geography and power: the work of Michel Foucault. Unpublished typescript of paper read at the Annual Conference of the Institute of British Geographers, University of Leeds, January 1985.

Evans, D. M. 1978. Alienation, mental illness and the partitioning of space. *Antipode* 10, 13–23.

Evans, R. 1971. Bentham's Panopticon: an incident in the social history of architecture. *Architectural Association Quarterly* (July), 21–37.

Field, J. 1858. *Prison discipline and the advantages of the separate system of imprisonment.* London: Longman, Brown, Green & Longmans.

Foucault, M. 1967. *Madness and civilization: a history of insanity in the age of reason.* London: Tavistock.

Foucault, M. 1972. *The archaeology of knowledge.* London: Tavistock.

Foucault, M. 1979. *Discipline and punish: the birth of the prison.* London: Penguin.

Foucault, M. 1982. Space, knowledge and power (interview with P. Rabinow). *Skyline* (March), 17–20.

Giddens, A. 1973. *The class structure of the advanced societies.* London: Hutchinson.

Gregory, D. 1987. Postmodernism and the politics of social theory (editorial). *Environment and Planning D: Society and Space* 5, 245–8.

Gregory, D. 1988. Areal differentiation and post-modern human geography. In *Horizons in human geography*, D. Gregory & R. Walford (eds.). London: Macmillan (forthcoming).

Himmelfarb, G. 1968. The haunted house of Jeremy Bentham. In *Victorian minds*, G. Himmelfarb (ed.), 32–81. New York: Alfred A. Knopf.

Lalor, J. 1861. Observations on the size and construction of lunatic asylums. *Journal of Mental Science* **7**, 104–11.

MacDonald, M. 1981. *Mystical Bedlam: madness, anxiety and healing in seventeenth-century England.* Cambridge: Cambridge University Press.

Mundy, J. 1861. Five cardinal questions on administrative psychiatry. *Journal of Mental Science* **7**, 343–70.

Neaman, J. S. 1975. *Suggestion of the devil: the origins of madness.* New York: Anchor.

Palmer, E. 1854. Description of the Lincolnshire County Asylum. *Asylum Journal* **1**, 73–5.

Philo, C. P. 1987a. "The Same and the Other": on geographies, madness and outsiders. *Loughborough University of Technology, Department of Geography, Occasional Paper No. 11.*

Philo, C. P. 1987b. Fit localities for an asylum: the historical geography of the nineteenth-century "mad-business" in England as viewed through the pages of the *Asylum Journal. Journal of Historical Geography* **13**, 398–415.

Philo, C. P. 1988a. Thoughts, words and "creative locational acts." In *The behavioural environment: essays in reflection, application and criticism*, F. W. Boal & D. N. Livingstone (eds.). London: Croom Helm (forthcoming).

Philo, C. P. 1988b. The space reserved for insanity: studies in the historical geography of the English and Welsh "mad-business." PhD thesis in preparation, University of Cambridge, Department of Geography.

Report from Select Committee on the Better Regulation of Madhouses in England (Parliamentary Reports, Vol. IV, 1814–'15) (including a preamble and four separate reports of evidence collected before the Committee).

Report from Select Committee on the Better Regulation of Madhouses in England (Parliamentary Reports, Vol. VI, 1816) (including three separate reports of evidence collected before the Committee).

Ricks, C. (ed.) 1969. *The poems of Tennyson.* London: Longmans, Green.

Sankey, W. H. O. 1856. Do the public asylums of England, as at present constructed, afford the greatest facilities for the care and treatment of the insane? *Asylum Journal of Mental Science* **2**, 466–79.

Sibbald, J. 1861. The cottage system and Gheel. *Journal of Mental Science* **7**, 31–61.

Sibley, D. 1981. *Outsiders in urban societies.* Oxford: Basil Blackwell.

Sibley, D. 1987. Out damned spot? the purification of space. *University of Hull, Department of Geography, Working Paper No. 1.*

Stevens, H. 1858. Insane colony at Gheel. *Asylum Journal of Mental Science* **4**, 426–37.

Tomes, N. J. 1981. *A generous confidence: Thomas Story Kirkbride and the art of asylum-keeping.* Cambridge: Cambridge University Press.

Tyerman, D. F. 1857. On the association and classification of patients at Colney Hatch, and appropriation of the recreation room as a dining hall. *Asylum Journal of Mental Science* **3**, 114–16.

Wolpert, J. 1976. Opening closed spaces. *Annals of the Association of American Geographers* **66**, 1–13.

Wolpert, J., M. Dear & R. Crawford 1975. Satellite mental health facilities. *Annals of the Association of American Geographers* **65**, 24–35.

Wolpert, J. & E. Wolpert 1974. From asylum to ghetto. *Antipode* **6**, 63–76.

Part V

Civil society

13

Incorporation theory and the reproduction of community fabric

GERALDINE PRATT

The role of homeownership in stabilizing the capitalist political economy is widely acknowledged. It is a source of academic and political debate and the rationale/legitimation of housing policy in many societies. In Canada, for example, the Progressive Conservative Party justified their proposed introduction of mortgage interest and property tax credits on the grounds that, among other things, "People who own homes feel a greater stake in the community and country in which they live" (Conservative Party of Canada, 1979, 3). The psychological impact of homeownership has been taken so seriously in the United States that in 1971 the National Institute of Mental Health convened a Working Conference on the Behavioral Effects of New Opportunities in Extended Property Ownership to determine "whether, besides the behavioral attributes associated with being a renter, those associated with being poor might be expected to change upon becoming an owner" (quoted in Perin, 1977, 61). As a final example, since 1978 the South African government has expanded homeownership opportunities among urban Africans. Mather & Parnell (1986) argue that government policy is an explicit reaction to the Soweto riots of 1976. Government officials believe that expanded homeownership will stabilize African society, with African homeowners unlikely participants in riots, boycotts, and strikes.

Social scientists have also pursued the ways that homeownership sustains capitalist social relations. Their understanding and explanation of this process have changed over the last decade, reflecting the influence of broader epistemological debates. There has been a shift from functionalist explanations drawing on social control metaphors to more nuanced explanations of social reproduction that acknowledge the mutually reinforcing influence of structural forces and individual interpretations, preferences, and actions. In the first section of this chapter, the main lines of incorporation theory, the argument that homeownership incorporates homeowners into the capitalist social order, will be outlined. The evolving nature of explanations of the ideological and political effects of homeownership, away from social control theory to a richer understanding that admits the role of multiple agents, unintended consequences of action, and contingent factors will be stressed. In a second section, the theoretical themes will be examined in context, through an empirical study of political incorporation and homeownership in Vancouver, British Columbia.

Incorporation theory – from social control to social reproduction

There are numerous strands to incorporation theory. First, most households require a mortgage, often amortized over 25 years, in order to purchase a house. Long-term indebtedness is thought to effectively trap workers in their jobs and reduce the likelihood of labor agitation. As early as 1871, Engels cautioned that mortgage debt would have this effect. Writing with reference to workers in the United States, Engels argued that "the workers must shoulder heavy mortgage debts in order to obtain even these houses and thus they become completely the slaves of their employers; they are bound to their houses, they cannot go away, and they are compelled to put up with whatever working conditions are offered them" (1935, 15). Obliged to meet fixed monthly mortgage payments on the threat of losing one's house and equity, workers with mortgages may be less likely to endorse strike action and other forms of collective resistance that would jeopardize their pay packet in the short term.

A second strand of incorporation theory stresses the divisions that homeownership creates within the working class (Harvey 1975, 1976). Working-class homeowners come to identify with their housing status group and fail to see the class–based interests they share with working-class renters. This can happen at work and within the community. In so far as access to homeownership depends on well-paying, secure employment, homeownership reinforces occupational divisions marked by differences in income and job security. Renters and homeowners may also oppose each other in the community, especially over development issues (Heskin 1983, Pratt 1982, Saunders 1979). For example, homeowners may wish to control the development of multi-family rental accommodation in opposition to renters who would benefit from its construction. There has been considerable debate concerning whether such conflicts reflect real differences in the material interests of homeowners and renters or simply reflect perceptions of status differences (Dunleavy 1979, Edel 1982, Pratt 1982, Saunders 1978, 1979, 1984). The details of this debate need not concern us here; the more general point is that housing tenure distinctions introduce divisions within the working class, both at work and within the community, and this is thought to weaken working-class struggle.

The view that homeowners and tenants are divided by perceptions of status rests on the supposition that homeownership has considerable psychological significance; this is a third strand of incorporation theory. The ownership of domestic property gives homeowners a psychological stake in the system. It legitimizes, through active participation, a central legal structure in capitalist society: the ownership of property. Noting this effect, Castells (1977, 161) comments that the suburban development program initiated in France in 1928 was "[i]n perfect accordance with the ideology of integration which wished to bring class struggle to an end by making each worker an owner – outside of his [*sic*] work of course." Many critics believe that the legitimation of private property is won at a small cost because widespread homeownership does not alter the basic (unequal) distribution of wealth in capitalist societies, and it does not free the worker from the necessity of doing waged work (Angotti 1977, Edel *et al.* 1984, Forrest 1983, Gray 1982, Luria 1976). In Luria's view, "homeownership is the reward of the defeated" (1976, 98).

Another dimension of the psychological significance of homeownership centers on the private satisfactions and sense of security and personal control which can be attained through owning a home. In so far as workers create satisfying private worlds in and around the home and community, they may more readily accept the increasingly impoverished nature of work (Agnew 1981a, 1981b, Arendt 1958, Brittan 1978, Handlin 1979, O'Connor 1984, Rakoff 1977, Saunders 1984, Sennett 1974). Work is accepted as a means to an end, a way of sustaining a pleasurable and fulfilling homelife. This changed set of priorities can be read into transformations in workplace struggles, away from attempts to alter the qualitative characteristics of work to an exclusive concern over wage rates and job security. The latter is a narrowed frame of workplace resistance that leaves the social relations of capitalism unchallenged. To the extent that homeownership fosters this narrowed vision, it functions as an agent of social integration.

These are the major threads of incorporation theory; the specifics of the thesis have changed considerably over the last ten years, responding to sustained epistemological and substantive critiques. Three criticisms will be reviewed. First, it has been recognized that the category, housing tenure, must be conceptually unpacked. Second, the conspiratorial interpretation of the process of incorporation has been rejected. Third, a one-dimensional causal account of the ideological effects of homeownership has been abandoned for a more fully contextualized understanding of the interrelationships between ideology, politics, and housing tenure.

A number of authors have noted that homeownership has been "fetishized" (Ball 1986, Gray 1982, Hayward 1986, Sullivan 1986). Homeownership is often ascribed a set of characteristics that are contingently but not necessarily related to the tenure form. For example, Angotti (1977, 43) claims that: "Individual homeownership, along with the private automobile, has reinforced the . . . isolation of working people from one another. It brutally separates workplace from residence, it reinforces family instead of factory relations, and isolates that worker as worker from the worker as consumer." A common problem, exemplified by Angotti's claims, is that the experiences of suburban homeowners are generalized to all homeowners. A home-centered lifestyle divorced from workplace concerns and relations may be typical of suburban but not inner-city homeowners. So too, in several countries, such as Britain and the United States, suburban homeownership is often a better investment than inner-city housing (Gray 1982, Williams 1984). It is likely that the meaning of property ownership varies across the metropolitan area as well, with suburban homeowners more likely to view their domestic property as a form of capital. This has implications for incorporation theory for, as Agnew has shown in the United States (1978), an investment orientation to homeownership is associated with a more privatistic orientation to urban development. It is also typically in suburban areas that a large proportion of homeowners hold mortgages and are therefore vulnerable to the types of pressures outlined by incorporation theory. There is a danger of reifying homeownership and overgeneralizing incorporation theory.

Incorporation theory has also been criticized because it oversimplifies the processes that led to the expansion of homeownership in many countries after World War 2. Homeownership has been presented as a vehicle for social control, engineered by finance capital, the state or capital-in-general (see Edel *et al.* 1984, 185–9, Saunders 1979, for critical reviews). It is often suggested that workers embraced the suburban vision against their own real interests, that suburban homes were, in Edel, Sclar and Luria's (1984) terms, "lawns for pawns."

It is now recognized that social control explanations radically simplify the processes that led to mass homeownership. An adequate explanation requires an understanding of how structural forces combine with individual actions and how the latter can unintentionally, but nevertheless actively, reproduce capitalist social relations.

One cannot underestimate the economic forces that structured the suburban expansion associated with mass homeownership. Over the last 50 years, many national governments have viewed expanded homeowner-ship, not simply as a means of political incorporation, but as a key means of regulating and stimulating the economy. One reason for this is that homeownership and the single-family house are effective vehicles for stimulating consumer demand (Aglietta 1979, Harvey 1982, 1985, Hayden 1981, Mackenzie & Rose 1983, O'Connor 1973, 1984). North American single-family housing developments are distinguishable by their absence of communal facilities, necessitating the purchase by individual households of a considerable array of consumer durables, such as washing machines and recreational equipment. The increased spatial separation between home and work, in fact often associated with homeownership, requires the purchase of transportation; in the North American context it typically creates the need for the purchase of at least one automobile by each household.

This economic explanation for expanded homeownership complements incorporation theory in the sense that both stress the importance of expanded homeownership for the stability of capitalist society. It is increasingly recognized (e.g., Mackenzie & Rose 1983), however, that the form of suburbanization occurring in North America is not a simple reflection of capitalist needs. There is a meaningful distinction between cause and effect; the causes of suburbanization and expanded homeowner-ship cannot be understood only in terms of their success in extending the demand for consumer goods (in other words, by their effects). At the very least, the pattern of suburbanization was the outcome of a contested struggle.

In the United States, the National Association of Real Estate Boards (NAREB) and the National Association of Home Builders (NAHB) lobbied hard for the "suburban solution": the homebuilders spent over $5 million in their efforts to shape the 1949 Housing Act. In fact, their tactics in promoting their suburban vision prompted a congressional investition (Checkoway 1980). The transport policy that led to the expansion of car ownership also resulted from intense lobbying by a coalition of auto-oil-rubber and construction interests, dating from 1916 (O'Connor 1973, Yago 1978). Thus, while suburbanization and expanded homeownership effectively extended mass consumption (and from this perspective can be seen to stabilize the capitalist economy), their development was not an automatic reflection of capitalist needs. Congressional reactions to the lobbying efforts of NAREB and NAHB also suggest that it is a real conceptual mistake to ascribe a single mentality to "the state" or capitalist class.

Further, the economic pressures for expanded homeownership must be balanced with an appreciation that homeownership was also sought after by the working class (Edel *et al.* 1984, Mackenzie & Rose 1983, Rose 1980, 1981). Even if one wishes to maintain that expanded homeownership is not in the long-term interests of the working class, either historically (Edel *et al.* 1984) or in the current context of rising house prices (Ball 1983, Hartman 1983), it is a violation of the historical evidence to suggest that homeownership was pressed on the working class by the state in the interests of the capitalist class. There is now a move towards a more nuanced understanding of the processes that led to expanded homeownership among working-class households, including an appreciation of how working-class households may have unintentionally, but nevertheless actively, reproduced capitalist class relations through attempts to improve their living conditions via homeownership (Edel *et al.* 1984, Mackenzie & Rose 1983, Rose 1980, 1981). In other words, though incorporation theory may correctly outline the *effects* of homeownership, it has overemphasized the extent to which those effects were the *causes* of expanded homeownership.

There is also considerable doubt, however, concerning the adequacy of early accounts of the socially integrative effects of homeownership. One of the problems is that there has been a tendency to overgeneralize claims about the effects of homeownership. A related issue is that the "effects" have been conceived of as a product, an end-point, rather than a process that must be conceptualized in, and likely varies across contexts. Homeownership has no one political effect that is constant across different places and times. In a very real sense, homeownership is a different tenure form in countries where patterns of subsidies, legal arrangements, and alternative housing tenures vary (Kemeny 1981). Edel (1982) questions the contemporary relevance of facets of incorporation theory on the grounds of historical contingency. He doubts the current applicability of Engels's thesis that mortgage debt chains workers to their jobs. Edel reasons that innovations in commuter transportation, the greater diversity of employment opportunities available within modern metropolitan areas, and the development of more efficient resale and mortgage markets weaken Engels's assumption that ties to a residence restrict employment opportunities. (Edel may be overgeneralizing his case as well. Engels's argument has contemporary relevance in a metropolitan area such as Vancouver where regular cycles of boom and bust, reflecting the resource-based regional economy, reach across industrial sectors and occupational groups.) Agnew (1981a,b) questions the transnational generalizability of the incorporation thesis. He argues that specific conditions in the United States fostered the links between homeownership, privatism, and possessive individualism. These links are less well developed in Britain, where a stronger working-class culture works against their establishment. The contingent and variable relationship between homeownership and political ideology is a point that will be pursued in the Vancouver case study.

The relationship between homeownership and consciousness is also distorted by the theory of ideology that incorporation theory has drawn upon. It is not only inaccurate to suggest that the desire for homeownership is implanted in a top-down manner. Incorporation theorists have often missed the reality of ideology as lived experience. People do not swallow ideology whole, they absorb contradictory ideas and often, at least partially, penetrate ideology and develop insight into the structural forces that constrain them (Lears 1985). Such partial penetration offers opportunities for resistance – this is the theme that will be explored in the Vancouver case study.

Blinkered by the social control metaphor, incorporation theorists failed to see the creative potential of "free spaces" existing within communities. In recent years, considerable attention has been given to the emancipatory potential of social movements that develop over consumption and reproduction issues (Castells 1978, 1983). Cox (1981) views community struggles aimed to stop or slow urban development as potentially progressive forms of resistance to the increasing commodification of all spheres of life and all places within the city. Incorporation theorists have tended to ignore these progressive side-effects of homeownership and attachment to community, both because their analysis is filtered through the lens of social control and because their conception of progressive political action tends to be fixed on traditional working-class politics within the production sphere.

Having established the general lines of incorporation theory and the criticisms that have been directed toward it, I now turn to demonstrate the dynamic relationship between politics and housing tenure in Vancouver, British Columbia. In this context, homeownership appears to have contradictory effects; homeownership opens the potential both for integration into and resistance to capitalist social relations.

Incorporation and housing tenure in Vancouver

Though the attitudes and actions of some homeowners in Vancouver are fairly accurately described by incorporation theory, on the whole the theory offers a static and partial assessment of the dynamic relationship between housing tenure and politics. Support for this claim will be drawn from a number of sources: from in-depth interviews with 100 renters and homeowners living in Surrey, an outer suburb in the Vancouver area,[2] from a quantitative analysis of a national survey,[3] and from an assessment of recent developments concerning illegal suites in Vancouver.

(a)   Evidence of incorporation

The interviews with Surrey households produced a wealth of statements that can be interpreted as supporting incorporation theory. Many Surrey homeowners spoke of their jobs, their feelings about renters, their views about community in exactly the terms suggested by incorporation theory. A selection of their comments conveys a sense of the meaning ascribed to homeownership and its influence on their lives.

A number of households were on strikingly regimented "consumption programs" and interpreted their jobs in terms of them. The experiences of one couple demonstrate the priority given to consumption, and the extent to which jobs are viewed as vehicles for consumption. These individuals summarized major stages in their life in terms of a savings and acquisition program. They had emigrated from Holland in 1970 and rented a house for the first four years of living in Canada. During this time they saved the downpayment to buy a house: "Ever since we got to Canada, this was the first thing on our mind." This is because they felt that they were "throwing money away" while renting. After four years of saving, they were able to purchase a townhouse in Surrey with a mortgage covering 89 per cent of the cost.[4] The mortgage was amortized over 25 years, but they paid it off in seven: "It took some doing but this was our objective." Then they started saving for a camper van. When interviewed, they had purchased the van and expected to pay it off within two years, at which point they planned to purchase a house. "One of these days," commented the woman, "I'd like to think about myself."

There were two particularly interesting aspects of their life-narrative. First, they had compromised their living conditions in order to own their own house. The house they rented upon moving to Canada had a large garden with fruit trees. The woman enjoyed gardening and still missed the rented house for this reason. Referring to the townhouse she had lived in for eight years, she stated: "I didn't want to live here." The townhouse was part of a strategy and, at the point of being interviewed, this couple planned to purchase a house within the next two years. Second, the woman worked as a sales clerk and clearly interpreted her job in relation to the purchase of consumption goods. "If I wouldn't have been working, the time for paying off [the townhouse] would have been longer. And we wouldn't have got the van for ten years."

Numerous people interviewed in Surrey commented directly on the influence of homeownership on one's commitment to a job and willingness to participate in collective workplace action. Their statements seem to support Engels's claims regarding the capacity of mortgage debt to trap workers in their jobs. I have presented this evidence in more detail elsewhere (Pratt 1986b). As an example, I report on the comments individuals made in response to a stereotypical statement about renters and homeowners. Respondents were asked whether they agreed or disagreed with a number of these statements, and were also encouraged to expand upon the basis of their opinion. Given the statement that: "Homeowners are more stable employees because they are responsible for their mortgages", 33 (or 66%) of the homeowners and 26 (52%) of the renters agreed. Comments included: "Yes, to the extent that if a person was in a job he didn't like, he'd probably be less prepared to quit. You have more of a stake and don't want to lose it." "Maybe . . . it creates a tendency to stick with a job no matter what." "Most of the yes-men at work own houses and they're afraid to lose them." "At work I see people over their heads with mortgages. They stay in a job when they could do better because they know that they can at least pay the mortgage at their present job."

In addition, a good proportion of those who were members of trade unions tended to acknowledge the influence of homeownership on labor activism. Of the 21 manual homeowners belonging to trade unions, 10 felt that the burden of mortgage debt had an immediate and direct impact on support of strikes and, more generally, contract negotiations. Representative comments included: "The young ones with the big mortgages are the ones who are worried. Last year our contract was up for negotiation and it came up when there were rumors of a strike. . . . And it was taken into account by the union leaders. There's no use going on strike if two weeks into it the members are mourning. You need a good morale on the picket line so you have to take it into account. . . . Sure it influences me. There's no catching up after a strike." As another example, a union official regretted that: "I don't know why it has an effect but it does. The ones I've talked to, it's hard to convince them that the banks can't get you for a year."

Many Surrey homeowners also voiced their perceptions of status differences between homeowners and renters. Perceptions of such differences were very common, and the following is a selection of comments that are by no means extreme. "In general terms, people who rent don't maintain their property. Their general living habits are lower. Their attitudes are different. You have to look at why they're renting. They're either kids or lower income people who don't care about their living surroundings. . . . If you don't have something to look after you tend to spend your money on alcohol or cars or trips." The theme of responsibility was a common one: "Homeownership creates maturity. You have to pull in your belt buckle and own up." Homeownership figured into perceptions of class: "I guess I'm a blend between working and middle class. You know, I'm a guy who worked in a mill. I'm not working in management. But I live in a house and don't rent." The depth of these cultural meanings was conveyed by a husband and wife who drew the distinction between owning and renting through their childhood memories. He recalled that "We always owned. We lived in the country. But I remember coming into town to see my uncle. They rented a house. Even then I knew the difference between renting the use of a house and owning a house. You always saw it on television . . . making rent payments and the worries, [and] the differences between renters and homeowners." His wife, on the other hand, "grew up in a town where 40 per cent of the labor force was employed in one industry. My parents had bought a house and had to sell it. I remember it hurt me as much as it hurt them. I don't think that the other kids knew but I did. Maybe I could feel the tension between my mother and father."

What was particularly interesting is that many respondents felt that these status differences should be acted upon, and reflected in public policy. Some felt that homeownership should be more heavily subsidized than the rental sector because, in the words of one respondent, "people are making an effort. Homeownership creates taxes for government." Others argued that homeowners should be bailed out when mortgage rates rise unexpectedly because "they're subsidizing the community. If they can stay in the community they'll keep putting money into it, at the local shops [such as lumber stores]." Some maintained that tenants should have restricted political rights at the local level: "I have no problem in terms of voting for elected officials, but for issues of zoning, development issues, people who own should have the say. Tenants can move if they don't like it."

All of these statements express facets of incorporation theory. And yet, there was a significant crack in the veneer of political incorporation; many homeowners felt strong antagonism toward either the government or the chartered banks on the basis of their personal experiences with housing programs or mortgage interest rates.

(b)   Means of incorporation as seed for resistance

One interview was conducted in a house literally barren of non-essential furniture. The living room and den were virtually empty. This was because the owners were concentrating their financial resources on paying off the mortgage. The wife was not at home. She worked evenings and nights. The couple had three children under the age of five but managed without day-care by working non-overlapping shifts. One of the main reasons that the respondent's wife was working was to pay off the mortgage. The respondent believed in self reliance and espoused a strongly individualistic philosophy. Government should "get out" of housing altogether. He believed that problems of housing affordability reflected the irresponsibility of individual homeowners who take on unrealistically high mortgages and then get into difficulties with repayment through lack of financial responsibility. This man's practical arrangements and political attitudes are compatible with incorporation theory, up to a point.

Questioned on the subject of mortgage rates, however, he outlined a conspiracy theory with great vehemence: "The high interest rates were manipulated by people who stood to gain. They realized that people in debt for things won't give them up. I know many people who are working for free. It's white slavery. . . . Once you have a big mortgage, there is no way of getting out of debt. I don't trust the system." The respondent himself had been forced to renew his mortgage at 18 per cent. It was this that had precipitated his wife's return to work and his commitment to rid himself of mortgage debt.

Mortgage interest rates was a subject on the minds of many of the Surrey homeowners interviewed in 1983. Interest rates for conventional mortgage loans rose dramatically through the 1970s, from an annual rate of 9.43% in 1971 to one of 18.15% in 1981, with a peak of 21.46% in September 1981. The majority of Canadian homeowners hold mortgages that are renewed at current interest rates on a periodic basis. The renewal period typically varies from two to five years, though at the height of the inflationary period in the early 1980s, a six month renewal period had become the norm. There were few mortgaged Surrey homeowners, therefore, who had not either anticipated the possibility or directly experienced a significant increase in mortgage payments.

This experience cultivated a critical discourse among mortgaged homeowners. It was particularly striking to hear this level of criticism voiced by otherwise conservative individuals. One such individual managed a large department for a multinational corporation, had been a regional director for the Jaycees, voted Social Credit (the right wing provincial party), and had recently attended a pro-life demonstration. He had signed only one petition in the last five years and this was to urge government to put a picture of Terry Fox on a stamp.[5] Asked about rent rebates for low income renters, he stated that "if people are renting they have to find a place they can afford and not spend their money on beer or whatever these people spend it on, and expect someone else to pay their rent." Yet he was extremely critical of the high interest rates: "They reflect manipulation by the money power brokers. It's a straight rip off." He was critical of the bank's unwillingness to renegotiate loans with homeowners who had been forced to renew their mortgages at the highest rates. Though he had been fortunate to obtain a 5 year mortgage at a 13.5 per cent interest rate in 1980, he had a friend who had lost his house, and he displayed a keen awareness of the exact expenses involved in renegotiating a mortgage loan to the lower rates available in 1983. He was galled by a comment recently made to him by a loans officer as he was arranging a consumer loan, to the effect that she had no sympathy for homeowners who were losing their houses. Criticism came from unexpected quarters; a couple in their early sixties who owned their house outright expressed resentment towards the banks, not because they were themselves affected by the interest rates, but because their children were having difficulty purchasing houses and their daughter felt constrained to defer her family so as to work full-time and save for a house.

Experiences with government housing programs also left some

homeowners embittered. One couple who had participated in the Assisted Homeownership Program (AHOP) felt that they had been misled about the terms of the subsidy and repayment. The husband repeatedly stated that he had wanted to send a bomb to the federal government in Ottawa: "I wanted to get out my old Abby Hoffman manual and send a bomb." Another couple had friends in the AHOP program who had lost their house, having to repay their loan at the same time as interest rates escalated: "That program messed people up."

In actuality, massive consumer revolts did not develop in Canada in the early 1980s. The mortgage interest rates had dropped by 1983. The most lasting effects have undoubtedly taken the form of individual accommodations, such as attempts to pay off mortgage loans over a shorter amortization period, to reduce the overall payments, and insulate oneself from fluctuations in interest rates. These individual accommodations can themselves be interpreted as instances of incorporation, in so far as they involve a more extensive commitment to waged work. Nevertheless, many Surrey homeowners felt very real resentments. The potential, under the appropriate conditions, for political organization of homeowners against the state or financial institutions should not be ignored. Homeownership has a more complex and contradictory impact than conceptualized by incorporation theory. It may tie workers to their jobs, create divisions between homeowners and renters, direct homeowners' political interests towards the defense of community, and provide them with a sense of belonging and a perception of vested interest in "the system." Simultaneously, however, homeowners may perceive that they have purchased certain rights along with their house, such as stable housing costs. Mortgage debt brings homeowners into a dependent, but also potentially conflictive relationship with the financial institutions. Government housing programs may generate criticism instead of lulling homeowners into political passivity. Castells (1977, 1978, 1983), among others (Bunge 1977, Fincher 1984), has emphasized the emancipatory potential of consumption-oriented community-based urban social movements. Because homeownership has been misleadingly labeled an "individualized" form of consumption (Dunleavy 1979), the possibility that homeowner revolts may challenge dominant societal institutions has not been adequately considered.

(c) Uneven incorporation

Incorporation theorists have typically not been sensitive to the historical (Edel 1982) and national (Agnew 1981a, 1981b) specificity of the meaning and circumstances of homeownership and its political impacts. Analyses of both the Surrey interviews and a national survey of political and social attitudes caution against overgeneralizing the political importance of homeownership, even within one region at one point in time. The results of these analyses have been reported elsewhere (Pratt 1986a, 1986b). A brief summary will substantiate the theoretical point.

Gray (1982, 270) has noted, "that owner-occupation [is presumed to act] as if it were a fairy godmother's wand. When waved . . . a previously ragged and unhappy Cinderella (a tenant) is changed into a beautiful and desirable person (an owner-occupier)." This supposition rests on the implicit assumption that homeownership outweighs the influence of other experiences, such as those at work. It matters not whether Cinderella is a sales clerk or an electrician.

Analyses of the Surrey interviews and the national data indicate that other life experiences, especially those at work, do influence the impact of homeownership. There are two groups for which homeownership tends to be unrelated to political attitudes and seems less strongly linked to social identity.

First, the political attitudes of skilled blue-collar workers are unrelated to housing tenure (Pratt 1986a). National survey data was analyzed to determine whether homeowners· are, in fact, more conservative than renters. Likely mediating factors were controlled, including occupational class, class (as defined in Marxist theory), household income, education, age, and stage in the lifecycle. The interaction between occupational class, housing tenure, and political ideology is particularly striking. Of the 16 questionnaire items examined, among skilled blue-collar household heads, there are only two items for which homeowners are statistically significantly more conservative than renters. Among household heads classified as skilled white-collar, and as managers and professionals, however, there are 13 and 10 items, respectively, for which homeowners are more conservative than renters. In other words, among skilled white-collar household heads, homeowners are more conservative than renters on almost every measure of political ideology examined. Among skilled blue-collar household heads, there is, in general, no difference between the attitudes of renters and homeowners.

This contrast between skilled blue-collar and skilled white-collar workers was evident among Surrey homeowners as well. Persons classified as white collar more often felt that housing tenure has a determining influence on one's economic and social standing. Asked simply whether they agreed or disagreed with the statement, "Home-owners have taken a step up the ladder of social and economic standing," 44 per cent of white-collar workers agreed. This compares to 22 per cent of blue-collar workers (chi square = 4.5, d.f. = 1, p. < 0.05). As well, 14 white-collar (28 per cent) compared to 6 blue-collar homeowners and renters agreed with the rather strong statement that "Homeowners are *only* concerned with their house and garden and tend not to be concerned with wider social issues" (although this contrast just approaches statistical significance: chi square = 3.06, d.f. = 1, p < 0.075). One white-collar homeowner explained that: "You get a comfort zone and spend your time cutting the lawn and tend not to be concerned with what is going on down the street." Another noted that: "You tend to live in your own little world. It's your castle. No – it's even more basic than that – it's security, and at an instinctual level, it's possessiveness." Further, it was particularly white-collar homeowners who felt that they had become more involved in their community since owning their homes (56 per cent of white-collar compared to 36 per cent of blue-collar: chi square = 5.19, d.f. = 1, p < 0.025).

Following from this, white-collar homeowners and renters tend more often to believe that renters should have less input into the local political process. Asked whether they thought that owners and renters should have an equal say in the future of the community, 16 (32 per cent) white-collar respondents compared to 8 (16 per cent) blue-collar respondents indicated that renters should have less input (although this contrast merely approaches statistical significance: chi square = 2.70, d.f. = 1, p < 0.10). In general, then, white-collar workers more consistently agreed with suggested links between housing tenure and social identity.

The different meaning and influence of homeownership seems to reflect the greater significance of collective workplace organizations for skilled blue-collar as compared with white-collar workers. Most (68 per cent) blue-collar workers interviewed in Surrey belonged to trade unions, while only a minority (14 per cent) of white-collar workers did. While the argument is speculative, it is possible that, in the absence of an identification with a production-based group, white-collar workers come to identify more fully with the home and community.

The diverse influences on the political values of skilled white-collar workers and managers may reflect another aspect of their class position. Many in these occupational classes are located in contradictory class positions, as defined by Carchedi (1977) and Wright (1979), in that their work time is evenly split between the supervisory function of capital and the productive function of workers. Located in a contradictory class position, they may not experience *a* class position as clearly, and the impact of variations in consumption relations may thus be stronger.

A second group for which incorporation theory seems not to hold is that of self-employed workers (Pratt 1986b). It can, of course, be argued that, as petty bourgeoisie, self-employed persons are already fully incorporated into capitalist society, and that incorporation theory has little relevance for this group. Nevertheless, the pragmatic and rational approach to homeownership and the housing market displayed by the self-employed is of some interest in relation to incorporation theory. It contradicts the theme of individual passivity, and the idea that home-ownership exerts a grand, politically soporific effect over the populace.

The Surrey sample was stratified by occupational class and housing tenure, into skilled blue-collar and skilled white-collar homeowners and renters. Individuals who identified themselves as self-employed in the City Directory were excluded from the sample. Nevertheless, almost half of those renting houses were found to be self-employed, either presently or within the last five years. Twenty-one of the 50 renters, as compared to 1 of the 50 homeowners, had been self-employed at some point between 1979 and 1983.

There are two ways in which this situation appears to fit with incorporation theory. First, it can be taken to demonstrate the way in which homeownership reinforces occupational and class distinctions: the incomes of those who are self-employed are often more variable than those of salaried workers. Access to homeownership may be closed to them. One quarter of the self-employed interviewed in Surrey had owned houses but had been forced to sell them because of financial difficulties. Second, self-employed renters expressed concerns that mortgage debt traps one in a job and saps one's entrepreneurial initiative (Pratt 1986b).

Nevertheless, a number of self-employed persons interviewed had not been trapped by homeownership and, in fact, had used the inflationary housing market of the 1970s as a means of amassing capital to start their business ventures. The housing market was viewed in a detached and calculated way. Homeownership seemed not to carry the wealth of meanings, such as affording personal control, privacy and security, that are commonly ascribed to it. Because few self-employed persons saw housing as a good investment at the time of being interviewed, given high interest rates and low rates of house price inflation, two-thirds preferred to rent. This contrasts sharply with the preferences of wage-earning renters: 82 per cent of employed renters stated that they would prefer to own a home but could not obtain the financing that would allow this.

For the self-employed, work was a central concern and homeownership was put to work in the service of their businesses. Experiences with homeownership did not draw their priorities to the community. The home was viewed pragmatically, not as a "haven in a heartless world."

Interviews with skilled blue-collar homeowners and self-employed renters indicate that the meanings attached to homeownership vary, depending on other influences in one's life. A synthetic approach that integrates individuals' multiple commitments and affiliations is required to understand the political significance of homeownership. A considerably more subtle argument than that which has tended to be presented as incorporation theory is necessary to describe adequately the socially integrative effects of homeownership.

(d) Homeownership and informalization of the housing market

In the latter half of 1986, illegal or secondary suites in single-family houses became a significant political issue in Vancouver. This issue highlights the contradictory role that homeownership can play in the process of social integration.

Local government reaction to secondary suites in Vancouver has oscillated over the last 47 years, from active encouragement to censure. In 1940, Vancouver homeowners were encouraged to relieve war-time housing shortages through provision of suites in single-family homes. In 1956, city council initiated a series of measures to close the suites. Invariably, widespread enforcement of closure has been stalled by recognition that the conventional rental sector is inadequate to the need for low-cost rental accommodation. In the fall of 1986, however, representatives of the conservative, business-oriented civic party, the Non Partisan Association (NPA), fought the civic election on a platform that included the elimination of secondary suites through the strict enforcement of the city's residential zoning by-laws (Krangle 1986a, 1986b).

Concerns about this issue had been spawned, in part, by the construction of increasing numbers of "jumbo" houses in Vancouver single-family neighborhoods. These houses are built to the maximum allowable floor space, frequently contain one or more suites, and impose upon their neighbors by blocking views and light and disrupting the visual continuity of streetscapes. Concerns also focus on congested parking caused by multiple use of single family houses, and questions of equity: homeowners without suites complain that those with suites are not paying their fair share of local taxes to compensate for the extra strains they place on municipal services. The NPA mayoral candidate identified the issue as one of neighborhood self-determination: "we should ask the neighbourhood's what kind of neighbourhoods they want. . . . We're not interested in putting people into the street. What we are interested in is maintaining . . . our neighbourhoods (Krangle 1987)."

The number of homeowners and renters directly involved with this issue is by no means insignificant. The city planning department estimates that there are, at present, between 21,000 and 26,000 illegal suites (MacAfee 1987). This means that, on average, one in four single-detached houses located in single-family zones contains this type of accommodation. The estimates vary by neighborhood; in some neighborhoods only 10 per cent of the houses are believed to contain a suite, but in others the estimates are as high as 59 per cent (Vancouver Planning Dept. 1986).

This phenomenon is certainly not unique to Vancouver (Baer 1986, Gellen 1985, Hare 1981, Rudel 1984). Slumps in the construction of new rental units throughout North America (Downs 1983), demographic shifts that increase the demand for housing, including the maturing of the baby boomers and an explosion in the number of one and two person households (Miron 1982), and reduced federal commitment to housing programs (Bourne 1986, Hartman 1983) have strained many housing markets in the United States and Canada and created pressure for private solutions through the informal housing market. A number of factors conspire, however, to make Vancouver an extreme case. Over 70 per cent of the City of Vancouver is currently reserved for single-family dwelling residential use. No other major North American city has as much of its area zoned in this category (McAfee 1987). At the same time, housing costs in Vancouver substantially exceed the national average (Canada Mortgage and Housing Corp. 1984), making many home-buyers willing accomplices in the informal housing market so as to afford the high costs of homeownership.

The situation is of considerable interest from the perspective of incorporation theory. The purchase of a house in Vancouver in many cases places the owners in an illegal relationship with local and federal governments. Secondary apartments violate the city zoning and building by-laws. Homeowners with illegal suites collect income that is not declared for tax purposes. A survey of classified advertizements in one of the major daily newspapers from January–February 1987 ascertained that illegal suites rent for 20 to 35 per cent less than comparable accommodation in approved multiple residences (*Vancouver Sun* 1987a). Even so, a one-bedroom illegal suite is likely to produce at least $4,500 of undeclared income annually. Two and three bedroom apartments produce considerably more. Along with initiation to the pleasures of property ownership, many of Vancouver's homeowners are schooled in the financial benefits of violating local by-laws and deluding the tax department. One can also speculate that, contrary to incorporation theory, the income obtained as a homeowner through illegal suites, *decreases* the homeowner's dependency on income obtained through waged labor.

The politicization of the issue reveals strongly-felt resentments among homeowners, between those who have illegal suites and those who do not. In the November 1986 civic elections, the NPA captured seven of the ten aldermanic seats on City Council, as well as the Office of the Mayor. Since this time, City Council has taken action against illegal suites, by amending the city's zoning to prevent approval on second kitchens, by initiating a neighborhood-by-neighborhood assessment process and by adopting an interim (pending neighborhood decisions regarding secondary suites) enforcement program. Due to a shortage of staff, enforcement is initiated by complaints filed by neighbors. The supervisor of property use inspectors has stated that the number of complaints increased five-fold as a result of the politicization of the issue. His department received over 300 complaints over just a five week period, between the end of January and the beginning of March 1987 (*Vancouver Sun* 1987b, A10). The situation has generated secrecy on the part of some owners of illegal suites. David Lane, the chairman of the Tenants' Rights Coalition, reports that he has received calls from illegal suite tenants who claim that landlords have requested that they keep their blinds down and park several blocks from their dwellings so that neighbors will not notice or complain about their secondary suites (*Vancouver Courier* 1987). The head of area planning for the City of Vancouver, Ron Youngberg, is quoted as saying, "Those who have suites are very quiet, of course they are, and those that don't have them say 'Hey, they're getting away with not paying their share of taxes' . . . so you can't deny it's an issue" (Vancouver *Sun* 1985, A6).

The presence of a lively informal housing market in Vancouver reinforces the claim that the ideological significance of homeownership must be studied in context. In Vancouver, purchase of a house may, in some sense, give people an ideological stake in the system. Left at that, however, this is a very "thin" reading of the situation. With the letting of suites to ease the costs of high mortgages, many Vancouver homeowners are no doubt breaking civic by-laws and federal income tax laws for the first times in their lives, comforted by the fact that they are doing so in the company of one in four homeowners. This is perhaps a small rebellion, but it is nevertheless a real violation of legal authority. Homeownership also offers access to income outside of the waged labor market. Finally, uneven participation in the informal housing market has generated tension amongst homeowners, reinforcing the necessity for greater clarity in conceptualizing housing tenure.

Conclusion

Incorporation theory is a noteworthy attempt to link community and social integration, attachment to territory and political consciousness. Incorporation theorists have been criticized, however, for reifying the category, homeownership; for assuming that the integrative effects of homeownership are constant across different housing markets and social classes; and for ignoring the possibility that homeownership has a multiplicity of effects, depending upon other factors in the housing market. Nevertheless, the central focus of incorporation theory, on the links between housing and social reproduction, is of continuing interest and a number of interesting research directions are suggested by critical reactions to incorporation theory.

The first emerges from a recognition that homeownership is a dynamic category; homeownership must be understood in relation to other housing tenures, and in the context of the total system of housing provision. This realization draws the focus away from homeownership *per se*, and leads one to consider the possibility of other housing-related processes of integration. Incorporation theorists have, for instance, tended to characterize renters as less constrained human subjects, and studies of renters tend to focus on renters' strikes and other forms of collective resistance (e.g. Castells 1983, Lowe 1986). The processes of integration related to the rental tenure are relatively unexplored.

Rent controls provide one interesting example of the ideological integration of renters as an unintended consequence of political resistance. Rent controls are typically introduced as the result of intense lobbying on the part of tenants, with the purpose of protecting tenants from the extremes of landlord avidity. Rent controls are, of course, typically opposed by landlords and developers, and shortages in rental housing are often blamed, in a simplistic way, on rent controls (see Marcuse 1981, for a critical evaluation of these claims). The supposed effects of rent controls are sometimes taken as evidence of the inevitable failure of government intervention in the "free" market (Hayek 1975). One can envisage a process by which political resistance on the part of renters has the unintended consequence of reinforcing the ideological legitimation of unconstrained private property rights in the eyes of landlords, developers, and renters alike. This is one example of a process of integration related to the rental tenure. The more general point is that the links between housing and social reproduction need to be explored in a broader context than an exclusive focus on homeownership allows.

Second, the importance of *local* variations in the relationship between housing tenure, political culture, and social reproduction warrants further attention. This is in line with the attention given to "locale" by a number of authors in this volume. Agnew (1981a, 1981b) has shown how national political culture mediates the meaning of homeownership in Britain and the United States. We have little understanding, however, of the way in which the integration and/or resistance effects of homeownership in turn shape local political cultures over time. As one example of such a process, one can speculate that Vancouver homeowners' resentment towards the chartered banks (institutions that, in the eyes of many Western Canadians, encapsulate the economic and political dominance of Eastern Canada) reinforced and exacerbated existing regional hostilities.[6]

Alternatively, the integrative effects of homeownership may cause communities of homeowners, especially in white-collar neighborhoods, to find political solutions through consumption issues, and may thus influence the *content* of local politics in some communities. Katznelson (1981) has highlighted the split between the politics of home and work in the US, with the politics of community focused around the provision of services and class-based politics localized within the workplace. He contends that this division contains and diffuses political conflict, playing a significant role in the social reproduction of capitalist relations in American society. Katznelson attributes the separation of issues addressed at work and within the community to the spatial separation between home and work, as well as to two features of the American political and institutional context: first, a comparative tolerance of trade unions by the courts in the 18th century allowed workplace grievances to be organized around the workplace itself; and second, all white male workers in American cities obtained political franchise relatively early and thus a class-based oppositional force did not develop against the state. Katznelson emphasizes the last two factors and leaves the effects of the spatial separation between home and work relatively unexplored. Research that documents the ideological significance of homeownership, especially for skilled white-collar workers, suggests that housing tenure must be considered in conjunction with the spatial separation between home and work for a fuller understanding of the role the community plays in structuring the content of urban politics. It also suggests that the content of local politics will vary by neighborhood, depending on characteristics of housing tenure, and the occupational class of residents.

Finally, the informalization of the housing market deserves more analytic attention and, in particular, should be considered in relation to the broader literature on the informal economy. Pahl (1984) has examined informal types of work in Britain, and come to the conclusion that access to informal employment is intrinsically dependent on participation in the formal labor market. In Britain, therefore, informal employment is unlikely to provide a buffer against unemployment. The informal housing market perhaps offers another means of "getting by," which is not dependent on the formal labor market (though highly dependent on access to housing). In such an instance, one is driven to ask the question that is frequently addressed in analyses of the informal economy: does the informal housing market offer a means of existence outside of capitalist social relations or is it, too, implicated in the extended reproduction of capitalist society?

Notes

I am grateful to the editors, and Jody Emel, for comments on a draft of this chapter.

1 Edel may be overgeneralizing his case as well. Engels's argument has contemporary relevance in a metropolitan area such as Vancouver where regular cycles of boom and bust, reflecting the resource-based regional economy, reach across industrial sectors and occupational groups.

2 The interviews in Surrey were conducted in the summer of 1983. They followed the quantitative analysis of the national survey data, and were structured so as to explore in more detail several interesting empirical regularities noted from the earlier analysis. They were meant to allow a more qualitative and contextualized exploration of patterns revealed and questions raised through the larger data set. The smaller sample was selected from households living in Surrey, a municipality in Greater Vancouver of relatively lower priced detached houses, duplexes, and townhouses. House-type was thus explicitly controlled. The suburban municipality was chosen to narrow household income and lifecycle to lower to middle income families with children at home, a cohort of some interest in the results of the analysis of the urban national sample. Also reflecting the initial quantitative analysis, the smaller sample was stratified by both housing tenure and occupational class, so that homeowners and renters were evenly divided into the occupational classes of skilled blue-collar and skilled white-collar on the basis of the occupations of the male household heads. The sample was selected from the *City Directory* so as to comprise a stratified random sample of skilled blue-collar and skilled white-collar homeowners and renters living in single detached houses, duplexes, and townhouses in a lower priced suburban municipality.

3 The national sample of 1984 respondents, 18 years old or more, is representative of the Canadian urban population, that is, persons living in census metropolitan areas (a continuous built-up region having a population of 100,000 or more). The survey data were gathered in 1979 as the second phase of the Social Change in Canada study (Atkinson *et al.* 1982), sponsored by the Institute for Behavioral Research (renamed the Institute for Social Research in 1984). The survey was used because it contains data for individuals on a wide variety of political attitudes and activities, as well as detailed information on housing and job-related characteristics.

4 In Canada, an 11 per cent downpayment is considered to be a small one by conventional mortgage lenders. In such circumstances, borrowers must obtain mortgage insurance and generally pay a higher mortgage interest rate. The couple's eagerness to own a house evidently outweighed this additional financial burden.

5 Terry Fox was a cancer victim, with one leg amputated, who ran across Canada to raise funds for cancer research.

6 The significance of this is highlighted by the fact that regions are thought by many social scientists (e.g., Elkins & Simeon 1980) to define the major divisions between Canadian "political cultures." For example, federal partisan affiliation is more strongly linked to region than class.

References

Aglietta M. 1979. *A theory of capitalist regulation: the U.S. experience*. London: New Left Books.

Agnew, John 1978. Market relations and locational conflict in cross-national perspective. In *Urbanization and conflict in market societies*, K. R. Cox (ed.), 128–41. Chicago: Maaroufa.

Agnew, John 1981a. Home-ownership and identity in capitalist societies. In *Housing and identity: cross-cultural perspectives*, J. S. Duncan (ed.), 60–97. London: Croom Helm.

Agnew, John 1981b. Home-ownership and the capitalist social order. In *Urbanization and urban planning in capitalist society*, M. Dear & A. J. Scott (eds.), 457–80. London: Methuen.

Angotti, T. 1977. The housing question: Engels and after. *Monthly Review* **29**, 39–51.

Arendt, H. 1958. *The human condition*. Chicago: University of Chicago Press.

Atkinson, T., B. Blishen, M. Ornstein & H. M. Stevenson 1982. *Social change in Canada*, National cross-section survey phase II, May–August 1979, Institute for Behavioural Research. Downsview, Canada: York University.

Baer, W. 1986. The shadow market in housing. *Scientific American* **225**, 29–35.

Ball, Michael 1983. *Housing policy and economic power: the political economy of owner occupation*. London: Methuen.

Ball, Michael 1986. Housing analysis: time for a theoretical refocus? *Housing Studies* **1**, 147–65.

Berry, Michael 1986. Housing provision and class relations under capitalism. *Housing Studies* **2**, 109–21.

Bourne, L. S. 1986. Recent housing policy issues in Canada. *Housing Studies* **1**, 122–8.

Brittan, A. 1978. *The privatised world*. London: Routledge & Kegan Paul.

Bunge, William 1977. The point of reproduction: a second front. *Antipode* **9**, 60–76.

Canada Mortgage and Housing Corporation 1984. *Housing in British Columbia: a statistical profile*. Ottawa: Canada Mortgage and Housing Corporation.

Carchedi, G. 1977. *On the economic identification of social classes*. London: Routledge & Kegan Paul.

Castells, Manuel 1977. *The urban question: a Marxist approach*. Trans. A. Sheridan. Cambridge, Mass.: MIT Press.

Castells, Manuel 1978. *City, class, and power*. Trans. E. Lebas. New York: St. Martins Press.

Castells, Manuel 1983. *The city and the grassroots*. Berkeley: University of California Press.

Chekoway, Barry 1980. Large builders, federal housing programmes, and postwar suburbanization. *International Journal of Urban and Regional Research* **4**, 21–45.

Conservative Party of Canada 1979. Homeowners deserve a break . . . economy recovers through stimulus [Canada].

Cox, K. R. 1981. Capitalism and conflict over the communal living space. In *Urbanization and urban planning in capitalist society*, M. J. Dear & A. J. Scott (eds.). New York: Methuen.

Downs, A. 1983. *Rental housing in the 1980's*. Washington DC: The Brookings Institution.

Dunleavy, Patrick 1979. The urban basis of political alignment: social class, domestic property ownership and state intervention in consumption processes. *British Journal of Political Science* **9**, 409–43.

Edel, Matthew 1982. Homeownership and working class unity. *International Journal of Urban and Regional Research* **6**, 205–22.

Edel, M., E. Sclar & D. Luria 1984. *Shakey palaces: homeownership and social mobility in Boston's suburbanization.* New York: Columbia University Press.

Elkins, D. & R. Simeon (eds.) 1980. *Small worlds: provinces and parties in Canadian political life.* Toronto: Methuen.

Engels, F. 1935. *The housing question.* New York: International Publishers.

Fincher, Ruth 1984. Identifying class struggle outside commodity production. *Environment and Planning D: Society & Space* **2**, 309–27.

Forrest, R. 1983. The meaning of homeownership. *Environment and Planning D: Society & Space* **1**, 205–16.

Gellen, M. 1985. *Accessory apartments in single-family housing.* Rutgers, N.J.: Center for Urban Policy Research.

Gray, Fred 1982. Owner occupation and social relations. In *Owner occupation in Britain,* S. Merrett (ed.), 267–91. London: Routledge & Kegan Paul.

Handlin, D. P. 1979. *The American home: architecture and society, 1815–1915.* Boston: Little, Brown.

Hare, P. H. 1981. Carving up the American dream. *Planning* **47**, 14–17.

Hartman, Chester (ed.) 1983. *America's housing crisis: what is to be done?* Boston: Routledge & Kegan Paul.

Harvey, David 1975. Class structure in a capitalist society and the theory of residential differentiation. In *Processes in physical and human geography,* R. Peel, M. Chisholm & P. Haggett (eds.), 354–69. London: Heinemann.

Harvey, David 1976. Labor, capital and class struggle around the built environment in advanced capitalist societies. *Politics and Society* **6**, 265–95.

Harvey, David 1982. *The limits to capital.* Oxford: Basil Blackwell.

Harvey, David 1985. *The urbanization of capital.* Baltimore: The Johns Hopkins University Press.

Hayden, Dolores 1981. *The grand revolution: a history of feminist designs for American homes, neighborhoods and cities,* Cambridge, Mass.: MIT Press.

Hayek, F. A. 1975. *Rent control: a popular paradox.* Vancouver: The Fraser Institute.

Hayward, D. 1986. The great Australian dream reconsidered: a review of Kemeny. *Housing Studies* **1**, 210–19.

Heskin, A. D. 1983. *Tenants and the American dream: ideology and the tenant movement.* New York: Praeger.

Katznelson, I. 1981. *City trenches: urban politics and the patterning of class in the United States.* New York: Pantheon.

Kemeny, J. 1981. *The myth of home ownership: private versus public choices in housing tenure.* London: Routledge & Kegan Paul.

Krangle, K. 1986a. City warned against illegal suite action. *Vancouver Sun,* October 4.

Krangle, K. 1986b. Civic election campaign. *Vancouver Sun,* November 14.

Krangle, K. 1987. City to go slowly to wiping out illegal suites, mayor says. *Vancouver Sun,* February 20.

Lears, T. J. Jackson 1985. The concept of cultural hegemony: problems and possibilities. *American Historical Review* **90**, 567–83.

Lowe, S. 1986. *Urban social movements: the city after Castells.* London: Macmillan.

Luria, Daniel 1976. Suburbanization, homeownership and working class consciousness. PhD dissertation. Amherst: University of Massachusetts, Dept. of Economics.

McAfee, A. 1987. Secondary suites: the issues. *Quarterly Review* **14**, 16–18. Vancouver: City of Vancouver Planning Dept.

Mackenzie, Suzanne & Damaris Rose 1983. Industrial change, the domestic economy and home life. In *Redundant spaces in cities and regions*, J. Anderson *et al.* (eds.), 155–99. London: Academic Press.

Marcuse, P. 1981. Housing abandonment: does rent control make a difference? Public Policy Report no. 4. Division of Urban Planning, Columbia University.

Mather, C. T. & S. M. Parnell 1986. Urban renewal in Soweto: 1976–1986. Paper presented at the International Geographical Union, Working Group on Urbanization in Developing Countries, Madrid.

Miron, John 1982. The two-person household: formation and demand. Research paper no. 131, Centre for Urban and Community Studies, University of Toronto.

O'Connor, James 1973. *The fiscal crisis of the state*. New York: St. Martins Press.

O'Connor, James 1984. *Accumulation crisis*. New York: Basil Blackwell.

Pahl, R. E. 1984. *Divisions of labour*. Oxford: Basil Blackwell.

Perin, C. 1977. *Everything in its place: social order and land use in America*. Princeton, N.J.: Princeton University Press.

Pratt, G. 1982. Class analysis and urban domestic property: a critical reexamination. *International Journal of Urban and Regional Research* 6, 481–502.

Pratt, Geraldine 1986a. Housing tenure and social cleavages in urban Canada. *Annals of the Association of American Geographers* 76, 366–80.

Pratt, Geraldine 1986b. Against reductionism: the relations of consumption as a mode of social structuration. *International Journal of Urban and Regional Research* 10, 377–400.

Rakoff, R. M. 1977. Ideology in everyday life: the meaning of the home. *Politics and Society* 7, 85–104.

Rose, Damaris 1980. Toward a re-evaluation of the political significance of home-ownership in Britain. In *Housing, construction and the state*, Conference of Socialist Economists, Political Economy of Housing Workshop. London: CSE.

Rose, Damaris 1981. Home-ownership and industrial change: the struggle for a "separate sphere." University of Sussex Working Papers in Urban and Regional Studies no. 25.

Rudel, Thomas K. 1984. Household change, accessory apartments and low income housing in suburbs. *Professional Geographer* 36, 174–81.

Saunders, Peter 1978. Domestic property and social class. *International Journal of Urban and Regional Research* 2, 233–51.

Saunders, Peter 1979. *Urban politics: a sociological approach*. London: Hutchinson.

Saunders, Peter 1984. Beyond housing classes: the sociological significance of private property rights in the means of consumption. *International Journal of Urban and Regional Research* 8, 201–27.

Sennett, R. 1974. *The fall of public man: on the social psychology of capitalism*. New York: Vintage Books.

Sullivan, Oriel 1986. Housing tenure as a consumption-sector divide: a critical perspective. Paper presented at International Research Conference on Housing Policy, Gavle, Sweden.

Vancouver City Planning Department 1986. Secondary suites in RS-1 single-family areas. Unpublished report to City Manager and Planning and Development Committee, August 20.

Vancouver Courier 1987. Illegal suites: a new wave? April 5, 6.

Vancouver Sun 1985. Illegal suites called epidemic. February 8, section A, 6.

Vancouver Sun 1987a. City Vacancy rate boon for renters. February 5, section A, 1.

Vancouver Sun 1987b. Neighbors clashing over big houses, broken views. March 21, section 1, 10.

Williams, Peter 1984. The politics of property: home ownership in Australia. In

Australian urban politics, J. Halligan & C. Paris (eds.), 167–92. Melbourne: Longman Cheshire.

Wright, Eric Olin 1979. *Class structure and income determination*. New York: Academic Press.

Yago, Glenn 1978. Current issues in U.S. transportation politics. *International Journal of Urban and Regional Research* **2**, 351–9.

14

Community exclusion of the mentally ill

S. MARTIN TAYLOR

Community exclusion was a predictable response to deinstitutionalized mental health care. The psychiatric hospital had served to separate the mentally ill from the public at large and thereby to sustain an out of sight and out of mind mentality. The quite rapid discharge of large numbers of patients, most of whom required continuing drug therapy, inevitably represented a threat to a long-standing and, for many, comfortable status quo. Territorial prerogatives were challenged as neighborhoods were selected as locations for the various types of facility needed to service the discharged client group. Public responses to these changed circumstances were not uniform however. Rejection of and opposition to clients and facilities quickly emerged among certain populations and in certain urban neighborhoods. The media were not slow to highlight resistance and effectively exaggerate both its prevalence and intensity. In contrast, the non-opponents, whether latent supporters or neutrals, remained largely invisible and unaccounted for by virtue of their passivity.

Systematic assessment of public response to community mental health care followed naturally in the wake of deinstitutionalization. The academic pedigree for the research comprised a marriage of perspectives derived from public facility location theory and social psychology, specifically attitude theory and measurement. The rationale for this partnership is succinctly stated by Dear *et al.* (1977, 139) in anticipation of the type of research which was shortly to follow:

Generally speaking, community perceptions of external effects determine the extent and intensity of opposition to a program and facility.

Two concepts emerge from this line of argument: spatial externalities and public perceptions. The first makes the link with public facility location theory in recognizing that facilities such as those provided to serve the mentally ill have effects which impact beyond the client population. These so-called external effects are typically unpriced and unintended, but none the less potentially crucial in determining public acceptance of a close-by facility. Facilities associated with negative effects are those which are generally regarded as having unwanted impacts on the host neighborhood

316

and are therefore likely to provoke territorial instincts to protect the status quo and to exclude uninvited intrusion. Mental health facilities are usually included in this category.

The identification of public perceptions as a key concept makes the link with social psychology and, in tandem, with behavioral geography given its mandate to investigate environmental perceptions and attitudes. In this field, terms can be used quite loosely with the potential for cloudiness in conceptualization, measurement, and inference. Fortunately, research on public attitudes towards community mental health care has been able to benefit from the work of Fishbein and others (Fishbein & Ajzen 1975; Ajzen & Fishbein 1980) and therefore to articulate more precisely the links between such interrelated mental constructs as beliefs, norms, attitudes, and intentions. The last ten years have produced several major studies of public attitudes towards the mentally ill and community-based facilities with the result that general conclusions are now emerging with respect to the prevalence of negative and positive sentiments and their correlates at both the individual and community scale of analysis (see Taylor 1987).

Beyond the proximal determinants of public attitudes, such as socio-demographic variables, the focus for much of the empirical research to date, lie more distal factors rooted in the social, political, and economic fabric of capitalist society which might well be regarded as more fundamental causes. The connections between these more general causes and specific attitudes and proximal antecedents are not easily drawn. Preliminary statements in this direction have been made (e.g., Dear 1981) which argue that social theory can infuse the empirical insights from behavioral analyses with interpretations of community exclusion as but one aspect of the forces at work in the larger social formation resulting in the social and spatial reproduction of the mentally ill. The argument is compelling as we witness the persistence of ghettoization, increased privatization of facilities and services, the rise of homelessness, reinstitutionalization, and failures to implement legislation designed to counter exclusionary practices. From this perspective, the experience of the mentally ill is paralleled by that of other marginalized groups. A fundamental change in their social and spatial status would require changes in the social formation which the dominant social, political, and economic groups either consciously or tacitly resist. The result is that improvements in the daily life environment of the mentally ill and others, whether in terms of housing opportunities, employment prospects, social support or indeed community attitudes and acceptance, are too often token. In this broader context, community exclusion of the mentally ill emerges as a manifestation of individual and collective desires to protect territory with the aims of maintaining the use and exchange values of home and neighborhood, and on a deeper, but related level, as a component within a process of reproduction which perpetuates the uneven distribution of life chances and advantages.

This chapter will not be the place to pursue these more macroscopic interpretations of events for the community-based mentally ill, their importance notwithstanding. Rather, consistent with my approach and

317

research in the field, the focus will be behavioral with the aims of synthesizing and discussing the empirical evidence on public opposition to community mental health care. Other chapters in this volume (e.g., Ch. 4) address social and spatial relations from a macroscopic perspective and thereby provide a broader context within which the discussion in this chapter can be usefully situated.

Three main issues are addressed. The first focuses on the prevalence of opposition to community mental health care. Is it as widespread and as strong as media reports of community conflict over facility siting might lead us to conclude? The second issue deals with the correlates of opposition. This is examined at the individual level in terms of the personal characteristics which have been consistently shown to relate to negative attitudes and also at the neighborhood level where census data can be used to profile rejecting communities. The third issue concentrates on the implications of the empirical evidence with reference to both the providers and users of facilities in the context of current redirections in community mental health care, particularly the dismantling of the inner-city ghetto and the decentralization of facilities and services. This leads on to some concluding remarks regarding the broader issue of the social and spatial reproduction of the mentally ill.

The prevalence of opposition

Research on opposition to community mental health care demands decisions with respect to indicators and methods of measurement. Current approaches in attitude theory suggest that we might choose to examine beliefs, attitudes, intentions or behavior or some combination. The selection has a strong bearing on the conclusions we might ultimately draw. For example, in our Toronto study (Dear & Taylor 1982) only a small fraction of those reporting negative attitudes toward facilities had translated their feelings into any kind of oppositional behavior. The implication is the need to collect information on multiple indicators of attitude and behavior and to examine the relationships between them guided by an underlying theoretical model. Our use of the Fishbein and Ajzen theory of reasoned action illustrates this type of approach. If attitudes are to be measured, there is the immediate question as to their target or object. Clients and facilities would both seem relevant and important to separate. It is quite feasible that individuals could hold different attitudes towards the mentally ill and mental health facilities. In the context of predicting behavior, Fishbein & Ajzen (1975) question whether we should be measuring attitudes towards targets or objects at all. Their thesis is that the failure of many past studies to show strong links between attitude and behavior is because of measuring attitudes towards general targets (e.g., the mentally ill) instead of attitudes towards specific behaviors.

The measurement issue is complicated still further when the question of labeling is considered. For example, in the case of measuring attitudes toward the mentally ill, how should this group be defined to survey respondents? The default option might be to provide no definition beyond the label and allow the respondent to answer based on some self-definition. Alternatively, a capsule description might be provided in an effort to achieve greater standardization. The same general options apply in the case of facilities. There are no simple answers to these issues, particularly in the absence of research directed towards assessing the effect of label on responses. Two implications follow however. The first is the importance of interpreting the results of attitude studies in light of how the values were defined. The second is the desirability of developing some standardized measures as the basis for allowing across-study comparisons.

With these methodological caveats in mind, attention turns to the evidence from recent studies of public attitudes towards the mentally ill and facilities. In this discussion, the term "attitude" is used as a generic consistent with the way it was typically applied by the original investigators.

The results obtained in recent studies using different measures of attitude towards the mentally ill have been consistent in showing quite a high level of public sympathy and tolerance. The most detailed findings are from our Toronto research (Taylor & Dear 1981). In the 1978 study, data were obtained from a stratified random sample of 1090 respondents, representative of different socio-economic status groups and of suburban and central city populations. Scores on four attitudes towards the mentally ill (AMI) scales, each of which comprised ten Likert statements, were skewed towards the positive end of the response scale. The "mentally ill" were defined as "people needing treatment for mental disorders, but who are capable of independent living outside a hospital." Mean scores on the 4 scales were: authoritarianism, 3.6 and social restrictiveness, 3.7 (maximum tolerance equals 5.0); benevolence, 2.2 and community mental health ideology, 2.4 (maximum tolerance equals 1.0). A subsequent survey (Dear *et al.* 1985) showed that similarly positive attitudes were shared by an Ontario-wide sample. In this case, a subset of 16 of the 40 AMI statements were used. The means on the four scales were: authoritarianism, 3.3 and social restrictiveness, 3.9; benevolence, 2.3 and community mental health ideology, 2.3. In short, the sample groups could be characterized as neither authoritarian nor socially restrictive, rather they appear generally benevolent and supportive of community mental health care.

Several studies in progress are using the AMI scales and the results should provide a standardized basis for a comparative analysis of attitudes towards the mentally ill in the community in a variety of urban and regional settings. Most notable among these is the Winnipeg Area Survey (Tefft *et al.* 1987) which was designed to replicate the 1978 Toronto study. Early results show remarkable consistency in AMI responses implying little variation over time and space in Canadian attitudes towards the community-based mentally ill.

Generally sympathetic responses to community-based mental health care are also evident in reported attitudes toward mental health facilities. Here again, there is an issue of definition in that response is likely to vary depending on such factors as the type of facility (e.g., group home or clinic), the type of client, and level of supervision. The tendency has been to elicit reactions to relatively unspecific facility types. For example, Rabkin *et al.* (1984) measured attitudes towards "a home for former mental patients" and a "psychiatric clinic." Dear & Taylor (1982) elicited attitudes towards "community mental health facilities" defined as including "outpatient clinics, drop-in centres, and group homes which are situated in residential neighborhoods and serve the local community" and as excluding "mental health facilities which are a part of a major hospital." This remains a very broad definition, especially as most respondents are reporting attitudes towards a hypothetical facility rather than one they are aware of in their own neighborhood. A more specific approach was adopted in the later study by Dear *et al.* (1985). Three facilities were presented as "vignettes" in which the number and type of clients were described as well as the level of supervision. Attitudes were measured for locations at three distances from the respondent's home: in the neighborhood, on the block, and next door.

The findings of the different studies of attitudes towards facilities are consistent in showing that opposition is limited to a minority. Rabkin *et al.* (1984) found that 75 per cent of respondents in their New York study would not object to having a mental facility "set up near home." Dear & Taylor (1982) showed that the percentage in their Toronto sample rating a community mental health facility as undesirable varied with the distance of the hypothetical location from the respondent's home. At seven to twelve blocks, 12 per cent rated a facility as undesirable. This rose to 24 per cent at two to six blocks and to 39 per cent on the same block. A major study by the Canadian Training Institute (CTI 1984) examined the attitudes of residents (N=1696) in three Ontario cities, Toronto, Ottawa, and London. Using Dear and Taylor's facility rating scale, it was found that the percentage rating a facility as undesirable rose from 8 to 15 to 23 per cent with increasing proximity to home. The Ontario-wide study by Dear *et al.* (1985) provides the most detailed results. Three different facilities, two group homes and an outpatient clinic, were rated, again for three distance zones. For all three facilities, there was a clear distance decay in opposition. The percentage rating a facility undesirable ranged from a low of 16 per cent for a small group home with 24-hour supervision located in the neighborhood to a high of 46 per cent for an outpatient clinic with about 200 clients located next door.

Additional evidence on reactions to facilities comes from the analysis of beliefs about facility impacts. In this context, data have been collected in the form of very general judgments as well as in terms of ratings on specific impact scales. In the former category, Rabkin *et al.* (1984) found that approximately half their sample perceived the presence of a home for former mental patients as having a bad effect on the neighborhood. A smaller fraction, approximately one third, believed a psychiatric clinic would have a negative effect. In neither case were there significant differences between residents in neighborhoods with and without facilities or between residents aware or unaware of the presence of a facility.

However, overall, the responses to various questions suggest that those aware of facilities had more positive perceptions of facility effects.

In our Toronto study (Dear & Taylor 1982), beliefs about facility impacts were measured by ratings on twelve semantic differential scales. For those unaware of facilities, the ratings were beliefs about the impacts a mental health facility would have on the neighborhood, and for those aware, they were beliefs about the impacts a facility had had. For all twelve scales the modal response was neutral, showing that many respondents perceived facilities having little or no effect. For the unaware group, the percentage perceiving negative effects exceeded the percentage perceiving positive effects on all twelve scales, ranging from a high of 46 per cent perceiving a negative effect on property values to a low of 20 per cent for effects on the visual appearance of the neighborhood. For the aware group, the percentage reporting negative impacts was higher on ten of the twelve scales, the exceptions being for the perceived effects on residential character and property taxes where opinions were evenly divided. In general, the percentage expressing negative perceptions was lower for the aware group suggesting that in so far as awareness had any effect on perceived impacts, it was in the direction of softening rather than hardening residents' views. However, cause and effect between awareness and facility perceptions could not be inferred because awareness covaried with other potentially salient variables, for example residential location within the city.

Taken together, these findings on attitudes toward the mentally ill and community mental health facilities imply a relatively high level of public tolerance and support. On first sight, this may seem difficult to reconcile with the apparent strong community opposition frequently encountered in attempts to locate facilities in neighborhoods. In part, an explanation may lie in the exaggeration of opposition due to conflict situations becoming media events. Situations in which facilities are introduced without conflict do not attract media and thereby public attention. Consequently, a disproportionate weight is given to the degree of opposition. On the other hand, the potential opposition to facilities represented in the various survey results should not be underestimated. The percentage rating facility locations as undesirable approaches 50 per cent in some instances. If we accept that this may be a somewhat conservative estimate due to a bias towards more positive ratings in response to social norms, then the data indicate quite a large proportion of the population is to some degree resistant to community mental health care. As we have suggested elsewhere (Dear & Taylor 1982, 117–18), even if opposition is limited to a small minority, we cannot conclude that facility locations will meet with easy acceptance in the community. We have to recognize that the opponents, though in the minority, can determine community response to facilities if they have political power and influence and if the non-opponents remain silent and voice no clear support. This combination of circumstances does not seem uncommon and makes sense out of assertions that as many as half of all psychiatric facilities planned for residential areas may have been blocked by community opposition (Piasecki, 1975).

While the evidence in the aggregate points to a relatively low prevalence of opposition, we need to recognize that there is a distinct "geography of intolerance," to borrow the phrase used by Dear & Wolch (1987). Consideration of spatial variations in opposition forms a suitable bridge between this section and the next in which profiles of opposition are examined.

Available data allow analysis of spatial variations in opposition at three geographic scales: intra-urban; inter-urban; and regional. Attention here is limited to the evidence provided by the three Ontario studies previously cited (Dear & Taylor 1982, CTI 1984, Dear *et al.* 1985). At the intra-urban level, the data from the 1978 Toronto study clearly show a strong distance-decay in tolerance away from the center of the city. For example, 38 per cent of the total sample rated the location of a community mental health facility on the same block as their home as undesirable. The percentages, however, varied considerably when broken down by census tract from a low of 12.5 per cent in an inner-city Toronto tract to a high of 86 per cent in a suburban tract in the borough of North York. This result is consistent with the history of community conflict over facility siting in Toronto which has typically been more vehement where a suburban location was proposed.

At the inter-urban scale, the study conducted by the Canadian Training Institute (Taylor 1986) provides data for three Ontario cities: Toronto, Ottawa, and London. Respondents were asked the same questions regarding the desirability of facility locations as used in the 1978 Toronto study, thus facilitating comparison of results. Again, focusing only on ratings of a facility within one block of home, the data show the per cent opposed as Toronto 26, Ottawa 21, and London 18. The difference between cities was not statistically significant. It does, however, suggest a somewhat higher level of opposition in Toronto where in the past conflicts over facility location had been more frequent and highly publicized. For the Toronto sample, the 12 per cent drop in those opposed between 1978 and 1984 begs an immediate explanation except to point out that the sample groups for the two studies are not strictly comparable, given that the CTI study focused attention on neighborhoods with existing correctional facilities and matched controls (CTI 1984).

The study conducted by Dear *et al.* (1985) to determine the effect of a province-wide advertizing campaign on attitudes toward the mentally ill provides data for examining regional variations in Ontario in opposition at two points in time, 1983 and 1984, the years of the baseline and follow-up surveys. Five regions were defined: Metropolitan Toronto, eastern, western, central, and northern Ontario. The sample design was such that, for each of the last four, data were collected from a comparable range of centers in terms of population size. As noted earlier, in the baseline survey respondents rated three facility vignettes (two group homes and an outpatient clinic) for three different locations: in the neighborhood, on the block, and next door. The strongest and most consistent regional effects were found for the second vignette which described a group home for 10–12 adult men and women with no live-in professional staff. For all three locations, regional differences in opposition were significant with the strongest opposition being expressed by the central and western region groups and the weakest by the eastern and northern Ontario samples with the attitudes of the Toronto region respondents approximately midway between. Although the regional variation was not as strong for the other two facility vignettes, one finding was consistent throughout: in all cases least opposition was reported by the eastern region sample.

The evidence therefore at these different geographic scales points to significant spatial variations in opposition to community mental health care. The logical next question is why these differences exist, which leads on to the next section where the individual and community level characteristics of opposition are examined.

Profiles of opposition

Early studies, largely predating the advent of deinstitutionalization, show significant relationships between various personal characteristics and attitudes towards the mentally ill. This research has been extensively reviewed in the several papers by Rabkin (1972, 1974, 1975) and requires only brief synopsis here.

The main variables examined in these earlier studies were age, education, occupation, social class, race, and ethnicity. In most cases their effects were considered separately even though they are interrelated. The outcome measures used varied considerably including patient vignettes, social distance scales, and Likert scales. Despite this methodological pot-pourri, some consistent findings emerged showing that those expressing more negative attitudes were generally older, less educated, and of lower occupational and social status.

Later studies conducted in the wake of deinstitutionalization have advanced this earlier work in several respects: by establishing whether the various correlates previously identified applied to attitudes towards the mentally ill in the community; by comparing the effects of personal characteristics on attitudes towards the mentally ill and attitudes towards facilities; by using multivariate methods to distinguish the separate and joint effects of the independent variables; and, perhaps most importantly, by embedding the empirical analysis within a theory of attitude formation.

The clearest attempt to date to embed the analysis of attitudes towards community mental health care within a theoretical framework is to be found in our work based on Fishbein & Ajzen's (1975) theory of reasoned action (Dear & Taylor 1982). Within their framework, attitudes play a central role linking external variables, beliefs, and subjective norms to specific social behaviors which can include in this context reactions to the mentally ill in the community. Adopting this approach, we attempted to incorporate the findings of earlier work in defining the specific components of a theoretical model of community attitudes to mental health care. Three sets of external variables were included: facility and user characteristics, personal socio-demographic, and indicators of neighborhood structure. These linked directly to three categories of beliefs regarding facility impacts, the mentally ill, and the neighborhood. These in turn linked to attitudes toward facilities and then to behavioral intention and to behavior, with the ultimate outcome variable being facility acceptance or rejection. The empirical test of the model using the Toronto data set incorporated all the links in the model up to and including behavioral intentions. The link to behavior was excluded by the absence of sufficient respondents reporting having taken action either in support of or in opposition to a facility. Within these limits, the findings confirm the relationships in the model between external variables and beliefs, between beliefs and attitudes, and between attitudes and behavioral intentions.

While grounded within a theoretical framework, and in that sense an advance over previous studies, the Toronto analysis was limited in furthering the explanation of attitudes because each set of relationships within the model was treated separately. Subsequent re-analysis of the same data (Hall 1980, Hall & Taylor 1983) goes a step further by testing an analytical model which examines the relationships simultaneously. This approach, using path analysis, permits estimation of the direct and indirect efforts of each independent variable and a more precise assessment of the overall power of the model to explain attitudes to facilities. The results show beliefs about the mentally ill and beliefs about facility impacts as having the strongest effects on attitudes towards facilities. According to the structure of the path model, more positive beliefs about the mentally ill lead to more positive perceptions of facility impacts, which, in turn, give rise to more favorable judgments of facility desirability. By comparison, both the direct and indirect effects of external variables, including facility, personal, and neighborhood characteristics, on attitudes to facilities were weaker.

The profile of opposition that emerges from the analysis differs with the outcome measure. For attitudes toward the mentally ill, more negative opinions are held by those who are: older, less familiar with mental illness, less educated, regular church attenders, and home owners. For attitudes toward facilities, opposition is more commonly reported by those who are: less familiar with mental illness, home owners, and residents of higher status neighborhoods. There is some evidence to suggest that the effects of socio-economic variables (education, occupation, and income) reverse such that higher status respondents express more positive attitudes toward the mentally ill but more negative attitudes towards facilities. Additional evidence of this reversal is emerging from analysis in progress using the data from the 1983 Ontario survey (Dear *et al.* 1985). A plausible interpretation of this result involves the conflicting concerns of higher status individuals. They tend to hold (or at least report) more liberal attitudes towards the mentally ill in general, but, at the same time, often have a greater stake in their home and neighborhood and as a consequence are more resistant to the perceived threat associated with the siting of a mental health facility.

Smith & Hanham (1981) also used path analysis in seeking to clarify the strength and pattern of effects of various socio-demographic variables on attitudes towards mental illness. The outcome measures, based on a ten item social rejection index, were the respondent's reactions to vignettes describing a serious and a moderate case of mental illness. Attitudes toward other social problems had the strongest effect on attitudes toward mental illness. Consistent with others' findings, homeowners were more likely to reject the mentally ill. Proximity emerges as a "two-edged sword." Residents in the neighborhood close to a mental hospital reported more positive attitudes toward serious mental illness, but this result was tempered by the finding that previous contact with mental illness and acquaintance with workers at the facility were not positively related to acceptance of the mentally ill. This implies that the effects of locational and social proximity are not necessarily consistent. As Smith and Hanham are careful to point out, the effects of contact may well depend on the characteristics of the individuals involved and the frequency with which contact occurs.

It is clear that the determinants of attitudes towards community mental health care clients and facilities are multiple and interacting and that studies to date have only begun to uncover the complex structure among the salient variables. Path analysis provides a powerful statistical tool for doing this. Its application, however, requires a theoretical basis on which to construct a path model. The statistical measures only form a meaningful basis for inference about the strength and pattern of effects when interpreted with reference to the theory used to develop the path model. It follows that combining path analysis with a theoretical model of attitudes, such as the Fishbein and Ajzen framework, holds the most promise for advancing understanding of the factors underlying public reaction to community mental health care.

To this point, the discussion of profiles of opposition has centred on disaggregate analyses of the effects of personal characteristics on attitudes toward community mental health care. A complementary approach involves aggregate, ecological analysis. In this context, the primary objective is to identify useful "markers" of opposition to clients and facilities in host neighborhoods. The emphasis is therefore strongly applied in seeking to aid and inform planning decisions regarding facility location and the assignment of clients to community settings.

The spatial unit usually chosen for this community level analysis is the census tract, given the immediate availability of census and other data by tract, and given its reasonable correspondence in area and population with urban neighborhoods and planning districts. Regression analysis can be used to predict aggregate measures of public attitude (e.g., the percentage in a census tract opposed to a facility) based on a set of ecological variables representing the physical and social characteristics of communities.

Several studies, including the disaggregate analyses previously described, inform the selection of ecological variables. The work of Trute & Segal (1976) is particularly relevant. They examined the characteristics of supportive communities for the severely mentally disabled and showed that support was strongest where there was neither strong social cohesion nor severe social dislocation. The former type of community tends to "close ranks" against the client; the latter tends to be too chaotic and threatening. Building on this and other findings, we developed a model of community reaction to mental health facilities comprising six dimensions of neighborhood structure: land use mix; demographic characteristics; socio-economic status; community homogeneity; community stability; and population density (Taylor *et al.* 1984). Indicators for each dimension were extracted from census and land use data for Metropolitan Toronto using the census tract as the unit of observation and analysis. The statistical reproducibility of these six dimensions was examined using factor analysis, and the results led to some revision and relabeling with the following seven neighborhood factors forming the basis for the subsequent regression analysis: neighborhood transience; scarcity of children; economic status; ethnic heterogeneity; sex ratio; residential land use; and institutional land use.

Multiple regression analysis was used to predict community reaction to facilities. Reactions were measured as the percentage of respondents in each census tract reporting negative attitudes to facilities. These data came from our 1978 Toronto survey. The percentage of variance explained ranged from 16 to 37 per cent and was highest for reactions to facilities closest to home (i.e., within one block). Three of the seven neighborhood factors were consistent and significant predictors: transience, scarcity of children, and economic status. Greater opposition to facilities was associated with stable neighborhoods with relatively large proportions of families with children and relatively high economic status. This profile of opposition corresponds closely with the social characteristics typically found in suburban neighborhoods and is therefore consistent with the results reported earlier describing the greater prevalence of opposition in suburban localities.

Implications

The implications of the evidence on community opposition to mental health care can be usefully summarized from the perspectives of the providers and users of services.

Superficially, the evidence presented seems to lead to a generally positive conclusion from the provider's standpoint. The prevalence of opposition is relatively low, pockets of strong resistance being limited to certain suburban localities. Facilities are often introduced into neighborhoods with minimal effect to the point that the majority of local residents remain totally unaware. Negative perceptions of impacts in advance of siting tend to diminish after the facility is introduced. The absence of any detrimental effect on property values (e.g., Boeckh *et al.* 1980) is perhaps the clearest example of the anticipation being worse than the event.

Despite this seemingly positive outlook, there may be storm clouds on the horizon because of the likely locations of future facilities. To this point in time, facilities have concentrated in inner-city neighborhoods. There are several reasons for this, including the availability of suitable properties, especially low cost housing for residential care services. Another factor, however, has been resistance on the part of suburban municipalities to share the burden of providing care for service-dependent groups in general, and the mentally ill in particular. In some cases, this resistance has translated into exclusionary zoning practices prohibiting the introduction of facilities. Such practices are now being removed and replaced by fair-share zoning policies which, in principle at least, open up the suburbs and permit a decentralization of services. The issue immediately arising from the adoption of a policy of decentralization is that of the potential reaction to facilities in communities which to date have not had to accept them. It is clear that the existing concentration of facilities corresponds quite closely with the inner-city zone of highest tolerance. Efforts to achieve a more equal distribution of facilities are likely therefore to create more opposition than has been experienced to date.

In this context, public education programs are often proposed as an important part of a strategy to promote greater acceptance of the mentally ill in the community. The effectiveness of these programs has, however, been largely assumed rather than demonstrated. Our recent evaluation of the Information and Action Program organized by the Canadian Mental Health Association in Ontario (Taylor & Dear 1986) provides some evidence to suggest that positive attitudes can be sustained in the face of potentially negative influences. The extent to which educational programs can offset and overcome existing negative attitudes remains uncertain, however.

With the prospect of increased decentralization of facilities, it is realistic to view educational campaigns as necessary but certainly not sufficient strategies for countering opposition. Anticipating greater community resistance to clients and facilities in the future, some have suggested that planning agencies develop a negotiating strategy with potential host neighborhoods. Such a strategy would include an agency commitment to an equitable share of the "burden" associated with different types of facilities and different levels of acceptance. If such assessments were possible, then an equitable allocation of facilities within a region could involve client/facility trade-offs whereby communities could negotiate with the planning agency to accept a larger number of low-burden facilities or a smaller number of high-burden units. In describing such a scheme, Dear & Wolch (1987) point out that this approach has already been attempted in some cities. Analysis of the CTI data on attitudes towards psychiatric and correctional group homes in Ontario cities (Taylor 1986) provides an example of how survey information might be used to obtain measures of the relative burden associated with particular facilities and client groups. The implications from this review for facility users, the mentally ill in the community, again on first sight appear to be generally positive. Historically, the gradual shift in public opinion away from authoritarian and restrictive beliefs about the mentally ill towards a more benevolent and community-focused view of mental health care implies a growing tolerance, sensitivity to needs, and perhaps willingness to support facilities and services. There remains a gap, however, between sympathy in principle and support in practice. The evidence presented would suggest that narrowing the gap involves a combination of positive first-hand experience with the mentally ill in the neighborhood, thereby reducing the fear of the unknown, and ongoing public education to promote and reinforce an understanding of needs. In this latter respect, the activities of groups such as the Canadian Mental Health Association are especially important, since they are well placed to initiate, coordinate, and evaluate efforts to communicate community mental health needs and priorities to the public at large; the recent Information and Action Program in Ontario is one example of the type of outreach that is possible.

Against this positive backcloth, we again need to consider the prospects for the future in light of the gradual decentralization of services. The current concentration of facilities of all types in inner-city neighborhoods correlates strongly with the distribution of clients. Clearly, there is circular causation involved, services are located to respond to need and users locate to have access to services. Decentralization implies some fracture of these close ties at least in the short run, with potential costs to users in terms of decreased access to care. If users are able to follow facilities and move out of the inner city, then other implications arise, specifically, the dismantling of the ghetto, which some regard as a supportive environment, not easily reproduced once the mentally ill are more widely distributed throughout the city. These concerns are, however, largely speculative and imply an understanding of the factors that influence coping and quality of life among the mentally ill in the community. This understanding is rudimentary at best and only now beginning to attract the research attention it deserves (Kearns *et al.* 1987). What is clear is that public attitudes are an important component in the complex web of factors affecting client wellbeing. This is poignantly expressed in the words of a client interviewed in our recent study in inner-city Hamilton:

> It's a subtle sort of thing. People aren't as understanding as they might be. . . . People give you funny looks. Neighbors on our street. Even here at the Center. It gives you a funny feeling, we're all isolated from the rest of society. They say "look at those guys in there. They're sitting around drinking coffee all day." You have to be sick yourself to know what it's really like.

In a broader context, one can question the implications of decentralization policy for the social and spatial reproduction of the mentally disabled. Is it likely to result in greater integration of the client group into the residential and social fabric and the gradual breaking down of the spatial and non-spatial barriers previously erected to maintain a separation from mainstream society? There is little evidence to suggest that such goals are in view as decentralization initiatives are devised and implemented. They seem to be prompted much more by immediate pragmatic political motives to relieve the social and financial costs of hitherto over-burdened inner cities. There is a point at which even the most tolerant of jurisdictions and neighborhoods demands a halt in the face of a seemingly ever increasing number of disabled, disadvantaged, and destitute.

Even though decentralization has not been motivated by a sense of need to better integrate the mentally disabled, might this yet occur as a fortunate by-product? Again, the prospects are not promising. If community resistance is overcome, host neighborhoods are more likely to manifest grudging acceptance than positive support. Planning and other mechanisms will be devised to maintain social and spatial distance with the result that we are likely to see the inner-city ghetto gradually replaced by smaller ghettos more widely scattered throughout the urban region. In which case, the client group might be relatively worse off, if such is possible. If the notion of the centralized ghetto as a coping mechanism has any validity, and admittedly the current evidence is not convincing, then removal to decentralized sites threatens the supposed informal support system presently experienced. Add to this the inevitable problems that might arise from decreased access to needed facilities and services and the related increased drain on already over-stretched financial resources, and one is left with a potentially daunting scenario for the client group. This may seem a pessimistic note on which to conclude but none the less realistic in a society in which self interest predominates over caring and the "landscape of despair" is thereby reproduced.

References

Ajzen, I. & M. Fishbein 1980. *Understanding attitudes and predicting social behavior.* Englewood Cliffs, N.J.: Prentice-Hall.

Boeckh, J. L., N. Dear & S. M. Taylor 1980. Property value effects of mental health facilities. *Canadian Geographer* 24, 270–85.

Canadian Training Institute (CTI) 1984. The effect of locating correctional group homes in residential neighbourhoods. Research report. Downsview, Ontario: York University, Canadian Training Institute.

Dear, M. 1981. Social and spatial reproduction of the mentally ill. In *Urbanization and urban planning in capitalist society*, M. Dear & A. J. Scott (eds.), 481–500. London: Methuen.

Dear, M., R. Fincher & L. Currie 1977. Measuring the external effects of public programs. *Environment and Planning A* 9, 137–47.

Dear, M. & S. M. Taylor 1982. *Not on our street: community attitudes to mental health care.* London: Pion.

Dear, M. J., S. M. Taylor, D. Bestvater & B. Breston 1985. Evaluation of the Information and Action Program. Research Report. Hamilton, Ontario: McMaster University, Department of Geography.

Dear, M. J. and J. Wolch 1987. *Landscapes of despair: from deinstitutionalization to homelessness.* Oxford: Polity Press.

Fishbein, M. & I. Ajzen 1975. *Belief, attitude, intention and behaviour: an introduction to theory and research.* Reading: Addison Wesley.

Hall, G. B. 1980. A causal model of individual responses to community mental health care. PhD dissertation. Hamilton, Ontario: McMaster University, Department of Geography.

Hall, G. B. & S. M. Taylor 1983. A causal model of attitudes toward mental health facilities. *Environment and Planning A* 15, 525–42.

Kearns, R. A., S. M. Taylor & M. J. Dear 1987. Coping and satisfaction among the chronically mentally disabled. *Canadian Journal of Community Mental Health* **6**, 13–24.

Piasecki, J. 1975. Community response to residential services for the psychosocially disabled. Paper presented at the First Annual Conference of the International Association for Psycho-social Rehabilitation Services, Horizon House Institute, Philadelphia.

Rabkin, J. G. 1972. Opinions about mental illness: a review of the literature. *Psychological Bulletin* **77**, 153–71.

Rabkin, J. G. 1974. Public attitudes toward mental illness: a review of the literature. *Schizophrenia Bulletin* **1**, 9–33.

Rabkin, J. G. 1975. The role of attitudes toward mental illness in evaluation of mental health programs. In *Handbook of Evaluation Research*, M. Guttentag & E. L. Struening (eds.), vol. 2, 431–82. Beverly Hills: Sage Publications.

Rabkin, J. G., G. Muhlin & P. W. Cohen 1984. What the neighbours think: community attitudes toward local psychiatric faciltiies. *Community Mental Health Journal* **20**, 304–12.

Segal, S. & U. Aviram 1978. *The mentally ill in community-based sheltered care*. New York: John Wiley.

Smith, C. J. & R. Q. Hanham 1981. Proximity and the formation of public attitudes towards mental illness. *Environment and Planning A* **13**, 147–65.

Taylor, S. M. 1986. The lesser of two "evils": community attitudes toward correctional and psychiatric facilities. *Ohio Geographers* **14**, 115–26.

Taylor, S. M. 1987. Community reactions to deinstitutionalization. In *Location and stigma*, C. J. Smith & J. A. Giggs (eds.). London: Allen & Unwin.

Taylor, S. M. & M. Dear 1981. Scaling community attitudes toward the mentally ill. *Schizophrenia Bulletin* **7**, 225–40.

Taylor, S. M. & M. Dear 1986. Changes in attitudes towards community mental health in a deteriorating social climate. In *Proceedings of the 2nd International Symposium in Medical Geography*. New Brunswick: Rutgers University.

Taylor, S. M., G. B. Hall, R. C. Hughes & M. J. Dear 1984. Predicting community reaction to mental health facilities. *American Planning Association Journal* **50**, 36–47.

Tefft, B., A. Segal & B. Trute 1987. Neighbourhood response to community mental health facilities for the chronically mentally disabled. *Canadian Journal of Community Mental Health* **6**, 37–50.

Trute, B. & S. Segal 1976. Census tract predictors and the social integration of sheltered care residents. *Social Psychiatry* **11**, 153–61.

15

Mirrors of power: reflective professionals in the neighborhood

DANA CUFF

Of all the territory that makes up the urban landscape, the most central in life is the home and its neighborhood. The neighborhood is the smallest political unit that is defined by physical space, whether that be the houses along a street, around a park, between some natural boundaries, or within a township. It is in the neighborhood, then, where the interaction between space and society goes public. Thus, it is not surprising that environmental planners and designers are involved in neighborhoods, guiding the everchanging system toward often ambiguous desired ends. Their participation, however, does not always lead to the desired ends, even when those can be determined. Here, I tell the story of one neighborhood in Houston, Texas, which is worth studying for several reasons. First, this neighborhood is interesting because it is so obvious an example of physical space which not only reflects but also renews the social order. That is, the case reveals the interactions between environment and society. Second, the case exposes the intrinsically conservative characteristics of design professions, which themselves reflect and renew the dominant socio-spatial patterns. And third, I relate this story because I have first hand experience in it, as a community advocate and design professional.

The neighborhood in question, Fourth Ward, has two parts: historical Freedman's Town and Allen Parkway Village, a public housing development. The neighborhood is the oldest black community in Houston, Texas, and the poorest neighborhood in the city. It is presently threatened by public and private redevelopment interests that would raze the area and scatter its residents. Indeed, a historical survey shows that approximately every 20 years outside agents with federal and municipal funds assert their power over the neighborhood, and the neighborhood both resists (by its mere survival) and succumbs (through its many losses). Now, as the intended final blow is readied to level the neighborhood, community action is sporadic, desperate, and factionalized, but there is resistance.[1]

The forces behind the assaults on the neighborhood, and behind the community's resistance to those assaults, are numerous and familiar. They concern labor, capital, urban restructuring, discrimination, demographics, land values, development as a growth industry, and class (see Ch. 4). While an important paper remains to be written explaining the interweaving of these factors into the history of Fourth Ward, my

objectives in the present work are more modest. With the physical artifact as evidence, I first show that the evolution of Fourth Ward simultaneously embodies and promotes the disadvantaged status of the poor, black community. My second objective is to examine the participation of environmental design professionals in this process of reflection and renewal of the larger social order. Although a great deal has been written on the positive and negative roles of certain agents in this process, the most important being the community and the state, we have few first-hand accounts of those actors who deliver design services, thereby producing a physical form for a community's sensibilities. Moreover, to understand the underlying structure of the institutions and ideologies that guide design professionals, the case in which the professional works for the disadvantaged party – the neighborhood – is most revealing. In such a case, underlying limitations meet self-stated goals to create unintended outcomes. Here the role of the design professional in social reproduction comes to the surface.

To put it another way, I explore the role of two other forces in the lives of neighborhoods: the physical environment itself and the design professional. I contend that each of these plays a part in social reproduction, that is, in the reflection and renewal of the dominant economic, cultural, and ideological structure of society. Fourth Ward is a unique neighborhood in Houston, Texas, but at the same time, the nature of its problems, the actors, and the events bear a fundamental likeness to those of other neighborhoods threatened by redevelopment (see for example Fainstein *et al.* 1983). Unique to my own analysis, compared to previous work on advocacy planning and design, is the attempt to unravel

332

Figure 15.2 Allen Parkway village, view to downtown. *Photo: Phyllis Moore*

the means by which the structure of professionalism interacts with community advocacy. From this analysis, there are two lessons to be learned. The first is a useful self-criticism that any community advocate-designer should undertake. The second is a more theoretical understanding of professionalism, in which community activism becomes the vehicle to uncover the underlying structure and limits of the design professions.

The neighborhood as force and artifact

Fourth Ward sits cheek by jowl with Houston's central business district, separated by only a freeway (see aerial photographs and Figs 15.1 and 15.2). In Fourth Ward is Allen Parkway Village, 1,000 units of public housing on 37 acres, and historic Freedman's Town, approximately 80 blocks of old, privately owned, primarily rental housing. The history of this place tells the unfolding story of what it means to be poor and black in Texas. Particularly in a neighborhood like this one, in which few established institutions exist that have maintained historical records, the environmental artifact imparts the untold story.

When two real estate speculators from New York, Augustus C. Allen and John K. Allen, laid out Houston's original townsite on Buffalo Bayou in 1836, they made the first in a long sequence of powerful development decisions about the city. Their grid, that potentially infinite blanket for land platting, embodied their willful control over the land, and the unlimited expansionism that has characterized Houston's growth. The Allen brothers' townsite was to become downtown, and just upstream there were low lying, insect infested lands that were subject to flooding and undesirable for development. Prior to Emancipation, in this area there lived a small number of free blacks along with some slaves who rented housing away from their owners and hired out their own labor. Thus began a long symbiotic relation: an oppressed community with its cheap labor source, bound to its advantaged neighbor, the central city. Already in the 1850 map of Houston, a small portion of the community has been platted. Although Fourth Ward was one of the largest black communities, it should be noted that the area in question was somewhat multi-racial, and that blacks were living on the city's fringes in all the wards (Wintz 1982).

With Emancipation, large numbers of freed slaves came to this nascent community in Fourth Ward, known then as Freedmantown. Between the 1860 and 1870 census, the black population of Houston nearly quadrupled, causing extreme housing shortages. Many of the newly arrived freed slaves came to Freedmantown to live in makeshift quarters of stables and shanties, and to worship in brush sheds. As early as 1880, all the current streets in the heart of Fourth Ward were platted (Breisch 1984). Its grid was a replica of the Allen brothers' work, only at a smaller scale and at an opposing angle. Here, quite literally, the environment reflects the dominant social group's patterns and interests, as if to say that Fourth Ward was inferior and did not belong to the city's core. By 1890, two street car lines moved through the area into the downtown, rendering the population of cheap labor both convenient and mobile.

In the earliest visual records dating from the 1890s, small frame cottages appear eight to ten to the block (see Fig. 15.3). Few of these have survived, due to their insubstantial construction, but at this time more substantial and elaborate structures begin to appear, some of which remain today (Breisch 1984). By the early 1880s, about 25 per cent of the black population owned land which may have been sold to them on credit by white speculators who subdivided low-priced farm land at what was then edge of the city. Exactly who built the hundreds of cottages and shotgun houses is not clear, but it is likely that white speculators filled tiny lots with inexpensive housing. The shotgun, a house type associated with early black settlements in America and perhaps originating in Haiti, suited the interests of both developers and residents. Its long, narrow footprint could be efficiently placed on small lots tightly organized into blocks, or placed in between larger structures, or at centers of blocks (Breisch 1984, 3). It was inexpensive to construct and easy to move if necessary. Outside, fronting the street is the narrow end of the building with its porch where much of the life of the community transpires. Inside the small house has a series of usually three to five equal sized, relatively ill-defined rooms, in which a variety of living arrangements can be accommodated. Even today, for example, residents use the front room as a sitting room by day and a bedroom at night.

From the 1860s until the Great Depression, the spatially confined community grew in population so that by the end of the 1920s, Fourth Ward was six times denser than the city-wide average. By the first decade

Figure 15.3 Bird's eye view, Houston, 1891 (partial view). In this map, looking due south, Freedmantown is just above the meandering bayou, where streets run true to the cardinal directions. A darker line plots the street car route from the neighborhood to downtown at the left. *Courtesy Houston Public Library*

of the 1900s, Fourth Ward had become a bona fide neighborhood with schools (including a college), hospital, churches, ball park, and commercial activities owned and operated by the black community (see Fig. 15.4). Up until the Depression, Fourth Ward was the economic heart of black Houston. By 1930, there were 11,500 residents, the lion's share of whom lived in rental housing (approximately 85 per cent; Breisch 1984, 5). The black population grew from 3,700 in 1870 to over 63,000 by 1930. Dense, substandard conditions in inner-city neighborhoods like Fourth Ward (which received less than their share of city services) produced an outward migration of higher income black families. This left a very low-income population of renters, a highly vulnerable group, on the most desirable land for development.

When the Depression struck, the vitality of the black community in Fourth Ward was abated. The weakening of the area left it prey to numerous attacks from parties who were themselves looking to regain a foothold in a shaky economy. The pivotal blow came from the city and the federal agencies in 1939, just two years after the United States Housing Act established the means for local public housing authorities to finance

Figure 15.4 Alexander map of Houston, 1912 (partial view). This map shows the radial organization of the wards, with the downtown at the center. In Fourth Ward, the ball park is shown at the south-eastern corner of the community. The area just below the bayou, called the reservation, was razed in the 1940s for the Allen Parkway Village public housing development. *Courtesy Houston Public Library*

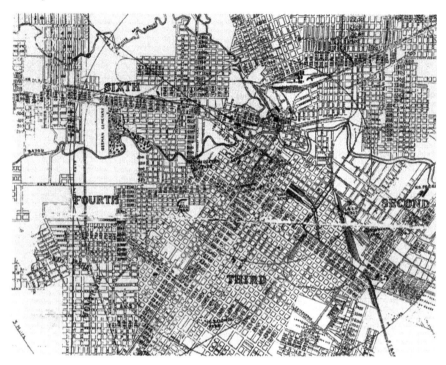

the construction of low-income housing. At this time, the oldest section of Freedman's Town was destroyed to make way for a large public housing project for white families only. Ironically, the 17-block area demolished was that part of Freedmantown which had been the site of the city's vice district, called the Reservation, between 1908 and 1917. Along with slum clearance, then, the new housing project would wipe away evidence of the city's past sins. Despite protest from the black community, 1,000 units of low-income housing were completed and surrounded by fences between 1942 and 1944.

The neighborhood's second major defeat came in 1953 when the Texas Highway Department, armed with federal highway funds, forced interstate Highway 45 through the neighborhood, severing east from west and again destroying blocks of homes. This effectively killed the portion of the neighborhood that remained on the downtown side of the freeway. As can be seen in the aerial photographs, one taken in 1955 (Fig. 15.5) and the other in 1980 (Fig. 15.6), this annexed to the central business district the commercial heart of Fourth Ward, its Carnegie Library and high school. Only Antioch Baptist Church, historically the strongest black

Figure 15.5 Houston, 1955. The freeway, still under construction, can be seen bifurcating Fourth Ward in the lower right quadrant of the photograph. The white roof of Antioch Baptist Church stands out in the area above the freeway. *Photo: Harper Leiper Studios*

congregation, remains as evidence of the old neighborhood while much of the rest became parking lots. Even the Fourth Ward street patterns were realigned and blocks enlarged, erasing all seams where traces of the historical black community might be detected.[2]

While it can be misleading to attach intentionality, meaning, and symbolism to every physical act, the events in Fourth Ward history can hardly be viewed otherwise. That public housing and highway construction were used as "slum clearance" programs is clear. (The importance of these programs and their federal dollars is exaggerated in Houston, where urban renewal funds were not available since the city has no zoning ordinance.) At the same time, paradoxically, both the public housing and the highway have helped forestall the demise of the remainder of Fourth Ward, called Freedman's Town by its current residents. The public housing development dominated the area, discouraging real estate speculation nearby and requiring a lengthy public process for demolition. For its part, the highway created a physical wall between two uneasy neighbors. Downtown has grown within its circumscribed boundaries, and will leap the barrier only with difficulty. Thus, both the public housing and the highway have had unintended consequences beneficial to the community.

This history sets a pattern, maintained until this day, of large scale government initiatied intrusions, with eventual unintended consequences,

337

Figure 15.6 Houston, 1980. The spreading downtown has eradicated the portion of Fourth Ward within the encircling highway. Just above the new, massive parking structure, Antioch Baptist Church is barely visible. In this view, the row upon row of housing at Allen Parkway Village is clearly shown in the lower left corner. *Photo: Harper Lieper Studios*

followed by smaller scale efforts by local agencies to reassert control. The first efforts to reflect a dominant social order do not also renew that order as expected, requiring continual environmental tinkering. In the 1950s and 1960s, the public housing was underpopulated since blacks were not admitted, and whites did not want to live surrounded by a black neighborhood. After being forced to integrate the public housing project in the late 1960s, eventually the development became dominated by blacks, and the housing authority began to worry about the area's growing political strength. Conveniently, a new population of low-income housing tenants became available in the 1970s: southeast Asians. The housing authority routed these immigrants to Allen Parkway Village, many of whom could not speak English, nor were they eligible to vote. Within the decade, the majority of APV's tenants were Indochinese who had less political power and greater vulnerability than American ethnic minorities. By controlling the public housing, the local agencies also controlled the neighborhood, and quite consciously manipulated the strength of the community's political voice.[3]

The housing authority now has applied for, and may be granted the right to demolish Allen Parkway Village. Only 150 households still reside in the 1,000 unit development, and those are being encouraged and harassed to move out. What is remarkable today is that the public housing is, for the second time, being used as a tool to wipe out undesirable neighborhoods. In Houston, Allen Parkway Village's 37-acre site is seen as key to redevelopment of the entire neighborhood. Once the public housing has been eliminated (and, incidentally, replaced by mid- to high-rise office buildings and upper-middle income housing), development interest will grow in adjacent Freedman's Town. Like the catalyst for a chemical reaction, the sale of Allen Parkway Village is expected to trigger a chain of events desired by the city, not the least of which is the demolition of historic Freedman's Town.[4]

The history of the place indicates that this effort is the third attempt to gather strength from federal subsidies to erode or eradicate Fourth Ward, and that unintended side effects will be produced. The public housing and the highway programs were used to accomplish, with only partial success, what this new redevelopment effort may achieve. Within Freedman's Town, landbanking has increased along with deterioration of the housing stock for several decades. Up through the early 1980s, Houston's aggressive economic expansion made this third and supposedly final assault seem inevitable once the aid of private developers had been enlisted. For the time being, the current economic recession has thwarted those plans. Political representatives of the area who previously took pro-development stands are now supporting community interests, as the political and economic tides change.

The design professionals

"What I'm trying to do is hold the front line with the help of outside professionals until I can get the community to come along with me." Lenwood Johnson, President, Residents Council of Allen Parkway Village.

About two months after moving to Houston in 1983, I found myself completely immersed in the activities surrounding Fourth Ward. Before long, I was interviewed on the local evening news arguing to save the community from redevelopment and by the local newspaper and radio journalists. Soon thereafter, an individual with tangential links to the mayor's office informed me that I was "ordering up a pair of concrete boots" if I got involved in this issue. I did get involved, and thus far, I have been spared the Jimmy Hoffa treatment. By now, there are new architects and planners who have stepped in to replace me and my colleagues from the earlier phases, just as my own participation followed that of numerous design professionals before me. By designers or design professionals, I mean all those trained specialists who create plans for shaping the physical environment, that is, architects, city planners, landscape architects, civil engineers, and urban designers.

I propose that physical artifacts reflect and renew the social order, and that design professionals contribute to that process – even in those circumstances when it is not their intention. This leads to their *production of unintended consequences*. To explain, I refer to my own case as just one typical example among many. After several years, I left Houston to take another teaching position, and since my own finances were bound to the university and not to compensation from my clients (the community) as in the traditional client-professional relationship, there were fewer strings attached to my departure. Even though I have continued to work with Fourth Ward neighborhood advocates, effective communication has been diminished by this move. To overcome some of the discontinuity in professional services that the community experienced (and expected from the outset), I explicitly sought the involvement of other design professionals. This new generation, however, did not utilize much of the preceding effort as a foundation for their undertakings, nor did the community's role in the design process remain as strong as in earlier work. For example, several design professionals began working with the city councilman for the Fourth Ward district, who had never supported neighborhood efforts. He selected a talented and well-intentioned but politically naive designer to assist with a plan for only partial rehabilitation of the neighborhood, with absolutely no resident participation. Allied with a group of design professionals concerned with preservation issues, a scheme has developed behind closed doors, if in fact it has developed at all. When challenged about why no community members have been involved in the planning, a number of telling arguments were raised:

(a) The study will be presented to the community when it is complete.
(b) Nothing has been done, so there has been no reason to involve others.
(c) The study will be limited to fairly technical issues, so it will not be political.
(d) The task group does not want to rock the boat, particularly with the councilman, preferring to stay clear of controversial individuals, and to stay out of politics.
(e) Adding community members will make the process more cumbersome.

Such arguments point out the inherent discrepancies possible between implicit intentions to assist the neighborhood and actions, where the latter contribute to the dominant power structure rather than to the community. Most clearly, the group views its position as "technical," while any position other than the one held by the group and councilman is considered both "political" and "controversial." Indeed, the group's proposal to preserve Fourth Ward may never come to light, for when oil prices dropped and the current economic crisis hit, private and public funds for implementation dried up.

An unintended consequence of our early involvement in Fourth Ward was to pave the way for later designers as well as policy makers to work in the community – including those who desire different ends. Early, more radical efforts can be followed by liberal or even conservative community assistance that contradicts the residents' position. In addition, some fundamental characteristics of design professionals (discussed later) worked against one party "carrying on" for a predecessor. Since our earlier work with the neighborhood groups (design proposals and cost estimates) was not well documented for a variety of reasons, the problems of transition were exacerbated.

If the recorded history of Fourth Ward is slim, records of the involvement of design professionals are virtually non-existent. Likewise, case study materials from similar neighborhood struggles do not typically follow the activities of the design consultants over time. Looking back to the neighborhood's history, there have been few clear roles for designers except for the civil engineers involved in the highway program and the architects and engineers for the public housing development. The original architects for Allen Parkway Village were concerned with creating livable shelter, albeit as a temporary residence since families were expected to "get back on their feet." Specifically, public housing was not to compete in terms of cost or amenity with private housing (Kravitz 1970). The lead architect, in fact, stated in a sworn affidavit his outrage at the city's current intentions to demolish Allen Parkway Village. The highway program, so monstrous in scale, swallows the roles that individuals such as civil engineers and designers may have played in the strategic location of the roads. These designers worked for the state, serving the state's interests and so it should not be surprising that their physical solutions often conflicted with residents' goal of keeping their neighborhood intact. But when designers working for the community fulfill goals of their adversaries, then we learn something about the limits of professionalism, and we learn why community organizations are rightly skeptical of professional assistance.

As Friedson argues in his excellent book, *Professional powers*, the actual knowledge and therefore activities that professionals bring to bear in their everyday worlds is different from the formal knowledge that is documented in texts and theory (1986, xi). As I have already stated, in the case of design advocacy projects for urban neighborhoods, we commonly find that professional conduct leads to unintended consequences, and to some form of conflict. The following analysis is based upon the Fourth Ward case, along with several other urban redevelopment efforts for which extensive documentation is available (see for example, Keyes on Boston, 1970; and Hartman on Yerba Buena, 1974). Basically, I present five fundamental realms where design professionals find themselves in conflict with communities – even those they are trying to serve: project orientation, specialization, patronage, autonomy, and locale. These five realms correspond to the differing constructs that designers and communities hold about time, knowledge, economics, and values, all of which are central to the ways problems are formulated:

(a) project orientation – time
(b) specialization – knowledge
(c) patronage – economics
(d) autonomy – values
(e) locale – space

Project orientation: the designer's time

Design professionals and community residents hold contrasting notions of time which confound priorities about short term, incremental change and long term plans. The fundamental difference is that the professional tends to perceive the community as a project, but for the residents the community is a way of life. So, while neighbors recognize that seemingly small issues may need to be resolved over long periods of time, professionals are more oriented toward goals than processes (a point made by Goodman 1971). Likewise, standards of professional service can stymie a neighborhood's response in a crisis. Professionals adhere to funding deadlines, the academic calendar, or public hearing schedules that may have little relevance from the community's perspective. One negative consequence of this temporal differential is that professionals "complete" projects, extracting themselves from neighborhoods whose projects are always incomplete in some sense. This creates discontinuities, periods of vulnerability, and so on.

To take one example from the Fourth Ward case, I return to the design professionals mentioned earlier who back the idea of a partial rehabilitation scheme in which a portion of the most significant historic buildings in the neighborhood and a portion of the public housing would be preserved and the rest sacrificed. They see this compromise as a way of saving at least part of the area. The leadership in the neighborhood, however, want nothing less than total rehabilitation, since they forsee the sacrifice of one portion now as the precursor to later sacrifices, and finally the loss of the entire neighborhood. At the least, the residents recognize they could lose the neighborhood in another way with this plan: to other social groups through gentrification. In other words, to create a preservation district from the residents' perspective is the first step to eradicating all but that one section of the neighborhood, which could easily become a cafe district for downtown business people rather than the heart of the black community. According to the designers, partial rehabilitation offers a project that could be realized, and realized within a reasonable time period. According to the neighborhood, such a "project" would unleash a force that, in the long run, would destroy their area. At the risk of overstating my point, the designers' plan to save part of historic Freedman's Town reflects the dominant ideology (e.g., that owners, not renters, should decide a neighborhood's fate; that a historical neighborhood is its buildings rather than its people) and, in turn, promotes that ideology (e.g., by breaking up a resistant political enclave; by taking control of a neighborhood through the vehicle of an historic district).

The designer as specialist: knowledge

Although the expertise that designers claim concerns physical structure, the form-related organization of spaces, and land uses, these affect the full spectrum of the community's existence. As rehabilitation and revitalization plans are proposed, they will inevitably embody social, political, and symbolic solutions as well as the explicit physical solution. As such plans are developed, social and economic planning must be incorporated. Yet, these dimensions are outside the designer's scope, and "responsible" professionals avoid stepping beyond the bounds of their expertise. Each professional working with the community has something less than the whole picture (e.g., housing, economic development, job training, political representation, legal action). Yet without comprehensive understanding, the physical designer can unwittingly work against the community's interests. For example, urban neighborhood revitalization without economic development will only pave the way for gentrification.

The trouble is, we do not have solutions to some of these problems; for example, we do not know how to stop gentrification. Even if we did, it is unlikely that the solution would be within a single profession's domain. Like the physician who is unwilling to prescribe anything but drug therapy for problems as unwieldly as depression or stress, the designer will try to find physical solutions to solve what are essentially non-physical problems: to halt gentrification, the designer proposes a building like a wall or a greenbelt like a moat. As professionals, we are reluctant to admit our ignorance or to give work away that we might do ourselves.

In a discussion of the planning professional's role, Norman Beckman notes that idealism and a focus on the future lead to intellectual leadership, but also to difficulty in rational integration (see Webber 1963), and by extension, to pragmatic implementation. This holds true for all design professionals. A disadvantaged community's first priorities are to improve basic conditions, which may be outside the professional's expertise if not interest. Even though the pragmatic requirements are many and the opportunities for design freedom are limited, nevertheless, there remains a tremendous need for creative solutions. The designers can be called upon, for example as we were in Houston, to determine the least costly, most dramatic rehabilitation proposal with a non-disruptive implementation plan. Few designers have the diverse skills required to perform well at such a task.

To take a simple example that makes the same point: in Fourth Ward, one of the community's strongest arguments is that the cost estimates for modernizing the public housing are highly inflated. To demonstrate this, the design professionals came up with alternatives that met all requirements yet were less costly. To get reliable cost estimates, however, required extensive knowledge in areas not traditionally belonging to design: public housing regulations and funding mechanisms, cost estimation, housing rehabilitation of distressed properties, upgrading subterranean utilities, and so on. And this knowledge was simple to grasp in comparison with the knowledge needed to understand the cost implications of revitalizing the historic part of the neighborhood. All the work for this undertaking was volunteered, which created difficulty in finding willing individuals and coordinating their efforts in a timely fashion. The less-than-ideal estimates that resulted did not have the persuasive power needed, creating a weakness in the community's stance on several occasions.[5]

The tradition of patronage: economics

The professional's traditional patrons have been the elite and the state, not the poor. Indeed, designers who are community advocates are generally asked to battle their own future clients. When the community's needs clash with the city's goals, for example, the designer-advocate will face a conflict of interest since the next job or negotiation may involve the city. Like the warning about concrete boots, Fourth Ward has had difficulty attracting even sympathetic professionals, who would not publicly take sides against powerful Houston developers and city officials. Exactly who is the community designer's patron? The community rarely pays for professional services (or pays minimal fees), but every professional must have an income to continue the delivery of professional services. The designer's real and prospective sources of income may be individuals or institutions that hold positions contrary to the community's goals. Thus, the university where the planner teaches, the city from whom the architect seeks the next job, and so on, are actually subsidizing the professional's community involvement. This can create a clear conflict of interest.

Traditionally, patrons have been single individuals or hierarchic entities which the designer can identify. By contrast, "the community" is a convenient fiction. Rather than one homogenous, static body, a community is dynamic and multifaceted, with factions and changing sets of actors. When such diversity is simplified into a one-dimensional entity, it is only a matter of time before the actual complexity erupts in conflict. Designers are rarely prepared to respond to such a dynamic situation, nor to analyze it so that it becomes intelligible. The idealized patron, on the other hand, gives direction to the professional, pays for the services rendered, and receives honor for serving that role.

The design professionals working on a partial rehabilitation proposal with the city councilmember have created a patron of sorts. They have found in the councilman an individual whose support will be beneficial over time, and a single individual to act as client. They have avoided the factionalized neighborhood interests and the problems such work might create with regard to their professional careers. The councilman, for his part, would perhaps increase his status by finding a more acceptable solution than that proposed by the planning department and the housing authority. Still, he prefers to act without community participation. Very clearly then, the design professionals are acting in a way to promote the existing social dominance of the local state over economically insecure citizens. Likewise, the spaces they design will inevitably reflect their own and the councilman's interests, since the residents and their advocates have been denied the right to participate.[6]

The bias for autonomy: values

The tradition of patronage evolved hand in hand with the professional bias for autonomy. The designer who grows frustrated with the community, an unwieldy client with a lack of clear direction, many faces, and many factions, is likely either to step back or to take the lead. Non-directive participation is more difficult, time consuming, and seemingly unproductive. Elsewhere, I have discussed the origins and implications of professionals' bias toward autonomy and away from collaboration (Cuff 1984). In a community project, that bias can be particularly detrimental. Designers who distance themselves from the community will have to rely on their own knowledge and objectives, which are likely to reflect their own roots in the establishment. In the Houston neighborhood, those well-intentioned professionals planning to save a part of historic Freedman's Town find it easier to work within the local government's hierarchy than within the neighborhood network. But in addition, working with the city permits greater autonomy for the designers. The residents want an active role, guiding the work that concerns their own territory, unlike the city councilman who can stand back.

Locale: space

The last fundamental discrepancy between a community and its professional design consultant concerns locale. Locales, according to Giddens (1984), are the physical regions that provide the settings for social interaction, ranging in scale from a room to a nation. Locale then is part of the context of situated interchange, giving meaning and definition to human activity. The concept is important for understanding the unintended, undesired consequences wrought by professionals attending community problems. The community's locale is its neighborhood, within which an individual's house is the personal nucleus. In turn, the neighborhood is perceived as a fixed and distinct context within the larger urban region. The professional community, however, has its own locale which overlaps only slightly with the neighborhood in that both are tied to their urban regions. A design professional's immediate context is the city where she or he practices, which is, in effect, a network of interconnected commissions. In addition, a designer belongs to a professional community which is national in scope via professional organizations, publications, conferences, and so on. Giddens makes the point that locales need not be spatially fixed (as, for example, with nomadic cultures) (1984, 164). This is applicable to professionals who are relatively mobile, establishing a context for action that is regional if not national.

In the Houston case, there are different spatial boundaries drawn around the significance of actions taken in Fourth Ward. The community members, in spite of being renters, are typically third generation residents who have lived in their present dwellings about 15 years. These people definitely see their homes and neighborhood as their territory. The design consultants' territory is Fourth Ward, but also Houston and beyond. In fact, this chapter is itself evidence of the extensive professional boundaries that I – and the reader – can draw around Fourth Ward.

From design to professionalism

In most cases, design professionals are rarely if ever the key decision makers in neighborhoods like Fourth Ward. Property owners, residents, regulatory agencies, water and sewer districts, city councils, and developers are the ones to set the highly delineated frame within which designers must act. Still design professionals do contribute with some freedom; that is, their actions are not wholly determined by the context, and so they must be viewed as active participants in shaping urban space. From the preceding discussion, there appear to be at least two nested ways that participation by design professionals promotes the existing dominant social order. First, the very structure of a profession embeds, tacitly and explicitly, the social structure of which it is a part. Second, the designer's specific actions, which stem from the structure of the profession, further the existing patterns of ideology and power.

While analyses of various professions have demonstrated their tendency to reproduce the social order (e.g., Becker *et al.* 1961, Bledstein 1976), the design professions have particular attributes relevant to this inquiry: their services are themselves public artifacts that have spatial location. Unlike the health professions, the design professions create something new that has a life of its own in the public arena. In this regard, law is the closest relative through its creation of precedent (Collins 1971), yet design is distinct because of its high visibility in the public realm and its territoriality. As the survey of Fourth Ward makes apparent, the physical artifact is a public display, there for any careful observer to read. Like the legal decisions that establish precedent, designed environments put forward in their own language statements about the conduct of society.

With other professions, the design professions also share the fundamental attributes which form the structural context for social reproduction. In the sociology of work, the definition of a profession is debated but there is agreement about some key characteristics of professions in contrast to occupations. In fact, these are the qualities that occupations try to acquire in their movements to professionalize (Cullen 1978). These are pertinent to the conflicts described earlier which community advocates face. A profession generally involves lengthy training and licensure, which means that it can regulate who enters the ranks and what they are expected to know. In general, professions seek self-regulation and peer evaluation. Each profession carves out a territory of expertise for which it claims to be the sole proprietor, and this expertise involves both mysterious, abstract knowledge and scientific, technical knowledge. Both of these types of knowledge are obscure to those that the professionals serve. Professions, including architecture and planning, tend to emphasize a service ideal while downplaying the profit motive. The rendering of services affects the public welfare, so that professionals assert their larger social role beyond any individual client's needs, thereby disclaiming economic ends.

The self-regulatory aspect of professions is part and parcel of the practitioner's bias for autonomy and the practitioner's determination of what constitutes a reasonable project and timetable. The sole proprietor-ship of professional expertise promotes the problem of "ignorant specialists" when facing complex problems like community revitalization. The distancing of professional knowledge from lay understanding promotes the practitioner's independence, and reinforces the ideal of elitist patronage. The service ideal veils the practitioner's financial needs and motives, and the strong ties to political webs of influence. Thus, all of these fundamental attributes of professions played a specific and significant role in the delivery of professional services to Fourth Ward.

Given my analysis, it is not possible to separate design services to disadvantaged communities from politics. In the early days of advocacy planning (e.g., Wilson 1963), this was a somewhat unsettling prospect. Those design professionals who refused to confront the political implications of their actions were certain to serve the objectives of the predominant power structure. This position was most strongly put forth by Robert Goodman in *After the planners* (1971). In his biting critique of the "elitist nature of environmental professionalism" (210), he called for a demystification of professional expertise, an increased role for the dweller, and a form of community socialism in which people can begin to free themselves from dependency on experts. While we may not agree with Goodman's broad social reforms, we cannot avoid the links between the environmental professions and the larger social order. While some less radical writers on the subject promote methods of comunity activism, such as negotiation and conflict resolution (see, for example, Cunningham & Kotler 1983), I want to avoid normative arguments stating the right and wrong ways for design professionals to engage in community undertak-ings. Rather, I want to promote an awareness of the unintended conse-quences of professional action, so that choices about advocacy methods are better informed. It is perfectly conceivable to me, for example, that the promotion rather than the resolution of conflict may be necessary at certain points in the design process and in the political process – indeed this was the case in Houston.

Conclusion

In the preceding analysis of Fourth Ward and the design professions, I have explained how design professionals, even community advocates, participate in social reproduction. The environment as artifact and tool, as well as the professions (by their very structures), further the political, social, economic and ideological status quo. Now the question arises, are there ways to break out of these patterns, to achieve intended rather than unintended consequences?

The answer to this question is tied to the vast literature on social change, which I will not attempt to summarize here. Instead, I would redirect our focus with regard to the topic of social change. It is well known that within neighborhoods, communities of residents have effectively organized grassroots political action. There is also something that could be called a professional community. The "turf" of the professional community is its knowledge and expertise, which is as dynamic as neighborhood territory. Current trends within the design community may hold implications for breaking that cycle of reflection and renewal. For example, in architecture where the production of professionals has increased dramatically in recent years, a number of new specializations have arisen, one of which is the "people industry" (Montgomery 1988). These new "specialists" may be far better equipped than their traditional predecessors to deal with the profession's changing landscape of more complex clients, increases in regulatory constraints and the political processes they engender, greater competition from engineers, and so on.

Looking back on Fourth Ward, there is another lesson about the context within which change may occur, that is, when the results of actions need not renew the previous situation. At the largest scale first, we need fissures in the system (the state, institutions, the economy) such as a recession or a change in the federal administration. When such a fissure occurs, as in a rock formation, a small amount of energy can loose an enormous force. In current parlance, a "window of opportunity" is opened, through which an alternative future can be envisioned. To take advantage of the fissure, the community can respond with grassroots action. If there is also to be a professional response, I propose it will come from those who have gathered a fundamental understanding of the everyday qualities of the neighborhood. Only through an intimate acquaintance with and appreciation for the everyday, can the designer overcome the differential between his or her project orientation and the community's way of life, between autonomy and participation, between specialized practice and holistic problems.[7] The patronage issue remains problematic. Perhaps this is why university professors, who are less interwoven into the patronage web than their freelance counterparts, are more likely to work successfully in community projects.

Notes

My thanks to Stephen Fox, Jennifer Wolch, and Ken Breisch, who read and thoughtfully commented on early drafts of this chapter, and to Walter Zisette for research assistance.

1 Present community action is organized by Lenwood Johnson, President of the Allen Parkway Village Residents Council, and by the Freedman's Town Association headed by Gladys House. To the best of my knowledge, the chapter describes the state of affairs at the time of writing, in late 1987.
2 For an overview of the history of Houston's black population, see the excellent monograph by Cary D. Wintz (1982).
3 In a study of the future of the Fourth Ward, the Housing Authority of the City of Houston acknowledged directing southeast Asians into the public housing in order to break growing political power among black residents.
4 In an unprecedented move, the Director of City Planning helped to organize Freedman's Town's absentee landowners to sell jointly to a single developer who will receive what no other developer in Houston has received: city-provided infrastructure such as utilities and streets. These efforts were unsuccessful in part because of the economic decline in Houston and in part because black landowners (especially the churches) scattered throughout the neighborhood would not participate in the scheme.
5 The worst example of this problem can be found in the GAO report (1986), where our cost estimates are called into question. The only individual willing and able to construct these estimates literally disappeared and thereafter we were without the documentation to support our figures.
6 For example, at an early stage an architect invited me to a meeting with the city councilman. I accepted, indicating my primary concern was that residents be involved in the negotiations and suggesting that the leader of the public housing tenants also attend. Shortly thereafter, the invitation was rescinded with the explanation that the councilman wanted to "keep the meeting small."
7 This corroborates and explains the more ideological arguments for community activism, as exemplified by John F. C. Turner, and the more scientific rationales for participant observation.

References

Becker, H. S., B. Beer, E. C. Hughes & A. Strauss 1961. *Boys in white*. Chicago: University of Chicago Press.
Bledstein, B. J. 1976. *The culture of professionalism*. New York: W. W. Norton.
Breisch, Kenneth A. 1984. National Register of Historic Places [NRHP]. Nomination form for Freedman's Town Historic District. November.
Collins, Peter 1971. *Architectural judgement*. Montreal: McGill-Queens.
Cuff, Dana 1984. Collaboration and the ideal of individualism in design. In *Architecture and the future*, P. Heyer & S. Grabow (eds.), 188–95. Proceedings of the Association of Collegiate Schools of Architecture.
Cuff, Dana 1985. Beyond the last resort: the case of public housing in Houston. *Places* **2**, 28–43.
Cullen, John 1978. Structural aspects of the architectural profession. *Journal of Architectural Education* **3**, 18–25.

Cunningham, James V. & M. Kotler 1983. *Building neighborhood organizations.* Notre Dame: University of Notre Dame Press.

Davidoff, Paul & L. Davidoff 1978. Advocacy and urban planning. In *Social scientists as advocates.* G. H. Weber & G. J. McCall (eds.), 99–120. Beverly Hills: Sage.

Fainstein, Susan S., Norman I. Fainstein, Richard Child Hill, Dennis Judd & Michael Peter Smith 1983. *Restructuring the city: the political economy of urban redevelopment.* New York: Longman.

Friedson, Eliot 1986. *Professional powers.* Chicago: University of Chicago Press.

Giddens, Anthony 1984. *The constitution of society.* Berkeley, CA: University of California Press.

Goodman, Robert 1971. *After the planners.* New York: Simon & Schuster.

Government Accounting Office 1986. Public Houston. Proposed sale of the Allen Parkway Village Project in Houston, Texas. GAO/RCED–86–160, September 16.

Hartman, Chester 1974. *Yerba Buena: Land grab and community resistance in San Francisco.* San Francisco: Glide Publications.

Housing Authority of the City of Houston 1983. *Technical Report: Allen Parkway Village/Fourth Ward.* Houston.

Keyes, Langley O. Jr. 1970. *The Boston rehabilitation program: an independent analysis.* Cambridge, Mass.: The Joint Center for Urban Studies of MIT and Harvard University. (Distributed by Harvard University Press.)

Kravitz, Alan S. 1970. Mandarinism: planning as handmaiden to conservative politics. In *Planning and politics: uneasy partnership.* Thad L. Beyle & G. T. Lathrop (eds.), 240–67. New York: Odyssey Press.

Larson, Magali 1983. Emblem and exception: the historical definition of the architect's professional role. In *Professionals and urban form*, J. Blau, M. E. LaGory and J. S. Pipkin (eds.). Albany: State University of New York Press.

Montgomery, Roger 1989. Architects' people: the conjunction of social science and American architectural practice over twenty-five years, 1959–1984. In W. R. Ellis & D. Cuff (eds.), New York: Oxford University Press.

Peattie, Lisa 1968. Reflections on advocacy planning. *Journal of the American Institute of Planners* **34**, 80–8.

Webber, Melvin M. 1963. Comprehensive planning and social responsibility: toward an AIP consensus on the professional role and purpose. *Journal of the American Institute of Planners* **29**, November.

Wilson, James Q. 1963. Planning and politics: citizen participation in urban renewal. *Journal of the American Institute of Planners* **29**, 242–9.

Wilson, James Q. (ed.) 1966. *Urban renewal: the record and the controversy.* Cambridge: MIT Press.

Wintz, Cary D. 1982. *Blacks in Houston.* Monograph prepared for Houston Center for the Humanities and the National Endowment for the Humanities (Fred R. von der Mehden, series editor).

16

The grassroots in action: gays and seniors capture the local state in West Hollywood, California[1]

ADAM MOOS

Introduction

This chapter examines the incorporation of the City of West Hollywood. The incorporation effort was the focus of a great deal of national and international press attention owing to the prominence of gay and lesbian persons living in the community. Many in the media considered the incorporation issue one of creating a "gay camelot" in Southern California. At the local level the incorporation effort was considered one of a tenants' movement to create a city where a strong rent control ordinance could be enacted. The aim of this chapter is to demonstrate that the city's successful incorporation resulted from a coalition of gays and tenants to create a city where these traditionally disenfranchized groups would have the ability for self-determination.

This study employs a manner of structuration based on hermeneutic inquiry similar to the one used by Dear and Moos in their examination of the ghettoization of ex-psychiatric patients in Hamilton, Ontario, Canada (cf. Moos & Dear 1986, Dear & Moos 1986). This method of inquiry was found to enable an analysis that considered the role of structure and agency in the creation of the ex-patient ghetto. The data sources for this study include information gathered from government agency reports, municipal and private files, as well as interviews with key agents. Additionally, my analysis relies heavily on my personal knowledge, since I served as campaign manager for Mayor Alan Viterbi in the 1984 incorporation election, and then served the city as its first Director of Rent Stabilization, a position I held from the opening of City Hall until mid 1986.

My examination into the incorporation movement begins with a consideration of West Hollywood's history and demographics and how these become implicated in the community's eventual incorporation. A discussion of the dynamics involved in getting the West Hollywood incorporation petition approved for ballot consideration follows. The chapter goes on to examine the politics of the incorporation effort, focusing both on the question of cityhood and on the races of the 44 candidates that filed for election to the first city council. Next, I assess the coalition's effectiveness in practice, through a discussion of the city council's legislative accomplishments during its first year of operation. Then, I examine the coalition's endurance via an analysis of the two city council elections that have taken place since the 1984 incorporation. The final section of this paper provides a reconsideration of West Hollywood's incorporation experience and suggests implications for the analysis of urban social movements which capture the local state.

What is West Hollywood?

Prior to incorporation, West Hollywood was under the jurisdiction of the Los Angeles County Board of Supervisors. The area's residents made a conscious decision to remain unincorporated in a 1924 vote when they defeated a proposal to become part of the city of Los Angeles by 810 to 754. The net effect of choosing to remain unincorporated was that West Hollywood would escape the stricter laws of Los Angeles and be under the more benevolent and uncritical eye of the County government.

The laxness with which the County governed the area was evident throughout the 1930s as speakeasies and purported underworld figures flourished in the community. The area became home to many counter-culture movements including the 1960s hippie movement which flourished on the Sunset Strip. And, in the mid 1970s, the gay sub-culture shifted into West Hollywod to escape the harassment of the Los Angeles Police Department who were consistently raiding gay bars and bath houses. The migration was not surprising since West Hollywood, under the jurisdiction of the Los Angeles County Sheriff's Department, had been home to many gays and their businesses previously without encountering any problems from law enforcement officials.

While the County was lax in controlling the social climate in West Hollywood, it was also very permissive concerning the community's zoning. The area's zoning map from the period prior to incorporation showed a predominance of R–4, multi-family high-density residential, throughout, with only two small blocks in the city having an R–1, single-family home zoning. Developers took advantage of the high-density zoning of West Hollywood and turned it into a community of apartment buildings.

The City of West Hollywood today very definitely reflects the County's governing of the area. This 1.9 square mile area is home to 35,703 (US Census 1980) persons making West Hollywood the densest area west of the Mississippi River. The housing stock in West Hollywood consists primarily of 20–30 unit apartment buildings, condominiums, and small single-family dwellings that often appear as two or three to a parcel. The predominance of rental housing within the city is evidenced by the fact that 88% of the community's residents rent their accommodations (Levine & Grigsby 1985). Senior citizens make up a large number of the city's renters and account for 28% of the area's population (US Census 1980). The continual migration of gays and lesbians into the area has provided the community with a gay and lesbian population estimated at 35–40 per cent of its residents.

Since the area was tolerant of the gay lifestyle, the community became the center of gay commerce as well. There are well over 100 gay-owned and -operated establishments, including two gay-run banks. Other businesses include interior design, for which West Hollywood is the industry's center, bars, boutiques, and restaurants. Gay capital is also present in the area as many gays have extensive residential or commercial property holdings, or both.

A move toward independence

The state government in California encourages incorporation for those areas which are able to be economically viable cities. The state considers County governments necessary to support areas until they are able to demonstrate that incorporation and hence self-government is practicable and approved by the voters of the area.

Incorporation of communities in California is governed by a Local Agency Formation Commission (LAFCO). Each county in California, except San Francisco,[2] has a LAFCO. The Los Angeles County LAFCO includes two members from the County Board of Supervisors, a Los Angeles City Councilmember, two mayors from other county cities, and a citizen member appointed by the Board of Supervisors. The role of the LAFCO is to assess the economic feasibility of an area wishing to incorporate, and to insure that the proposed municipality is not being formed to benefit solely a special interest.[3] As long as the LAFCO determines that an area passes on these criteria, it places the issue of incorporation on an upcoming ballot to allow the area's residents to decide whether to incorporate the area.

Prior to the successful 1984 effort, area residents had tried on three prior occasions to incorporate. These all failed and the most recent effort in 1980 collapsed when LAFCO prepared a very unfavorable report on the area's economic viability as a separate city.

In the Fall of 1983 community resident Ron Stone, frustrated by the unresponsiveness of the County Board of Supervisors to the area's needs, began to consider incorporation as a real possibility. Stone researched the issue and determined that the area would be economically viable within the criteria of LAFCO. He met several times with LAFCO staff to persuade them to look into West Hollywood as a viable incorporation effort. He presented the staff with economic information he had collected from the various county agencies that served West Hollywood, and convinced the staff to revise the 1980 report that had killed the prior effort. LAFCO produced an initial report which indicated that West Hollywood's tax base at its present level of services would leave the city with a $5 million surplus.

Stone's next step was to build the West Hollywood Incorporation Committee (WHIC) which would oversee the political effort necessary first to put the issue on the ballot and then work toward the passage of the area's incorporation. Stone recognized three issues that were of primary concern to West Hollywood residents. First, he realized that Los Angeles County's rent control law would expire in December 1984, and that extension of the law beyond that date was not likely, given the conservative majority that made up the board of supervisors. Second, as a gay man he recognized the importance of self-determination for the area's large population of gay and lesbian residents. Third, he knew from prior association with the area's community organizations that local control over development, parking, and service provision was a concern to many. Thus, he saw the need for a coalition of residents in order to take the incorporation effort to the next stage.

He attempted to enlist the help of gay community activists. Initial contacts to the Stonewall Gay and Lesbian Democratic Club and the Harvey Milk Lesbian and Gay Democratic Club proved fruitless. The gay establishment which had for years operated with relative success in West Hollywood under the County government opposed Stone's efforts. Stone's initial success with bringing the gay community into the incorporation effort happened through a very unlikely source. Bob Craig, publisher of the gay periodical *Frontiers*, was trying to improve magazine's circulation and enlisted the services of advertizing executive Art Guerrero. Guerrero had become aware of Stone's work to incorporate West Hollywood and suggested to Craig that he should make the issue the central focus of *Frontiers* in the coming months as a gay cause which would not only aid the incorporation effort, but also increase the magazine's circulation. Guerrero would later sit on the board of directors of the incorporation committee, and Craig would eventually replace Stone as head of the committee when Stone began his individual campaign for the city council.

Stories began to appear in *Frontiers* and Stone was able to engender the support of Valerie Terrigno, then the president of the Stonewall Gay and Lesbian Democratic Club, as well as the Harvey Milk Gay and Lesbian Democratic Club. Additionally, with the issue appearing frequently in *Frontiers*, as Guerrero predicted, incorporation became a major gay interest.

Stone also sought the area's tenants as part of the coalition and turned to the community's main tenant organization, the Coalition for Economic Survival, to sell them on the idea of incorporation as a way of bringing rent control to the community. The Coalition for Economic Survival (CES) had been politically active in the area for years and provided a political voice for tenants, most of whom were senior citizens. The relationship between seniors and CES is very evident in that 20 persons of the 30 member CES steering committee are seniors. CES played a role in the passage of the County's rent control ordinance in 1980. CES was also the driving force to gain passage of Proposition M,[4] a county-wide measure that would not only have extended the life of rent control within the county, but also would have greatly strengthened the law. Although Proposition M failed at the county level, it passed overwhelmingly in

West Hollywood by a 5:1 majority, demonstrating CES's strength at organizing their senior citizen dominated membership of 2,000 West Hollywood residents, and making them the recognized leader of the area's tenants.

The third group Stone sought to bring into the coalition was the community's establishment. These residents had been involved in the limited local control that County Supervisor Ed Edelman had allowed for the area. This group included persons who helped write the community plan and sat on Edelman's community advisory committee. The majority of persons in this group were initially against cityhood, primarily because they both feared and respected Edelman. However, after a series of meetings with Edelman, this group became more and more convinced of the necessity for local control and cityhood and became active in the incorporation movement.

With the findings of the initial LAFCO study favorable to the incorporation effort and the beginnings of the coalition in place, Stone began the next phase of the effort which was to meet LAFCO's requirement of having 25 per cent of the area's registered voters sign petitions indicating that they supported the issue of incorporation being placed on the ballot. With its membership base and its political savvy, CES provided Stone with the organizing ability to collect the necessary number of signatures for the issue to proceed to the next step. The CES volunteers, primarily senior citizens, and several members of the gay and lesbian community who were interested in the movement, provided the "legwork" necessary to gather the required number of signatures to further the incorporation effort.

During this initial phase of the incorporation effort, the only opposition to incorporation came from County Supervisor Edelman. Edelman's office released a report which questioned the accuracy of the LAFCO study and claimed that the area would not necessarily be economically self-supporting. The pro-cityhood forces questioned the accuracy of Edelman's report and decried it as an attempt to stop cityhood. Edelman claimed that he was "neutral" with regards to the issue of incorporation and was only trying to ensure that a "correct" analysis of the area's revenue and expenses was accomplished.

The Edelman study failed to halt the success of the Cityhood movement. The necessary number of signatures was gathered, and the LAFCO hearing to determine the feasibility of the area to incorporate was set. LAFCO approval was not a foregone conclusion for the cityhood proponents. Los Angeles County's LAFCO had a majority of conservative members who were neither favorable to rent control nor to gay and lesbian issues. However, the strength of the LAFCO staff's report attesting to the economic viability of West Hollywood as a city meant that LAFCO could not deny the application on this basis. At the LAFCO hearing, opposition to the Cityhood question was raised by the West Hollywood Study Group, an association of residents and area landlords, who argued that not enough information had been gathered even to allow the issue to appear on the ballot.[5] Despite this opposition, the LAFCO

voted unanimously to place the issue on the ballot forcing an incorporation election. Following the LAFCO approval, the County Board of
Supervisors placed the issue on the November 1984 ballot.

Opponents to cityhood turned their attention towards the courts in an
attempt to keep the issue from reaching the ballot. The West Hollywood
Study Group filed a lawsuit claiming that an Environmental Impact
Report (EIR) would have to be done for the incorporation to proceed. The
Taxpayers Association of West Hollywood, which was financed by one of
the area's larger developers, Lorraine Howell and Arthur Laurence, filed a
lawsuit against LAFCO in an attempt to halt the election. The Howell and
Laurence suit was based upon an "independent financial analysis" they had
commissioned; it predicted a $2.5 million deficit for West Hollywood.
LAFCO, Los Angeles County, and incorporation committee attorneys
argued against these claims and were successful in getting the suits
dismissed by demonstrating the flaws in the Howell and Laurence study,
and in arguing that changing the administrative nature of governing the
area did not require an EIR be done.

LAFCO's approval of West Hollywood's incorporation was the effort's
single most critical event, save for the ultimate vote by the residents. Two
elements combined to make the approval possible. First and foremost,
there was the report prepared by the LAFCO staff. Since the formation of
LAFCO one incorporation has taken place in Los Angeles County every
two years. The LAFCO staff have definite desire to assist areas to
incorporate in order to perpetuate the agency's existence. The actions of
Ron Stone in getting the LAFCO staff to consider West Hollywood were
critical, but once the LAFCO staff decided to handle West Hollywood as
their 1984 incorporation effort, their actions were aimed at making sure
the LAFCO board approved it for the ballot, since the staff would not
want to be embarrassed by placing before the commission an incorporation effort that would not be viable enough to pass. Thus, LAFCO staff,
aware of the explosiveness of the rent control and gay issues in West
Hollywood, took every precaution to ensure that the commission could
not reject the proposal on the basis of the area's lack of economic viability.

A second key ingredient to the LAFCO process had to do with its
chairman, County Supervisor Peter Schabarum. Schabarum is a very
conservative member of the board of supervisors, and to that end had a
political stake in West Hollywood's incorporation. Liberal supervisor
Edelman stood to lose a good deal of political clout if the area
incorporated. This aspect along with Stone's lobbying of Schabarum for
the standard conservative arguments of citizens' rights to local control lead
to Schabarum supporting the effort. And, with Schabarum's support, the
remaining members of the LAFCO, who were as conservative (if not
more so), ended up supporting the effort.

Incorporation politics

With the incorporation issue determined to be on the ballot, the stakes involved in it altered dramatically. Prior to the LAFCO hearing the focus of the debate around cityhood concerned local control and the economic viability of the area to exist as a city. Proponents of the incorporation issue did not want to change the nature of the debate from the "good government/self-determination" issue toward specific "post-cityhood" issues and thereby jeopardize receiving LAFCO approval for the effort. Following the LAFCO vote this focus shifted in two ways. First, the issue of rent control, which had been treated publicly as a footnote up to that point by many of the pro-cityhood organizers, was pushed to the forefront of the incorporation question. Second, the gay establishment which had not initially favored incorporation embraced the idea.

The initial members in WHIC included CES, the tenants and seniors group, and several representatives of the gay community. However, the segment of the gay community present was not its traditional leaders. Stephen Schulte, who was eventually elected to the first West Hollywood City Council, was not working for the area's incorporation, but instead was preparing to run for the Los Angeles City Council against an incumbent. Major West Hollywood landowner and Democratic Party activist Sheldon Andelson, possibly the most powerful member of the gay community, and other power brokers of the gay community gave neither money nor time to the effort.

The reasons for the apathy of the gay establishment rest in the history of the gay community in West Hollywood. Gays had become successful in West Hollywood by virtue of the fact that they were quietly active. Unlike gays in San Francisco and New York who were militant in their political activities (see, for example, Castells 1983), the gay community in Los Angeles preferred to politic behind the scenes. Edelman was perceived by many in the gay community as a friend, and many were concerned that supporting cityhood could put them out of favor in the event the measure failed.

This aspect of the incorporation effort changed, however, when the incorporation issue was approved for the ballot. Now, the gay community saw that it had the potential to elect openly gay and lesbian candidates to seats of power. The question at this point was no longer one of turning the community's back on its friend, Edelman, but one of "this is our chance to determine our own fate." In this sense, the question for the gay community became one of local control – control of issues that affected and concerned gays – and having a direct voice in how those issues would be handled.

The shift in focus towards the rent control issue following the LAFCO approval was not nearly as dramatic. Rent control was a major issue from the outset, and the fact that the incorporation effort had reached the LAFCO approval stage was due in large part to CES who had organized the signature gathering effort. However, up to the LAFCO approval opponents of cityhood (and for that matter rent control) could only

357

articulate their objections in terms of tax base, expenditures, and service delivery since cityhood activists had given rent control the same status as other issues of self-determination such as zoning and traffic control. The objective of this strategy for cityhood proponents was to avoid rent control becoming the central issue of the incorporation effort. While rent control was critical to the creation of West Hollywood in the final outcome, too much emphasis on the issue could have killed the incorporation effort at a very early stage in the LAFCO process.

Once the LAFCO approval was granted, the anti-cityhood forces altered the focus of their campaign and decided to attack the two issues driving cityhood other than local control. Cityhood opponent and major West Hollywood landowner Frances Montgomery commissioned a survey to determine if West Hollywood's senior citizens could be swayed from supporting cityhood by the realization of the potential for a gay takeover of the city council if the incorporation were to be successful. The survey's only impact was further to damage the credibility of anti-cityhood forces and leave them open to charges of extreme homophobia.[6] If anything, the net impact of the survey was to further the perspective within the gay community that cityhood for West Hollywood was necessary for them to overcome these kind of gay-baiting tactics. In order to attack rent control, landlords and developers formed the organization, West Hollywood Concerned Citizens. This group became the landlord equivalent to CES in terms of mobilizing political support against cityhood around the rent control issue.

Cityhood foes led by the area's landlords attempted to thwart the incorporation drive in the last week of the campaign by having the county government agree to create a Special Rent Control District for West Hollywood for the purposes of extending the County's rent control law in the area. The landlords, who at this point realized Cityhood was very likely to pass, feared the kind of rent control law that the new city council would put into place and preferred instead the weaker county law. Two of the three conservatives on the board of supervisors were opposed to rent control and were not going to extend it in the county. Edelman, and the board's other liberal member, Kenneth Hahn, were joined by conservative Peter Schabarum, who was also the chair of the County's LAFCO, in voting to create the special district for West Hollywood. Cityhood activists attacked this act as a mere landlord ploy to thwart the incorporation effort.

Schabarum's vote in favor of the special district creation came as a result of a petition signed by West Hollywood's landlords claiming that they were in favor of extending rent control in West Hollywood. The motivation for this approach by West Hollywood's apartment owners was to avoid cityhood and the rent control law that would follow the city's creation. For Schabarum, the vote in favor of the district's creation represented a "flip-flop" from his original support of cityhood. Given Schabarum's role as chair of LAFCO, it is unlikely that the LAFCO staff would have proceeded with a project that would not have had Schabarum's support. Schabarum's initial favorable leanings towards West

Hollywood's incorporation probably included a desire to remove a very valuable area from Edelman's control, as well as a traditional conservative leaning towards a community's ability for self-determination when viable. His vote to create the rent control district, which was considered by cityhood proponents to be a move against the incorporation, reflected Schabarum's conservative stance against stringent rent control laws such as the one West Hollywood would be expected to adopt. Although this vote to create a special rent control district had no impact on the incorporation election, it does lend credence to the notion that if rent control, as opposed to local control, had been the focus of the debate concerning cityhood during the LAFCO approval process it is unlikely that the city's residents would have ever had the opportunity to vote on incorporation.

Alongside the City of West Hollywood's incorporation campaign were the campaigns of the persons hoping to serve on its first city council. The city council's five open seats were the focus of 44 candidates of whom 40 ran active campaigns. Given the importance of rent control and the gay and lesbian issues in creating the city, it is important to note that all candidates supported some form of rent control, and were supportive of gay and lesbian issues as well. The distinction between candidates occurred in the language each employed to explain his/her stand on the issues. Rent control advocates argued for "strong rent control," while the many candidates with landlord/developer interests and backing talked of "fair and equitable rent control." Similarly, several candidates maintained gay and lesbian issues as a central focus to their campaign while others stressed "equality for all."

The large number of gay and lesbian candidates seeking a seat on the city council became a significant factor in the election. Nineteen openly gay and lesbian candidates mounted campaigns for seats on the new city council. The candidates differed greatly in their approaches to rent control, development, fiscal concerns, and other incorporation issues. It meant that for the gay community support of a gay or lesbian candidate became conditioned on his/her stands on particular issues and not solely on her/his sexual orientation. Thus, while being gay was important, there were too many gay candidates in the race to make that a sole criterion for selection. This experience is quite unlike prior political experiences where the gay community would unite behind a single gay candidate in a race for public office (for example, consider Harvey Milk's attempts in San Francisco). The accessibility of West Hollywood to gay politicians desiring to hold public office was at once both a source of strength and weakness. The strength came from the ability of many openly gay and lesbian candidates to seek office without having to worry that their sexual orientation would negatively impact their chances of getting elected. The weakness arose from the splitting of the gay community's vote amongst standard issues and an inability to rally around a slate of candidates who would best represent the gay community and the furthering of its political agenda.

The rent control advocates, on the other hand, provided their constituency with only five *real* choices for the five council seats. Very early in the campaign, CES had picked a slate of five candidates that it backed for the city council under the premise that these candidates would pass the strongest rent control law possible. Three of the five CES candidates were members of that organization, while the other two were endorsed by the organization based on their rent control stands. While CES had selected candidates based upon their strong support of a strict rent control law, they also understood the importance of reaching out to the gay community. The five CES endorsed candidates included a gay man and a lesbian as well as a senior citizen female, and two straight males.

The cityhood vote passed with a 2:1 majority. The five members of the city council were, in order of election: Valerie Terrigno, Alan Viterbi, John Heilman, Helen Albert and Stephen Schulte. Terrigno, Viterbi, Heilman and Albert were all endorsed by CES. Schulte, while not receiving CES support, did receive a great deal of backing from the gay community.

The elections of Terrigno, Heilman and Schulte provided the first city council with a gay majority. The elections of Terrigno, Viterbi, Heilman and Albert provided the city council with a pro-rent control majority. The newly elected council represented the coalition of seniors and gays who supported them. The test for the coalition would now be in the manner in which the council dealt with each group's agenda.

Campaign assessment

An analysis of the election results provides an understanding of the coalition that made incorporation a reality and placed the inaugural council in their position. Prior analysts of the city's incorporation have been quick to focus on the rent control aspect to explain cityhood (see, for example, Ufkes-Daniels 1985). These studies have noted the large numbers of tenants residing in the community and claimed that West Hollywood's cityhood grew from an activist tenant movement. While there is a great deal of validity to this perspective, it keeps too narrow a focus and denies the richness of the social movement that took place within the community.

The incorporation effort was about more than rent control. Local control and gay rights have already been shown to have been a major motivating force behind the persons who were involved in the effort, behind those whose only involvement in the campaign was to vote on the issue, and for the future councilmembers. Additionally, it was demonstrated above that the local control aspect of incorporation was critical for it to achieve LAFCO approval. Another measure of the greater than rent control aspect of the cityhood effort can be found by comparing the results of the cityhood vote with results from the aforementioned Proposition M campaign from the year before. Proposition M concerned rent control only and passed by a 5:1 margin in West Hollywood. Cityhood, on the other hand, passed by a 2:1 margin. The reasons for this difference can be found by examining first the Proposition M campaign and then the cityhood effort. Proposition M was a county-wide measure placed on the ballot in an "off year" election. Although West Hollywood provided a strong pocket of support, landlords invested great sums of money throughout the county to ensure the measure's defeat. There was little interest in West Hollywood to organize an anti-Proposition M election, since those against the measure had a much greater area to focus upon with their efforts; and those areas provided a greater possibility for support of their position than would be the case in West Hollywood. The incorporation effort involved a stake greater than rent control, and therefore generated a larger interest which resulted in a higher voter turnout. Although cityhood's 2:1 majority was overwhelming, it was much less than the 5:1 result from the year before. In order to understand the importance of this difference, we need to turn our attention to the results of the council race.

I noted earlier that only four of the five candidates on the CES slate were elected to office. If rent control had been the only pervasive issue in the campaign towards incorporation, we could have expected CES to sweep all five spots on the council. In addition, we would also have expected to have the top three vote getters in the race be the three who were members of the organization, and therefore who should have received solid and unwavering support from the tenant populace that CES represented. However, as I indicated above, this was not the case. While the CES slate did take four of the five seats, the top two finishers in the race were the additionally endorsed persons, Terrigno and Viterbi, and the member of the slate who did not win was one of the members of CES. That seat was the one won by Schulte.

What the results of the council race indicate is that a coalition was responsible for the cityhood victory, and that this coalition's influence is evident in the final finish of the council race. Terrigno, a lesbian, was an early member of the cityhood effort. Her success was tied both to her extensive recognition among gay (and especially lesbian) voters, as well as to her endorsement by CES. Schulte won a seat on the council without the CES endorsement because he was by far and away the most visible member of the gay community running for office. Prior to his election, Schulte had served as the Director of the Gay & Lesbian Community Services Center which is just outside the city of West Hollywood. Within the gay community the Director's position is one of power, prestige, and ultimately recognition. Thus, despite the large number of gay candidates in the race, Schulte's recognition within the gay community made him the only gay candidate that most gay individuals felt deserved solid support and could therefore transcend the problem of gay vote splitting. Additionally, by virtue of his position, Schulte received the solid backing of the gay political establishment which further benefited his chances. Heilman's ascension to the city council reflected the solid support he received from the tenants in West Hollywood. Additionally, Heilman was not without support in the gay community as he was (and the other CES candidates were) supported by the Harvey Milk Lesbian and Gay Democratic Club.

Although it was a coalition of renters, gays, and local control advocates that gave cityhood its overwhelming victory, the local control contingent did not gain a seat on the city council. Most notable in his absence was Ron Stone without whom the cityhood effort would never have happened. Stone's defeat was attributable to his failure to broaden his appeal beyond the local control issue and to become readily identifiable with either the rent control issue or the lesbian and gay issue. Thus, while the local control argument was critical to getting cityhood approved by LAFCO, it was not considered critical by the voters in electing a council.

A further analysis should be made of Stone's failure to obtain a seat on the council. Stone did not run a campaign of any sort, and instead tried to appeal to an element in the community that would recognize his efforts and abilities and elect him to the council. However, the first council campaign was not a simple affair by any stretch of the imagination. With 40 candidates mounting active campaigns, and several of the candidates spending over $20,000 to obtain a seat, the volume of campaign literature distributed dwarfed those candidates who did not have a large campaign war chest. Thus, while Stone's name recognition was very high during the LAFCO approval stage, he quickly fell into the shadows once the candidate campaigns began.

While cityhood's passage reflected a definite coalition of all three groups, the city council representation of the coalition included only the rent control/senior citizens and gay community. A clearer understanding of the effect of only two-thirds of the cityhood coalition can be seen by examining the actions of the city council after it took office. The following paragraphs briefly detail some of the key legislative actions the council took in support of the coalition that placed them in office.

Opening night

The City of West Hollywood was officially born on November 29, 1984. Much hoopla surrounded the initial meeting as residents wanted to be part of the historic event. National and international media were there to cover the first meeting of a government body that had a majority of gay and lesbian elected officials.

The first meeting, though, was not merely ceremonial. After the city council had established itself as a legal entity and handled the traditional tasks, it passed three pieces of legislation that reflect back on the heart of the incorporation effort. First, the council passed a rent moratorium that included a rent roll-back, prevented any rent increases, and protected against arbitrary evictions. Second, they enacted moratoriums on both building and conversions within the city. Third, the council passed an anti-discrimination ordinance that made discrimination of all persons illegal, and articulated that a person's sexual orientation can not be a valid basis for discrimination of any kind.

The nine months following that inaugural meeting saw the council dealing with the items of concern to all three segments of the initial cityhood effort. The rent control issue became the first major piece of legislation the council tackled. The council spent six months accepting testimony, debating the issues, and eventually drafting one of the country's strictest rent control laws. While the council was focusing on rent control, several of its other actions were directed at satisfying the gay agenda. The council passed a domestic partnership ordinance that provides gay couples living together with a sanctioning of their relationship. The council also passed an ordinance requiring that all contracts the city enter into have language in them that is consistent with the language in the anti-discrimination ordinance in order to ensure that contractors could not discriminate on the basis of sexual orientation either. The council's commitment to the gay community extended beyond legislative action. The council, after much debate, agreed to pay for additional law enforcement costs associated with the annual Christopher Street West Gay Pride Parade when that group threatened to move the parade from West Hollywood, where it had been for the previous ten years, to another location if the council did not provide the funding. The council also dealt with two issues that were problems within the gay community. The council passed ordinances preventing discrimination by gay bar owners of gay minorities, preventing discrimination against persons who dress opposite their gender. Finally, the council enforced the anti-discrimination ordinance by threatening to charge a local restaurateur who for many years had posted a sign above his bar saying "Fagots Stay Out!". The sign was removed.

Another major accomplishment of the council during this initial period was to fund several social services important to the community in general, but to the city's senior citizens and gay and lesbian residents in particular. The city's first operating budget (City of West Hollywood 1985) included $1,331,369 in social service grants. The vast majority of this money was earmarked for senior services particularly public transportation discounts, housing programs, kosher meals for the elderly, and legal services; and for gay community programs sponsored by the Gay and Lesbian Community Services Center and Aids Project Los Angeles. The social services grant both augmented services that the county had funded previously, but more importantly was used to establish new services that were important to each group.

The council's focus on rent control, the gay agenda, and social service provision during its first nine months in office left very little time available for dealing with standard local control issues such as the general plan or traffic problems or even the setting up of various commissions. The council did, however, maintain a very open and receptive forum to the concerns of the city's residents, and was able to deal with some of their concerns in a much more rapid fashion than would have been possible when the area was unincorporated.

Following these initial stages, the city began to deal with both the general plan and with developing the local bureaucracy. The council's approach to developing the general plan was consistent with the aims of local control. The council directed the city's consultant to employ a heavy public participation model to ensure that all interested residents would have the ability to let their concerns be known. With the emphasis on the public participation element, the general plan was going to take more than two years before it would be in place.

The second time around

In April 1986, the city's residents had their first referendum on the newly formed government. A field of ten candidates ran, including three incumbents who were elected just 16 months earlier. The candidates in this race represented many different elements within West Hollywood – tenants, gay men, landlords, local control advocates, conservatives, and liberals. CES endorsed only two candidates for the three spots, incumbents Heilman and Albert. CES did not endorse Schulte, but neither did they proffer a third candidate to take his spot. While not supporting all three incumbents, CES did nothing outright to threaten the incorporation coalition. Heilman and Albert were dedicated to strong rent control and were perceived to be supportive of gay and lesbian issues, and both had been open to concerns of residents about local issues in general. In this manner, CES had set up Heilman and Albert to receive a very wide spectrum of support from the community.

Another community group, West Hollywood Good Government (WHGG), offered a slate of candidates who wanted to be perceived as more moderate on rent control than the CES backed candidates in the race. This group also endorsed Schulte. This endorsement would prove a probem for Schulte in gaining re-election. WHGG was strongly perceived by renters as having one mission – gutting the rent control ordinance that the tenants in West Hollywood had fought for following incorporation. Schulte's association with this group did not help him with the tenant portion of the original coalition, and when WHGG attacked Heilman and Albert, Schulte became in danger of losing whatever support he had left in the tenant world. Schulte salvaged some tenant support and retained office when he, at the last minute in the campaign, disassociated himself from WHGG because of these attacks. Thus, the re-election of the three incumbents reflected the incorporation coalition, but at the same time strains in that coalition were visible as Schulte was perceived to be against one element of it.

The coalition of gays and seniors remained intact at the expense of the community's local control advocates. Ron Stone attempted to gain a seat on the council during this second election with arguments focusing on the inability of the council to govern effectively. He argued from a local control perspective claiming that while such items as the anti-discrimination ordinance were important, they were not the primary job of local government. He ran on a platform that focused on the council's inability to deal with many of the problems that went unsolved by the county prior to incorporation and were still neglected by the current city council 16 months later.[7]

The second election signaled the end of West Hollywood's "honeymoon period." Up to this point, criticism of the council and staff was minor and concerned pet grievances of individual citizens and/or businesses. There had been no major group challenges to council actions from the coalition that passed incorporation. Two groups within the coalition intensified their scrutiny of the council. Gay leaders were angry that West Hollywood, while having a council with a gay majority, had a staff without a gay or lesbian department head. They coupled this issue with charges of homophobia against key city staff members. Local control advocates were even more restless and angry. The local control aspect of the coalition had not received priority from the council who had first focused on rent control and gay/lesbian issues. Charges from this group were focused on the perceived waste in government as the city's first year operating budget was $13 million higher than the $11 million figure LAFCO had indicated it would cost to run the city. Other concerns were that city staff were not members of the community, but career bureaucrats who were not sensitive to the needs of the residents. This part of the coalition had yearned for local and responsive government and did not want an overly bureaucratized environment that would make West Hollywood as unresponsive to their concerns as the County had been previously.

One more time

A unique opportunity existed in West Hollywood for residents to partake in another referendum just eight months after the election where the three incumbents had received approval. The conviction of the city's first mayor, Valerie Terrigno,[8] on federal embezzlement charges provided a vacancy on the council that was filled via a November 1986 special election.

Unlike the previous elections, this one had only three candidates, of which two were serious contenders. Gene LaPietra represented the gay establishment. He was similar to Andelson in the power he wielded with elected officials and in his ability to achieve results. LaPietra is also very wealthy and would spend $350,000 of his own money in the campaign. His connection to West Hollywood had been through the gay community, and he only established residency in the city just prior to the deadline for filing his candidacy. Abbe Land represented what LaPietra did not. She is a straight female who has lived in West Hollywood for many years. She was chair of West Hollywood's Planning Commission and on the board of directors of CES.

Once again, the coalition was going to be severely tested. However, unlike the previous campaigns, there were only two candidates on the ballot enabling a clear choice by the voters on the issues that were of primary concern to them. However, during the campaign certain discoveries about LaPietra's past made him a less than solid choice for the gay community. LaPietra had previously been convicted for obscenity law violations leading many in the gay community to fear him as a potential negative standard bearer who could embarrass the community in the same manner Terrigno's conviction had done. Conversely, many who supported LaPietra did so because they did not want to lose the gay majority that had originally existed on the city council. Another factor that would influence gay community support of LaPietra was his endorsement by West Hollywood's landlords. Gay and lesbian renters were left to decide which issue was more critical to them in selecting the candidate for the city council. As for local control, the only endorsement of consequence that LaPietra received was from Schulte.

Land, on the other hand, received solid support from the community's senior citizens, the rent control advocates, and the local control segment of the community. Her major problem was that she was perceived to be weak on gay and lesbian issues, since she was not a member of that community, nor particularly active in it. However, while she was not the primary candidate of the gay segment of the coalition, her list of endorsements included solid backing from CES, the rent control constituency, and all the major players at the local level including councilmembers Heilman, Albert and Viterbi and a vast array of city commissioners and board members.

The results of this election were very similar to the original vote on incorporation. Land won receiving 62.1% of the vote. Land's victory was overwhelming, as she won every precinct in the city. The strength of her victory reflected the strength of the original coalition that passed cityhood. Although it is certain that she did not become the solid choice of the gay community, she received such a majority that it is safe to conclude that she won over the part of that community concerned with issues broader than the gay agenda, or who felt LaPietra to be a liability to the furthering of both gay and lesbian causes.

Conclusions

The City of West Hollywood's incorporation represents a successful grassroots political movement resulting from a coalition of gays and seniors. As I have attempted to indicate in the preceding paragraphs, it was necessary for this coalition to hold together in order to capture the local state via the political process, and then to promote the coalition's interests once the representatives of both groups attained city council seats. Although a local control contingent was important to the creation of the city, the agenda of those persons connected to local control interests has not been realized, since a local control candidate has not yet been able to win a seat on the council. Therefore, the initial actions of the city council reflected primarily the concerns of the city's senior and gay and lesbian populations and served to legitimate the creation of the city to them. The passage of many gay rights pieces of legislation satisfied the members of that community who desired to capture the local state in order to establish a territory where their sexual orientation could not be used as a weapon against them in gaining political, economic or social stature, or both. The passage of one of the country's strictest rent control laws satisfied those who wanted to establish a territory where tenants could control their own housing destiny.

While the creation of the local state resulted from a coalition of groups, it must be remembered that the actions of one individual were critical to the success of the social movement ever getting to the political arena. Ron Stone managed to succeed where many before him failed. The reasons for Stone's success stem from his knowledge of the incorporation process in California and from his ability to debate with politicians and bureaucratic staff about these issues in their language and at their level. Even though Stone lacked official power, he was able to keep the incorporation issue alive, and eventually see it through to its approval. Stone's foresight in building the original incorporation coalition proved critical both during LAFCO approval stage, and beyond as two-thirds of that original coalition eventually made up 80 per cent of the West Hollywood City Council. Yet, during the West Hollywood incorporation election politics, the nature of the debate changed and Stone's ability to influence the electorate was negligible in comparison to his ability to influence the politicians and bureaucrats in the pre-approval stage. Thus, the social movement could not have succeeded without the productive actions of Stone, even though he did not represent the coalition that eventually captured the local state.

Another factor to consider in assessing the social movement is the characteristics of the urban area which forms its locale. The importance of West Hollywood as the setting for the above drama can not be overlooked since it provided a discrete forum whereby political power could be realized by the coalition. The area's concentration of tenants/seniors and gays provided that opportunity for these normally powerless groups to form a coalition and capture the local state from the county government. If the area did not have the extreme mix of 88 per cent tenants and 35 per cent gays, then it is extremely unlikely that a coalition of these groups would have been victorious in capturing the local state. This notion is evidenced by the failure of West Hollywood tenants to have rent control extended in Los Angeles County prior to cityhood. Thus, the social geographic characteristics of West Hollywood enabled these two groups to achieve a political majority and therefore legitimacy, power, and control over the affairs in their community.

For West Hollywood the story is not over, but just beginning. The reality of the local state in West Hollywood has required that the council deals with more than just popular grassroots issues and broadens its focus to include the more traditional municipal tasks such as planning, development, traffic, services, and crime. Even though gay rights and rent control were found to be the battle cry in the first two elections, these other issues facing the council will most likely gain in importance as rent control and gay rights become institutionalized and "taken for granted" as part of everyday life in West Hollywood. In dealing with these other issues, the council will have a very potent new member in their coalition, the city bureaucracy. The professionals inside city hall continue to gain greater influence in the development and implementation of city policy. The potential exists for the council to find progressive solutions to these "standard" issues by keeping the grassroots involved in the governing process, so ensuring that staff actions are not completely insulated from community interests.

Notes

1 The author wishes to thank Mayor Alan Viterbi, Ron Stone, Larry Gross, Stephen Braun and Dori Stegman for allowing themselves to be interviewed for this article and for their generosity in allowing me access to their files. The author also thanks the city council, staff and residents of the City of West Hollywood who unknowingly contributed to this research.
2 The City and County of San Francisco are the identical geographic area, thus there is no unincorporated area in San Francisco County.
3 Prior to the State of California's creation of LAFCOs, a number of cities were created that benefited a specific industrial or commercial interest tax situation. LAFCOs were designed to eliminate the creation of these types of cities.
4 A proposition is a citizen-sponsored initiative that appears on the ballot via a petition drive to gather signatures of registered voters.
5 The West Hollywood Study Group enlisted the aid of a political consulting firm to gather signatures of residents opposed to the incorporation effort. While there is a legal basis for their effort, LAFCO regulations require an application be withdrawn if 50 per cent of the area's registered voters sign a petition opposing incorporation, fewer than 10 per cent of the signatures submitted were found to be of persons who either lived in West Hollywood or were even registered to vote.
6 Homophobia refers to a fear of homosexuals merely on the basis of their sexual orientation. It is a term reserved for the particularly unenlightened.
7 For example, prior to incorporation residents had wanted the County to install a signal light at the intersection at San Vicente and Cynthia and had no success in having this accomplished. The signal was installed on March 30, 1987, 2½ years following incorporation.
8 Prior to her election to the West Hollywood City Council, Valerie Terrigno served as Director of the Crossroads Counseling agency. In March 1986 she was convicted of embezzling funds from this federally funded agency. Her seat was vacated the following month, and the council scheduled a special election to fill it for the remainder of her term.

References

Castells, M. 1983. *The City and the grassroots*. Berkeley, CA: University of California Press.
City of West Hollywood 1985. *Preliminary budget revision: fiscal year 1985–86.*
Dear, M. J. & A. I. Moos 1986. Structuration theory in urban analysis: 2. Empirical application. *Environment and Planning A* **18**, 351–73.
Levine, N. & J. E. Grigsby 1985. *A survey of tenants and apartment owners in West Hollywood*. Los Angeles, CA. The Research Experience and The Planning Group.

Local Agency Formation Commission, Los Angeles County 1984. *Staff report on the proposed incorporation of the city of West Hollywood.*

Moos, A. I. & M. J. Dear 1986. Structuration theory in urban analysis: 1. Theoretical exegesis. *Environment and Planning A* **18**, 231–52.

Ufkes-Daniels, F. 1985. *The tenant movement in West Hollywood, California: an analysis of social theory and public policy.* Paper presented at Westlakes Division Meetings of the Association of American Geographers, October.

US Department of Commerce, Bureau of the Census, 1980. *Census of population.* Washington: US Government Printing Office.

West Hollywood Incorporation Committee 1984. *Cityhood.* West Hollywood, CA: West Hollywood Incorporation Committee.

17

Disability and the reproduction of bodily images: the dynamics of human appearances

HARLAN HAHN

Human beings clearly are not identical. People possess varying mental or emotional capabilities, but perhaps the most obvious evidence of individual differences is to be found in their physical appearances. This undeniable, but frequently overlooked fact has vast psychological, economic, social, and political implications. As Kolakowski (1974, 16) has noted, "We maintain that people should be considered as material beings, but nothing shocks us as much as the idea that people have bodies: . . . all these factors can play a role in social processes regardless of who owns the means of production and thus . . . some important social forces are not products of historical conditions and do not depend on class division." In this study, however, an effort is made to demonstrate that major differences or similarities in human appearance have political origins and have been used to serve social and economic purposes.

The segment of the population that is the principal focus of this analysis consists of people with visible disabilities. For many years, physical disability was regarded primarily as an individual problem stemming from functional limitations that represent a personal misfortune and an irremediable indication of biological inferiority. Recently, however, the traditional view has been challenged by the emergence of a "minority group" model based on a *socio-political* definition which regards disability as the product of the interaction between the individual and the environment (Hahn 1985a, 1985b, 1986b, 1987a). From this perspective, the major problems of a disability can be traced to a disabling environment rather than to personal defects or deficiencies. Perhaps even more importantly, this interpretation has indicated the possibility that widespread evidence of attitudinal aversion toward disabled people might be ascribed to "aesthetic" anxiety, or the fear of those who are seen as alien and strange, instead of to "existential" anxiety, or a perceived threat to the abilities of the observer (Hahn, forthcoming). In a basic sense, visible and severe disabilities seem to represent a more significant departure from the conventional human form than other attributes such as skin color, gender, and age that have also been associated with claims of biological inferiority and that have been used as a source of discrimination for centuries. As a result, the mounting attempts by people with disabilities to combat prejudice and discrimination may confront modern societies with a more pressing and compelling need to assess the impact of perceptibly different appearances in the development of the social structure than they have previously been prepared to conduct.

Ironically, perhaps a crucial explanation for the prior failure of visibly disabled persons to emerge as a prominent political group can be attributed to the stigma which has also been the basic source of this oppression. Unlike many other minorities, disabled individuals do not live in close proximity to one another; and their opportunities to interact either among themselves or with the nondisabled majority are sharply restricted by the spatial organization of communities and by architectural constraints (Hahn 1986a). The majority also lack a sense of generational continuity between parents and children. Perhaps most significantly, however, men and women with visible disabilities have not yet gained a positive sense of self-identity that might otherwise form a foundation for cohesive social and political action. Virtually every society has attempted to cope with the problem of disability by seeking maximal improvements in the functional capacities of disabled individuals, or by policies of eradication that range from infanticide or genocide to miracles, as well as the disproportionate emphasis on prevention and cure in contemporary telethons. Almost no attention has been devoted to proposals that would support or even permit the continuing presence of disabilities as a source of dignity and pride. Nor has any major effort been made to reverse the devaluating consequences of the stigma of unattractiveness usually associated with a visible disability. Yet, the legacy of human experience encompasses a wide range of principles and perspectives that might contribute to changes in the perception of visible disabilities and to a major redefinition of the identity of disabled persons. The exploration of these historical trends also may uncover important new understandings about the role of bodily differences in society.

There seem to be at least three distinct connotations that might be applied to the concept of social reproduction in this context. The first, of course, is strictly biological; and both the relative lack of genetically determined disabilities and the perceived unattractiveness of these traits may at least partially account for the neglect of disabled persons as a significant element in human geography. The second inference is primarily functional and concentrates attention on the presence of people with disabilities as a critical component of the industrial reserve army (Gough 1979). But, even in a capitalist economy that stresses the productive and efficient use of labor, an exclusive focus on functional considerations does not appear to provide a comprehensive means of explaining why groups such as disabled adults are relegated to the surplus labor force. Consideration must, therefore, be given to a third implication which suggests that social reproduction may also encompass the perpetuation of an imagery of acceptable and appropriate physical appearances. Although these images often seem to value bodily differences as well as similarities, they may be more restrictive and conformist in some eras than on other occasions. In addition, there are obviously important cultural variations in tolerance for deviant appearances or disabilities. This study focuses primarily on changing concepts of appearance and attractiveness that have emerged in western societies since the 18th century. Specifically, an attempt is made to glean from these trends insights and information that could promote the acceptance and the identity of persons with visible disabilities. In order to provide a broad cultural and theoretical background for this investigation, however, it is necessary to include a brief interpretation of significant developments affecting the treatment of human similarities and differences in earlier societies.

Politics and the emergence of physical differences

Although social commentators long have noted that the creation of methods of political control often seemed to reflect a prevalent human motivation to associate with people who *act* or *behave* like themselves, correspondingly less attention has been focused on the common desire to be surrounded by others who *look* like themselves. In a fundamental sense, however, this latter impulse seems to be intimately associated with one of the most difficult issues faced by any political regime, namely, the problem of succession. Whether initially chosen by social contracts, military conquest, or other practices, the creation of ruling families seemed to provide the (usually) male political leader with some assurance that the successor to his title (regardless of other temperamental or intellectual capabilities) would at least be "in his own image." For mere mortals, physical rather than mental or emotional perpetuation seems to represent the only means of ensuring posterity; and the propensity to transfer leadership in a manner that embodies this principle has been so strong that elaborate regulations were devised to prevent its violation. The necessity of at least one additional parent of the descendants of rulers seemingly contributed to the development of a complex structure of breeding populations, clans, kinship systems, tribes, and groups. The widespread adoption of the principle of patrilineal descent and the need for delicate marriage alliances to preserve political stability or tranquility in geographic areas also necessitated the formation of strict rules of sexual conduct that extended beyond the feudal era. A fundamental feature of early social structures, for both elites and masses, however, seemed to reflect a proclivity to be surrounded by people with similar physical appearances and to reject or avoid those perceived as alien or strange.

Although the exact origins of these patterns of human behavior remain a mystery, at least two separate lines of speculative inquiry seem especially worthy of further exploration. The first is represented by the concept of narcissism which, though it has been an important element of psycho-analytic theory, usually has been defined by that tradition as a form of self-love which may reflect either a developmental stage or a defensive mechanism. While this concept may be found along with Freud's descriptions of Oedipal strivings, as an example, he did not appear to infuse narcissism with the same meaning that it was given in the Greek myth: an obsessive preoccupation with personal vanity. In part, this reticence probably has been shared by many other observers of human behavior. While one of the most common forms of social research – eavesdropping – clearly reveals that a disproportionate number of comments overheard in everyday conversation refer to physical appearances, for instance, the profound implications of these facts seldom have been incorporated in scholarly investigations. Yet social theories also have reflected a steady expansion of the concept of the self to include personal feelings about the perceptions of others that contribute to a sense of individual identity. From this perspective, pathological connotations may be divorced from the concept of narcissism; and the term might be employed to refer to the common – though vain – human propensity to associate with others who possess similar appearances. While there are also grave dangers in extrapolating collective behavior from psychological predispositions, the popular impulse to seek neighbors who look like themselves may also help to account for the notion of territoriality and the definition of political jurisdictions. As a result of this tendency, of course, serious disadvantages may be imposed upon groups such as persons with visible disabilities who are perceived as unacceptable or deviant.

A second theoretical approach which might contribute to an explanation of the role of physical appearance in social and political development involves a materialistic investigation with major economic and political implications. If narcissistic principles provide a partial interpretation of the search for perceptible similarities in the organization of human communities, the latter orientation seems to suggest a means of understanding some of the differentiation that has emerged among separate groups. Since ruling families obviously did not display unique physical traits that would allow their progeny to be distinguished from others with similar attributes, some means may have been necessary to preserve distinctions in appearance. A major solution to this problem might have been found through the invention of clothing which has often been regarded as an extension of the body and which, unlike nudity, tends to accentuate rather than to obliterate physical differences. While the clothing worn by ordinary people remained essentially unchanged for centuries, the costumes of powerful leaders underwent rapid and increasingly extravagant changes – in part to separate themselves from common folk and from those who pretended to hold power. In fact, the rise of wealthy commercial interests prompted the enactment of numerous sumptuary laws which prohibited the wearing of clothes reserved for the ruling class (Ewen & Ewen 1982). The importance of clothing in molding perceptions of the human body is underscored by Hollander's (1978) study which shows that the changing proportions of nudes painted by artists in various historical eras are directly related to the contours of fashionable clothing of the time. Yet, these intriguing facts raise more fundamental questions. Has clothing and fashion been used primarily as a device in the stratification system to perpetuate distinctions based on appearance between the ruling classes and the remainder of the population? Or is it possible, from a perspective which recognizes the narcissistic instinct of human nature, that clothing actually represents the circumference of the outer boundaries of an acceptable human appearance, thereby excluding groups such as visibly disabled persons whose bodily deviance may not be effectively concealed by their apparel? To explore such questions, attention must be devoted to the influence that differences and similarities in appearances have exerted on the human psyche and behavior.

Differentiation and the shaping of human appearances

The constant historical preoccupation with physical similarities and differences naturally yields a corollary question. Is there an appearance which is generally recognized as non-human? Although most attempts to supply this type of image have been derived from supernatural sources that depict gods or devils with awesome powers, perhaps even more fruitful information about this subject can be obtained from the mythologies and folk or fairy tales that have often antedated recorded history. Many of these stories are populated by an extraordinary number of characters with disabilities. Furthermore, most of the disabled figures are cast in villainous or abominable roles. Often the transition from a "bad" to a "good" role is marked by a miraculous cure of the disability. Although it is nearly impossible to summarize the vast data about deviance and disability contained in these sources, at one level they seem to represent an attempt to furnish people with a model of the non-human consistent with Goffman's (1963) statement that, by definition, individuals with a stigma such as disability are considered "not quite human." Fisher (1973, 73) also has concluded, "Despite all the efforts invested by our society in an attempt to rally sympathy for the crippled, they still elicit serious discomfort. It is well documented that the disfigured person makes others feel anxious and he becomes an object to be warded off. He is viewed as simultaneously inferior and threatening. He becomes associated with the special class of monster images that haunts each culture." From a perspective that equates disability with the attempt to conceive of a non-human image, therefore, the proclivity to seek similarities in others may be so strong that those with prominent differences or disabilities may be relegated to the status of pariahs, to be shunned by the rest of society.

And yet there is another dimension to these phenomena. Myths and folk tales including characters with obvious disabilities have persisted for an extraordinary length of time, perhaps for an even longer period than most tales that contain an exclusively nondisabled cast. Accounts of extraordinarily different creatures have continued to reappear in legends ranging from the wild men of the Middle Ages to contemporary reports of gigantic beasts in the Himalayas and the Pacific Northwest. Even the modern popularity of monster or horror movies may indicate a reluctance by viewers to flee in terror from the sight of a character whose appearance may not be associated with the human species. People seem to be simultaneously disturbed and fascinated by extreme distortions of

374

conventional anatomical characteristics; and, at least in some cases, the intrinsic appeal of such differences might overshadow the fear they may evoke. The continuing depiction of fictional characters as well as persons with prominent disabilities in literature and the arts seems to signify an unwillingness to abandon an innate interest in human differences that may outweight narcissistic tendencies. The striking parallels between the visibly different or disabled human shapes exhibited both by early drawings of people in unknown primitive cultures and by extra-terrestrial beings in science fiction also may reflect a basic desire to exercise a far-ranging imagination about possible variations in the body (Renard 1984). Moreover, inhabitants of many cultures have actually sought to disable themselves in order to enhance their own appearance or attractiveness. In addition to efforts to reshape the head and to distort other parts of the anatomy, some societies have encouraged body scarification as a means of approximating their own images of beauty (Rudefsky 1971). Whether or not the conspicuous differences of disabled individuals and imaginary creatures represent the influence of a conception of what is non-human, they have also at least occasionally inspired a magnetism that reflects stimulation or attraction rather than revulsion or disgust.

Perhaps at least a partial solution to this puzzle is suggested by a cross-cultural examination of human physiognomy. After a comprehensive review of bodily characteristics, Brain (1979, 147) has concluded, "All human societies use body decorations to disguise their kinship with animals; clothing, ornaments and painting attempt to underline man's kinship with the world of culture, not the world of nature, with the gods and the spirit, not the animals. Decoration distinguishes us from the brutes." Since both adornments and distortion represent the transformation of the natural into the cultural body, this distinction seems somewhat more convincing as an explanation of the diverse ways in which humans have sought to alter their appearances than interpretations that have focused on the polarization of the sexes (Lakoff & Scherr 1984). Yet there is a critical dynamic process involved in both types of differentiation. In either case, the crucial component of bodily imagery revolves around the concept of differences. Human beings may not only wish to separate themselves from animals, but they are also attracted to and fascinated by animals. Obviously, similar statements can be made about men and women. Attempts to mold the presentation of the body cannot be explained merely as efforts to extend the distance between themselves and a specific reference group, whether it be animals or the opposite sex; they seem to reflect a duality that encompasses attraction as well as avoidance.

A major element of the persuasiveness of the human–animal dichotomy as one foundation for the investigation of physical differences, however, also seems to derive from connotations surrounding the description of these creatures. For many centuries perhaps the principal materialist struggle of men and women with their environments involved the effort to gain dominance over animals; and it seems likely that they might have wanted to display these triumphs in their appearance. In addition, the common use of "animalistic" to refer to base or unacceptable instincts may have prompted a desire to react against the symbolic vestiges of such traits in the decoration of their bodies. For similar reasons, the distinction between humans and animals seems to provide a more credible basis for assessing the role of persons with disabled or deviant characteristics than the differentiation between human and non-human. For most people, the concept of non-human represents a rather vacuous or amorphous category with which they have had almost no familiarity aside from their understanding of devils or deities. By contrast, nearly everyone has had experience with animals; and they may naturally become a primary focus as an inimical sign by which to gauge physical appearance. In fact, both the mythical creatures that have pervaded ancient and modern legends and persons with disabilities are more likely to be described or disparaged as animals than as non-human.

The crucial nature of the human-animal dualism, however, does not necessarily imply that attributes associated with animalistic features – such as visible disabilities – may be uniformly and invariably perceived as unattractive. Some evidence indicates that efforts to unravel the "human" or "animal" traits of the mythical hybrid figures of ancient cultures are affected primarily by characteristics of the legs and the head (Nash & Pieszko 1982). Art also reveals that aesthetic preferences about the proportions of other parts of the unclothed human anatomy, including the torso, chest, and trunk, have changed significantly over time (Clark 1956). The history of beauty, even in the western world, reflects a constant state of flux. While attractiveness has sometimes been described as symmetry, harmony, and proportionality, it has also been characterized by unevenness, discordance, and angularity. In fact, Lakoff & Scherr (1984, 48) have suggested that attraction "is the excitement of encountering opposites . . . and the beautiful object . . . is made of the mixture of opposing, even paradoxical, qualities. Beauty . . . offers the comfort of the familiar, coupled with the shock of the unexpected. . . . There is refinement and vulgarity at once in the truly beautiful, organization and chaos, uniqueness and sameness, wildness and placidity." The principles that shape changes in aesthetic tastes concerning the human body are complex. In part, they appear to be based both on a narcissistic search for perceptible physical similarities and on an interest in differentiation that may revolve around the animal-human distinction. Additional explanations of these trends and their impact on persons with visible disabilities, however, would seem to depend on a relatively detailed examination of a specific historical era.

Physical appearances in the 19th century

Despite their prominence in historical fables, persons with visibly disabled or deviant characteristics probably did not form a major portion of the population prior to the 19th century. In a harsh environment that few disabled individuals could endure, there may have been little need for most people to abandon the medieval assumption that disability signified supernatural intervention either as punishment or as a special form of divine providence. But significant changes were occurring in science and religion that threatened these superstitions. Turner (1984), for example, has shown that the medieval popularity of the ascetic body was reinforced by Protestant leaders such as John Wesley who, by stressing health and nutrition, may have promoted the trim anatomical contours which ultimately evolved as a symbol of modern hedonism. Similarly, the emerging field of medicine, which sought to substitute scientific for mythical beliefs before it gained effective remedies for most diseases, emphasized exercise and vigorous activity in a trend that was promptly amplified by 19th-century literature (Haley 1978). In fact, religion and medicine seemed to join in a moral hegemony that has had a continuing effect on scientific research and on human behavior. Illness was widely interpreted as a sign of moral laxity or indulgence; the cure was obviously the maintenance of a strong body controlled by a firm will and a disciplined mind. The difficulty was that this position made few allowances for persons with disabilities or other chronic health conditions. Disability seemed to be regarded as an unfortunate inability to meet the functional standards imposed by the environment, which required little more than efforts to seek the restoration of all possible physical faculties and possibly the extension of charity. No consideration apparently was given by medical or religious leaders to the possibility that society also might have a responsibility to alter the environment to accommodate the interests of disabled citizens. Nor were persons with disabilities granted any exemption by these authorities from the standards of bodily appearance that they sought to prescribe.

These trends were exacerbated by other scientific developments in the 19th century, including the discoveries of Darwin and the questionable social theories that they generated. In an age when scientists seemed determined to prove the biological inferiority of disadvantaged groups with identifiable physical characteristics such as gender or skin color, no one seemed to doubt that disabled persons could be classified among the "unfit" whose survival was doomed without extraordinary acts of benevolence. A powerful eugenics movement was launched to prohibit marriage by disabled adults through laws which seemed to exhibit little concern about whether or not the disability was genetically transmitted. Perhaps most significantly, there appeared to be a consensus that the state had no duty to promote equality between citizens by adopting policies to ameliorate clinically demonstrable biological deficiencies. In a plan that could be readily justified by functional criteria rather than by underlying aesthetic considerations, most people with visible disabilities were confined and concealed in homes or institutions.

Nonetheless, the presence of disabilities continued to have an impact on 19th-century society. Some of the most extremely disabled or deviant individuals joined freak shows where they often managed to form a subculture which, despite their exploitation, may have provided a place of refuge and status as well as a cathartic outlet for the curiosity of the nondisabled (Fiedler 1978). Epidemiological rates of disability also seemed to be associated with trends in beauty in oddly dissimilar ways. The increasing use of facial cosmetics, for example, apparently originated as a reaction against earlier smallpox epidemics (Banner 1983, 40). By contrast, the growing popularity of a palid complexion apparently was intended to emulate the appearance of tuberculosis patients who were reputed to possess unusual amorous energy and spiritual qualities (Sontag 1978). Furthermore, in the 19th century, many persons, especially women, continued to disable their bodies in an attempt to approximate prevailing concepts of beauty. Chinese foot-binding, which was apparently undertaken for erotic purposes, did not diminish until this era. In the western world, the principal form of body sculpting was the tight-lacing of corseted waists which, though roundly condemned as fetishistic and medically dangerous, has later been interpreted as an effort by socially mobile women to gain increased power by manipulating their sexuality in a manner that threatened the dominant patriarchy (Kunzle 1982, 299). Although the explanation of these developments – as well as almost all other trends in fashion – has failed to yield general agreement, they seem to imply dissatisfaction with existing configurations of the body even when encased in massive amounts of clothing. Such discontent may form a major impetus for the changes that constantly affect fashion. Human beings have always seemed to feel that they could improve upon the aesthetic appeal of their bodies, even through alterations that might otherwise be considered disabling or deviant.

Perhaps an even more significant influence on the assessment of the human figure, however, can be traced to fashion trends that began early in the 19th century. Emerging from an era in which the ornate costumes previously reserved for ruling classes had been discredited by a romantic interest in natural styles and by political revolutions in France and the United States, aristocrats joined ordinary citizens in adopting a costume that consisted primarily of white muslin dresses for women and simple suits for men (Lurie 1981, 61–2). Significantly, male clothing has remained essentially unchanged since that time. By the second and third decades of the 19th century, however, a new industry had arisen to encourage affluent American women to replace their simple hand-made garments with a rapidly shifting array of more fashionable apparel designed and produced for an expanding market (Banner 1983, 26). Eventually, perhaps both women's styles and ideas about beauty changed more drastically and more extensively in the 19th century than in any prior era. Yet, despite the changes in flourishes or accoutrements that reaped huge profits for this industry, the basic outline of dresses throughout the Victorian age continued to be characterized by a heavy bulging skirt that extended to the ground, occasionally embellished by bustles and padded corsets to produce a S-curved figure; a pinched waist; a puffed bosom; and a diminutive head topped by a horizontal bonnet. Perhaps most remarkably, to the uneducated eye, the silhouette of this costume might be misperceived as resembling a legless bird with a tiny sloping torso, a billowing chest, and a small tufted crown. In fact from a similar vantage point, the entire effect may be perceived as more animal-like than human. The tilted posture of these styles appeared to suggest a creature mounted on a ball-shaped dome prepared to use its upper appendages for movement or flight rather than for other purposes. These fashions seemed to betray fewer distinctively human attributes than comparable clothing such as simple muslin dresses or classical draped gowns. Although gentlemen had doffed their plumage by the 19th century, the look of the prosperous Victorian matron seemed oddly incongruous when compared with the dichotomy between humans and animals that has apparently exerted a continuous influence on efforts to mold the human anatomy.

The impact of the Industrial Revolution

Perhaps a cogent interpretation of Victorian fashion can be traced to one of the most significant historical developments that occurred during that era: the advent of the Industrial Revolution. By the end of the 19th century, for example, the United States and many other countries had been transformed from rural and agrarian societies to predominantly urban and industrialized nations. In this new landscape, the most prominent contrapuntal element that could be compared with the human body was no longer symbolized by animals. As millions moved from the countryside to the cities, experience with most animals steadily declined. In an industrialized environment, the dominant antithesis to the human form, especially for men who directed or worked in factories, was the machine. Most persons were no longer surrounded by animals; they were surrounded by machines.

The social changes spawned by the Industrial Revolution seemed to produce major modifications of human appearance as well as interesting distinctions between women and men. Like the attempt to differentiate themselves from animals, the polarization between humans and machines seemed to reflect a combination of repulsion and fascination. Initially, for upper-class males, the growing presence of machinery inspired a form of mimicry. The popularity of the stovepipe hat, which symbolized factory smokestacks, and the portly look which projected material success and stability, undoubtedly indicated the extent to which prosperous gentlemen of the mid- to late-19th century were prepared to embrace these unprecedented economic developments. As the disadvantages of capitalist enterprises became increasingly apparent, however, these fashions subsided.

By contrast, for the affluent 19th-century woman who spent her days in the home rather than in the factory, movement from the rural countryside may have decreased the need to emphasize her distance from untamed predators. Neither machines nor animals were an omnipresent feature of her surroundings. Perhaps even more significantly, her perception of animals had been transformed from fearsome beasts to domesticated creatures playing a role that, ironically, seemed to parallel her own position in the Victorian household. The middle-class woman in 19th-century urban areas had little need to dress in a manner that would stress the distinction between herself and antithetical aspects of her environment.

In these circumstances, she may have been able to project her underlying attraction to innocent animal qualities in a style that exemplified continuity with costumes worn by aristocratic females of earlier centuries who were similarly protected from ferocious creatures of the wild. Although the growing stress on athletic activities for both sexes began to promote a somewhat less stooped and encumbered posture for women, the basic configurations of the female form had not changed appreciably by the end of the 19th century.

As vast numbers of women left home for the workplace during World War 1, however, a dramatic change in women became increasingly obvious. The flapper of the 1920s looked nothing like her counterparts in previous eras. Dresses were hung straight on bodies with flattened breasts and buttocks, in a style that many critics described as "boyish." Perhaps most importantly, drastically shortened skirts revealed that the human female actually had legs. The style hardly resembled either the legless creatures of the 19th century or any known species of animals. While countless interpretations have been offered to explain the rapid transformation of modern women, surprisingly little attention has been given to perspectives based on the persistent effort by humans to stress the dissimilarities between themselves and animals as well as machines. The 20th-century woman increasingly found herself in a world dominated by machines; and, in an attempt to separate herself from these ubiquitous mechanical contrivances, she may have felt a strong need to accentuate her distinctively human attributes which overshadowed any secondary desire to identify with opposing symbols. As machines have become an increasingly inescapable component of modern society, this trend has shown few signs of abating. Both women and men of the 20th century have displayed a desire to expose increasing amounts of flesh, probably the most human of all anatomical traits, as visible proof of their detachment from mechanical or animal iconography. These patterns have not occurred without significant undercurrents. The upper-class penchant for furs prior to World War 2 and the contemporary popularity of leather, for example, seem to exhibit a continuing interest in the appearance of animals. Similarly, designers have sought to capitalize on the mesmerizing effects of machinery through the production of metallic dresses and other apparel; but, aside from drawings of futuristic garments, machine-like clothing has never captured the public imagination. Throughout the 20th century, the primary need to differentiate the human body from both machines and animals has seemed to overwhelm any secondary desire to copy them.

The dynamic interaction affecting the use of human, animal, and mechanical symbols in shaping the appearance of men and women in various historical periods does, however, contain important implications for the perception of physical deviance or differences. In view of the wide variations in bodily imagery that have both gained prominence and declined in a relatively short span of time, for example, there does not appear to be a uniform boundary defining the acceptability of human physiognomy that corresponds with the circumference of clothing, even the costumes of the Victorian era, or any other known parameter. Human beings have continued to disable themselves in order to fulfill cultural criteria of attractiveness. Changing fashions have periodically decreed that the human, and especially the female, anatomy assume different postures, different proportions, and different shapes. Instead of revealing a universal ideal by which all men or women are to be judged, the history of the fashionable body discloses a complex interplay between uniquely human features and other entities with which they might be compared. At various times, perhaps the relative disappearance of one of these symbols has permitted individuals to project the appeal of that image; this pattern may have emerged in the 19th century. On other occasions, a specific contrasting point of reference probably became so dominant that people have sought to stress their peculiarly human traits in opposition to it; this tendency seems to pervade the association between humans and machines in the 20th century. Perhaps, in another era, or culture, a decline in technology might prompt a renewed interest in animal images; or possibly the growing prominence of animals could promote increased enchantment with machines. Conceivably, at some time in the future, men and women may learn to integrate and to express all of these elements in a single identity that combines and amplifies the attractiveness of each dimension. In any event, the analysis of historical variations in concepts of beauty does not seem to be governed by mysterious forces beyond the realm of intellectual scrutiny; on the contrary, these phenomena are influenced by principles that can be verified and subjected to examination in an effort to uncover means of promoting the acceptance of an expanded range of physical differences.

Idealized standards of human appearance

Perhaps even more significant impacts on changing human appearances, however, were exerted by additional trends that emanated from the Industrial Revolution. Beneath external shifts in fashion, other dynamic processes both fueled and intensified these cycles. These processes were stimulated in part by technological innovations. The availability of sewing machines by the 1850s and paper patterns by the 1870s, for example, provided women with an expanded range of choice about the manner in which they could present themselves to others (Steele 1985, 83). Gradually, the clothing industry replaced home-made garments as the major source of the attire that people wore. The mass production of clothes and other goods generated a need for mass consumption that was promptly filled by mass advertizing (Hahn 1987b). In prior centuries, changes in style had been confined primarily to ruling classes that could afford their own seamstresses and individually tailored costumes; but the mechanization of production as well as improved methods of distribution and dissemination allowed prescribed modes of appearance to permeate the entire society. The primary role of fashion no longer focused on efforts to distinguish elites and masses. Swept along by images that they were scarcely able to resist, ordinary women and men would not remain impervious to these developments. Whereas changes in the clothing adopted by a ruling oligarchy previously had furnished evidence that may have been of principal interest to fashion historians, the ability to translate or to decode the messages communicated by variations in human appearance became essential to understanding the culture that emerged from the Industrial Revolution. Capitalist enterprises permitted the body to impose a massive imprint on all levels of socio-economic hierarchy and ensured that the definition of physical attractiveness would continue to exert a decisive effect on social behavior.

The growing prominence of fashionable images promoted by mass media since the 19th century has had several major consequences. Initially, there has been an erosion of the traditional demarcation between inner beauty and external manifestations. Steele (1985, 213) notes that, by the end of the century, "writers were much less likely to posit a rigid distinction between 'true' inner beauty and its 'false' and artificial counterpart." In a gradual but striking reversal of earlier moral attitudes, the growing acceptance of popular clothing and cosmetics has been steadily promoted by the belief that an appealing exterior was reflection of spiritual beauty. The element of superificiality seemed to disappear in the modern equation that what is beautiful is good (Dion *et al.* 1972). Conversely, of course, visible traits that denote a marked departure from conventional ideas about attractiveness have been increasingly interpreted as signs of moral as well as aesthetic faults.

Second, many critics of advertizing and the media have mistakenly concentrated on the argument that advertizing messages promote the conspicuous consumption of goods that buyers do not really need or want. Opponents often have replied that this claim unduly exaggerates the ability of the media to mesmerize consumers in a manner that is not directly reflected in sales records. But both sides of this controversy apparently neglected the fact that a principal effect of the media has not been to sell specific products; they have sold imagery and, especially, an idealized image of the human body. Through the persuasiveness of the media, the aspirations of elites and masses have been joined in narcissistic striving for a shared vision of a flawless physical appearance. The potential for class conflict that might otherwise result from reactions to the ubiquitous presence of machines in an industrialized milieu has been effectively defused by an hegemony of media representations fueling the relentless search for bodily perfection. The "capitalist realism" of advertizing has increased its power both by what it does not communicate and by the pictures that it furnishes the public (Schudson 1984). Conspicuous by their absence have been alternative or divergent models of attractiveness. Although most humans bear only a faint resemblance to the characters that dominate magazines, billboards, television, and movies, they have few other likenesses to emulate and hardly any optional means of discharging the frustrations that they are encouraged to feel about their corporeal selves.

Finally, the male and female figures depicted in the media represent a highly restrictive – and almost inherently unattainable – standard of physical appearance. Historically, these visual presentations have dis-proportionately excluded oppressed groups with deviant physical attri-butes such as skin color, age, or disability. By deprivng them of positive models and by making them virtually invisible to the general public, this omission has been a potent method of undermining their social and political influence. The women and men who appear in the media have been limited largely to young, white, unblemished, nondisabled, immacu-lately adorned, and perfectly proportioned types with extraordinary charismatic and sensual appeal. To an even greater extent than this description indicates, their features signify criteria of bodily perfectibility that few mortals even hope to approximate, let alone reach. The unachievable nature of these images, in turn, generates increased mass consumption resulting from anxiety produced by the apparent discrepancy between external characteristics of the average man or woman and the photogenic characters that they constantly confront in the media. Although there are few reliable estimates of the amount of money spent in the frenzied search for garments and products to enhance personal beauty, they probably exceed the resources allocated for food (Giddon 1983, 455). The body has been pressed into the service of capitalist expansion, and there seem to be few means of altering this pattern.

Expanding human attraction

Although often portrayed as a seemingly innocuous by-product of other events, the preoccupation with physical appearance bequeathed to the 20th century by the Industrial Revolution has produced serious social and political problems. Countless persons have been socialized to believe that crucial rewards such as jobs, prestige, and even social status are determined primarily by personal attractiveness; and abundant empirical evidence indicates that this possibly self-fulfilling prophecy might be true (Berscheid & Walster 1974, Landy & Sigall 1974, Schuler & Berger 1979). The neurotic compulsions engendered by these developments have yielded a host of complications including the creation of a new class of celebrity politicians or public leaders whose photogenic qualities seem to exceed their capacities to discover creative solutions to persistent social problems. Personal relationships frequently have been contaminated by the supposi-tion that overpowering attraction exempts individuals from responsibility for making conscious or thoughtful decisions. Similar tendencies have promoted medical difficulties that range from tension and eating disorders to the abuse of cosmetic surgery. Aerobic exercises have seemed to focus more on the sensual goal of reshaping the proportions of the body than on the health-related objective of gaining increased physical fitness, which was the original impetus of this fad. Some of these trends appeared to reach a peak in the 1980s. In fact, one observer has suggested that, whereas the 1960s was the age of collective action, in which citizens sought to create a better world through social movements, and the 1970s was the decade of individual introspection, in which persons sought to create a better world by changing their inner selves, the 1980s may be the era of the veneer, in which people are more concerned about the surface of their bodies than about either political or psychological issues.

The inescapable superficiality of modern perspectives of beauty may be exemplified by imagining a drab and boring world in which everyone was identical, or physically perfect. An intolerable blandness even might permeate a society confined exclusively to men and women whose appearance resembles the range of physical attributes usually presented in advertizing, television, and films. Perhaps the view of the human body as a *gestalt* is no longer appropriate. Modern men and women have seemed to display an extraordinary obsession with anatomical packaging, which may denote a latent desire to possess or control the entirety of a partner. Yet, even during moments of physical intimacy, sexual intercourse involves an essential connection extending from the genitals (which, oddly enough, are usually considered unattractive in this culture) through the heart (the emotional center of the body) and the mind (engaging the intellect too), rather than an act that merely involves the packaging. An enhanced awareness of the separate components of human beings may comprise a valuable antidote to the flimsy and artificial nature of personal relationships based on external appearances.

Perhaps the deleterious effects of these trends can be illustrated most graphically by conceiving of a bell-shaped curve that represents a continuum of human appearances. Whereas the modal point or peak of the curve indicates the physical characteristics of the average human being, the idealized attractiveness projected by the media would be located at the extreme end of one of the tails, perhaps on the right. The use of this idealized image as a yardstick for comparison might reveal that a "usual range of attraction" generally extends from this benchmark to include an area delineated by a point on the right slope of the curve. In statistical terms, this population may be roughly defined as ranging from $+4$ to $+1$ standard deviation to the right of the mid-point. Since this area does not even reach the average person, men and women may necessarily be required to expand their perceptions to a point on the left slope of the curve that represents "a frequently expanded range of attraction." Statistically, this arena probably can be delimited by $+1$ or -1 standard deviation. Finally, there are probably some rare individuals who have managed, for reasons that have not yet been investigated, to develop "an unusually expanded range of attraction" that may extend, say, to the second standard deviation to the left of the mid-point. Although such a graph merely portrays a heuristic attempt to illuminate the decisive influence of media images on modern perceptions and preferences, it could be compared with figures designed to depict similar ranges of attraction in prior eras that were marked by diverse cultural conditions such as less prevalent media images of the human physique, more variable standards of beauty, and heterosexual partnerships determined by arranged marriages rather than by romantic love. Perhaps the most significant conclusion that can be drawn from this illustration, however, is that, even when attention is focused on the area of the normal curve encompassed by an unusually expanded span of attraction, there still remain some individuals in the extreme left end of the tail of the curve, including perhaps a disproportionate percentage of people with visible disabilities, whose appearance may be considered so different that they might be rendered unacceptable in many kinds of social interactions.

There appears to be little doubt that visible disabilities are commonly perceived as culturally unattractive traits which often bar the individual bearing such stigma from jobs, opportunities, other benefits, and personal relationships. And yet attractiveness is a learned rather than an instinctual phenomenon. Moreover, it is acquired from a culture containing a rich and dynamic legacy that has often found a compelling appeal in human differences as well as similarities. Much of this tradition has apparently been stifled by the prevalence of restricted and conformist images of the body in advertizing and mass media which have acted as the instruments of capitalist development. And yet even a cursory review of historical trends affecting the presentation of the human physique indicates a fertile potential for regaining an appreciation of the fascination of physical differences, and for redefining the identity of men and women with disabilities.

Fundamentally, attractiveness can be defined as a multi-dimensional dialectical phenomenon encompassing traits that the culture defines as unattractive and characteristics that are popularly perceived as attractive or appealing. Attraction is essentially the excitement generated by the tension between these opposing elements. In this analysis, the antithetical symbolism of human, animal, and mechanical attributes has been identified as three fundamental dimensions of this process. Each of the symbols that is dissimilar to human attributes seems to induce a simultaneous aversion and fascination that is manifested by the modification of the body in various historical periods. From the perspectives of persons with visible disabilities, perhaps the most remarkable aspect of this dynamic is the fact that many such individuals combine all of these elements. Those who move in a seated or reclining position or who ambulate with canes or crutches that represent three or four appendages; persons with sensory or developmental impairments that may prompt tentative, awkward, or unanticipated postures or movements; individuals of short stature; people allegedly disfigured by amputations, burns, or scars; and many other children or adults with physical deviations or disabilities occasionally may be glimpsed as reflecting properties often associated with animals. Similarly, the prevalent ·use of technical aids including beds, respirators, catheters, prostheses, braces, wheelchairs, crutches, walkers, white canes, hearing aids, and dialysis units by persons with various disabilities may prompt a misperception that such mechanical contrivances comprise an integral part of their bodies. Amid all of these external trappings, people with obvious disabilities possess qualities that are unmistakably human. But the appearance of a disabled person need not entail an attempt to focus exclusively on these features or to obliterate the accoutrements that resemble animal- or machine-like components. Unlike the figure of the nondisabled individual, the common configurations of people with visible disabilities combine both the dominant human motif and secondary themes that reflect the fascination and appeal of animals and machines. Perhaps important advances toward recognizing the attractiveness as well as the acceptability of the appearance of disabled women and men can be made, therefore, by appreciating the stimulating and intriguing qualities of each of the separate dimensions that have defined and determined beauty throughout history.

The awareness that attractiveness is shaped by a combination of elements rather than by a single idealized standard comprises a crucial vehicle for redefining the meaning of a visible disability. From a multidimensional and dialectical perspective, disability *is* beautiful; and disabled citizens can find dignity and pride in bodily manifestations of disability as well as in other aspects of their appeal. As a result, they may gain a positive sense of personal and political identity. This consciousness does not entail a need to look beyond or to erase the overt signs of a disability in order to discern a beauty with extraordinary appearance. Nor is it necessary for disabled persons to dwell exclusively on inner qualities or personality characteristics to project a vision that is pleasing to the senses. There is an aesthetic of disability and anatomical differences that is both sensuous and artistic. Humans must learn alternative ways of identifying the attractive qualities of bodily variations both in themselves and others. This is a task encompassing vast political implications that can only be accomplished through intense social and political activity.

Modern images of the human body indicate a merger of narcissistic and economic motives supporting elite efforts to restrict active participation in modern society to persons whose appearance denotes physical similarities rather than differences. The hegemony of media images, which act like a distorting mirror that provokes the anxiety of its viewers, must be shattered to allow both disabled and nondisabled observers to gain a sense of comfort and satisfaction with their own bodies. As the group that has perhaps been most oppressed by prevailing standards of physical attractiveness, persons with visible disabilities might appropriately play a leading role in a countercultural movement to challenge this imagery. Denial and the emulation of nondisabled models no longer comprise a feasible strategy for disabled individuals seeking social acceptance and approval. Moreover, they may anticipate the support of nondisabled allies who also find existing criteria of physical beauty overly constricting or conformist and who are searching for optional means of appreciating the stimulation of human differences. Perhaps significant progress toward this objective can be achieved by recognizing that the body has been an important instrument of political communication throughout history. Physical appearance has been used as a device for promoting both stability and change. In much of the 19th and 20th centuries, the human anatomy has been shackled to mechanisms of capitalist expansion. But the dynamic principles of bodily attractiveness also encompass methods of disrupting the status quo and of unleashing the potential for revolutionary transformations.

References

Banner, L.W. 1983. *American beauty*. Chicago: University of Chicago Press.

Berscheid, E. & E. Walster 1974. Physical attractiveness. In *Advances in Experimental Social Psychology*, L. Berkowitz (ed.), **7**, 158–215. New York: Academic Press.

Brain, R. 1979. *The decorated body*. New York: Harper & Row.

Clark, K. 1956. *The nude: a study in ideal form*. Garden City, New York: Doubleday.

Dion, K., E. Berscheid & E. Walster 1972. What is beautiful is good. *Journal of Personality and Social Psychology* **24**, 285–90.

Ewen, S. & E. Ewen 1982. *Channels of desire*. New York: McGraw-Hill.

Fiedler, L. 1978. *Freaks: myths and images of the secret self*. New York: Simon & Schuster.

Fisher, S. 1973. *Body consciousness: you are what you feel*. Englewood Cliffs, N.J.: Prentice-Hall.

Giddon, D. B. 1983. Through the looking glasses of physicians, dentists, and patients. *Perspectives in Biology and Medicine* **26**, 451–8.

Goffman, E. 1963. *Stigma: notes on the management of spoiled identity*. Englewood Cliffs, N.J.: Prentice-Hall.

Gough, I. 1979, *The political economy of the welfare state*. London: Macmillan.

Hahn, H. 1985a. Toward a politics of disability: definitions, disciplines, and policies. *Social Science Journal* **22**, 87–105.

Hahn, H. 1985b. Changing perceptions of disability and the future of rehabilitation. In *Social influences in rehabilitation planning: blueprint for the twenty-first century*, L. G. Perlman & G. F. Austin (eds.), 53–64. Alexandra, Virginia: National Rehabilitation Association.

Hahn, H. 1986a. Disability and the urban environment: a perspective on Los Angeles. *Society and Space* **4**, 273–88.

Hahn, H. 1986b. Public support for rehabilitation programs: the analysis of U.S. disability policy. *Disability, Handicap, and Society* **1**, 121–37.

Hahn, H. 1987a. Civil rights for disabled Americans: the foundation of a political agenda. In *Images of the disabled/disabling images*, A. Gartner & T. Joe (eds.), 181–203. New York: Praeger.

Hahn, H. 1987b. Advertising the acceptable employable image: disability and capitalism. *Policy Studies Journal* **15**, 551–70.

Hahn, H. (forthcoming). The politics of physical differences: disability and discrimination. *Journal of Social Issues*.

Haley, B. 1978. *The healthy body and Victorian culture*. Cambridge, Mass.: Harvard University Press.

Hollander, A. L. 1978. *Seeing through clothes*. New York: Viking.

Kolakowski, L. 1974. Introduction. In *The socialist idea: a reappraisal*, L. Kolakowski & S. Hampshire (eds.), 1–17. New York: Basic Books.

Kunzle, D. 1982. *Fashion and fetishism*. Totowa, N.J.: Rowman & Littlefield.

Lakoff, R. T. & R. L. Scherr 1984. *Face value: the politics of beauty*. Boston: Routledge & Kegan Paul.

Landy, D. & H. Sigall 1974. Beauty is talent: task evaluation as a function of the performer's physical attractiveness. *Journal of Personality and Social Psychology* **29**, 299–304.

Lurie, A. 1981. *The language of clothes*. New York: Random House.

Nash, A. & H. Pieszko 1982. The multidimensional structure of mythological hybrid (part-human, part-animal) figures. *Journal of General Psychology* **106**, 35–55.

Renard, J. B. 1984. The wildman and the extraterrestrial: two figures of evolutionist fantasy. *Diogenes* **127**, 63–81.

Rudofsky, B. 1971. *The unfashionable human body*. Garden City, New York: Doubleday.

Schudson, M. 1984. *Advertising, the uneasy persuasion*. New York: Basic Books.

Schuler, H. & W. Berger 1979. The impact of physical attractiveness on an employment decision. In *Love and attraction: an international conference*, M. Cook & G. Wilson (eds.), 33–6. Oxford: Pergamon Press.

Sontag, S. 1978. *Illness as metaphor*. New York: Farrar, Straus & Giroux.

Steele, V. 1985. *Fashion and eroticism: ideals of feminine beauty from the Victorian era to the Jazz Age*. New York: Oxford University Press.

Turner, B. S. 1984. *The body and society*. New York: Basil Blackwell.

Subject Index

Milton Keynes UK
Ingram Content Group UK Ltd.
UKHW031140141024
449569UK00024B/1196